Complex Networks and Dynamic Systems

Volume 2

Series Editor

Terry L. Friesz
Pennsylvania State University
University Park, PA, USA

For further volumes:
http://www.springer.com/series/8854

Satish V. Ukkusuri • Kaan Ozbay
Editors

Advances in Dynamic Network Modeling in Complex Transportation Systems

Springer

Editors
Satish V. Ukkusuri
Purdue University
West Lafayette, IN, USA

Kaan Ozbay
Rutgers University
Piscataway, NJ, USA

ISBN 978-1-4614-6242-2 ISBN 978-1-4614-6243-9 (eBook)
DOI 10.1007/978-1-4614-6243-9
Springer New York Heidelberg Dordrecht London

Library of Congress Control Number: 2012956192

Printed on acid-free paper

Springer is part of Springer Science+Business Media (www.springer.com)

Preface

This book focuses on recent developments in dynamic network modeling (DNM) including aspects of route guidance and traffic control as it relates to transportation systems and other complex infrastructure networks. Dynamic network modeling is generally understood to be the mathematical modeling of time-varying vehicular flows on networks in a fashion that is consistent with established traffic flow theory and travel demand theory. Presently, we estimate that there are approximately 250 scholars around the globe actively involved in research, demonstrations, and applications pertaining to the body of knowledge related to various aspects of dynamic network modeling.

The area of DNM is a very active one. It is not only a purely research area but also an area with important social impacts because it can provide real-world solution that will improve the efficiency of our transportation systems without having to build new roads. Some of the important findings are:

1. There are many theoretical and practical aspects of the "Dynamic Route Guidance and Control" problem that are being addressed by academicians, private sector, and transportation agencies as evidenced from the list of presenters at the workshop.
2. Development of realistic and robust mathematical models of traffic flow, control, dynamic traffic assignment, and data processing needed for the development and deployment of effective route guidance and control systems remains as major challenges.
3. Different control strategies such as congestion pricing, traffic responsive signals, diversion during nonrecurrent congestion, and speed control for different transportation facilities are some important examples of real-time control strategies that are needed. However, it is important to ensure that the deployed control strategy would work in real time under real-world conditions. This requires extensive off-line evaluation of the capabilities of underlying control and guidance models and strategies.
4. Increasingly, the private sector is involved in the collection and dissemination of traffic information for real-time control of transportation systems throughout

the world. However, their current contribution to the research and development of advanced systems is rather limited. Better coordination between the private sector and academia will be very productive. Similarly, public agencies that are the owners of transportation systems also need to better interface with academia and the private sector to be able to expedite the deployment of these "Dynamic Route Guidance and Control" systems that will drastically improve the efficiency of their transportation networks.

5. Real-time collection of speed, travel time, flow, and other traffic data over very large transportation networks remains to be a major challenge that is being tackled by private companies and academicians. More work is needed to ensure reliability and accuracy of the collected traffic data.
6. This is a long-term and wide-ranging research area that spans between the control of individual vehicles and large multimodal transportation networks. Moreover, successful development of such real-time online control and guidance systems requires the interfacing of algorithms, software, and hardware in such a way that the resulting system is robust and reliable. Thus, major research, development, and deployment investment are needed to successfully implement dynamic network models.

Dynamic network modeling as a field has grown over the last 30 years with contributions from various scholars all over the field. The basic problem which many scholars in this area have focused is related to the analysis and prediction of traffic flows satisfying notions of equilibrium when flows are changing over time. In addition, recent research has also focused on integrating dynamic equilibrium with traffic control and other mechanism designs such as congestion pricing and network design. Recently, advances in sensor deployment, availability of GPS enabled vehicular data and social media data have rapidly contributed to better understanding and estimating the traffic network states and have contributed to new research problems which advance previous models in dynamic modeling.

This book mainly contains some of the papers presented at the National Science Foundation workshop on "Dynamic Route Guidance and Traffic Control" which was organized on June 7–8, 2010 at Rutgers University by Prof. Kaan Ozbay, Prof. Satish Ukkusuri, Prof. Hani Nassif, and Prof. Pushkin Kachroo. This workshop brought together various experts in this area from universities, industris, and federal/state agencies to present recent findings in this area. Various topics were presented at the workshop including dynamic traffic assignment, traffic flow modeling, network control, complex systems, mobile sensor deployment, intelligent traffic systems, and data collection issues. This book is motivated by the research presented at this workshop and the discussion that followed where a volume that summarizes recent advances in the aforementioned areas was seen as an important book. The organizers invited a select set of researchers to contribute chapters to this book. More than 15 scholars from U.S. universities and abroad have accepted to write manuscripts for this book. The book focuses on recent methodological advances and application of dynamic network modeling to transportation systems. The book is divided into four sections:

1. *Recent Algorithms in Dynamic Routing and Guidance*: A fundamental problem in dynamic modeling is to develop dynamic routing algorithms which consider various data sources and uncertainties in the system.
2. *Methodological Advances in Dynamic Network Assignment and Traffic Control*: In this section, various papers related to recent mathematical programming formulations for dynamic modeling will be compiled.
3. *Applications of Dynamic Network Modeling*: In this section, papers related to various applications from evacuation and simulation-based modeling will be compiled.
4. *Data Needs for Real-Time*: Dynamic Route Guidance and Traffic Control: In this section, papers related to data needs and availability for the successful implementation of real-time control and routing algorithms will be compiled.

The papers that were selected for this book were rigorously reviewed by various experts in this field. We thank all authors who submitted their work for consideration. In addition, we thank the dozens of referees for their important work in reviewing the papers. We would also like to acknowledge the financial support provided for the Dynamic Route Guidance and Traffic Control Workshop by NSF's Civil, Mechanical, and Manufacturing Innovation Division of the Directorate for Engineering under the award #0951147 and Professor Robert L. Smith who was the NSF program manager and made major contributions to the content and success of the workshop. Additional information about the NSF workshop is available at http://ritslab.rutgers.edu/agenda.html. Special thanks go to Prof. Terry Friesz and editors of Springer for graciously allowing us to edit this book.

West Lafayette, IN, USA — Satish V. Ukkusuri
Piscataway, NJ, USA — Kaan Ozbay

Contents

1 **Dynamic Traffic Assignment: A Survey of Mathematical Models and Techniques** 1
Pushkin Kachroo and Neveen Shlayan

2 **The Max-Pressure Controller for Arbitrary Networks of Signalized Intersections** 27
Pravin Varaiya

3 **Coordinated Feedback-Based Freeway Ramp Metering Control Strategies "C-MIXCROS and D-MIXCROS" that Take Ramp Queues into Account** 67
Ilgin Gokasar, Kaan Ozbay, and Pushkin Kachroo

4 **Solving the Integrated Corridor Control Problem Using Simultaneous Perturbation Stochastic Approximation** 89
Jingtao Ma, Yu (Marco) Nie, and H. Michael Zhang

5 **Analyses of Arterial Travel Times Based on Probe Data** 115
Isaac Kumar Isukapati, George F. List, Stacy Eisenman, Jeffrey Wojtowicz, and William Wallace

6 **A Multibuffer Model for LWR Road Networks** 143
Mauro Garavello and Benedetto Piccoli

7 **Cell-Based Dynamic Equilibrium Models** 163
W.Y. Szeto

8 **Information Impacts on Traveler Behavior and Network Performance: State of Knowledge and Future Directions** 193
Ramachandran Balakrishna, Moshe Ben-Akiva, Jon Bottom, and Song Gao

9 Modeling Within-Day Activity Rescheduling Decisions under Time-Varying Network Conditions ... 225
Yunemi Jang, Yi-Chang Chiu, and Hong Zheng

10 Dynamic Navigation in Direction-Dependent Environments ... 245
Irina S. Dolinskaya

11 An Approach to Assess the Impact of Dynamic Congestion in Vehicle Routing Problems ... 265
H.M. Abdul Aziz and Satish V. Ukkusuri

12 Incident Duration Prediction with Hybrid Tree-based Quantile Regression ... 287
Qing He, Yiannis Kamarianakis, Klayut Jintanakul, and Laura Wynter

Index ... 307

Chapter 1
Dynamic Traffic Assignment: A Survey of Mathematical Models and Techniques

Pushkin Kachroo and Neveen Shlayan

Abstract This paper presents a survey of the mathematical methods used for modeling and solutions for the traffic assignment problem. It covers the static (steady-state) traffic assignment techniques as well as dynamic traffic assignment in lumped parameter and distributed parameter settings. Moreover, it also surveys simulation-based solutions. The paper shows the models for static assignment, variational inequality method, projection dynamics for dynamic travel routing, discrete time and continuous time dynamic traffic assignment, and macroscopic dynamic traffic assignment (DTA). The paper then presents the macroscopic DTA in terms of the Wardrop principle and derives a partial differential equation for experienced travel time function that can be integrated with the macroscopic DTA framework.

1.1 Introduction

Traffic assignment is an integral part of the four-stage transportation planning process [see Gazis (1974) and Potts and Oliver (1972)] that includes:

1. *Trip Generation*: Trip generation models estimate the number of trips generated at origin nodes and/or the number of trips attracted to destination nodes based on factors such as household income, demographics, and land-use pattern. This data is obtained using surveys conducted periodically.

P. Kachroo (✉) • N. Shlayan
Department of Electrical & Computer Engineering, University of Nevada Las Vegas, 4505 S. Maryland Pkwy, Las Vegas, NV 89154-4007, USA
e-mail: pushkin@unlv.edu; neveenshlayan@gmail.com

S.V. Ukkusuri and K. Ozbay (eds.), *Advances in Dynamic Network Modeling in Complex Transportation Systems*, Complex Networks and Dynamic Systems 2,
DOI 10.1007/978-1-4614-6243-9_1, © Springer Science+Business Media New York 2013

2. *Trip Distribution*: From the total number trips generated and attracted at each node, trip distribution algorithms generate an origin–destination (O–D) matrix, in which each cell entry indicates the number of trips from one specific origin to one specific destination. Hitchcock model (Hitchcock 1941), opportunity model (Stouffer 1940), gravity model (Voorhees 1956), and entropy models (Wilson 1967) have been used for trip distribution algorithms.
3. *Modal Split*: Modal split analysis takes each cell value in the O–D matrix and divides it among various alternate modes of travel. The models are built based on performing discrete choice analysis on survey data [see Ben-Akiva and Lerman (1985)].
4. *Traffic Assignment*: This step assigns each O–D flow value onto various alternate paths from that specific origin to the destination node. Assignments are based on optimization, usually using either Wardrop's user-equilibrium (Wardrop 1952; Sheffi 1985) or system optimum.

This four-step process comes from the traditional transportation planning area and is not designed for real-time operations, such as traffic responsive real-time incident management. However, a lot of research has taken place in the area of traffic assignment, especially dynamic traffic assignment that enables researchers to study transient traffic behavior, not just the steady-state one which the static assignment is designed for. A survey paper (Peeta and Ziliaskopoulos 2001) provides an excellent survey for the research work that has been performed in the area of dynamic traffic assignment. This paper, in contrast to that survey work, provides a survey of the mathematical framework that has been used in this area and presents the results to enable the reader to grasp the various mathematical tools that have been used to study and analyze this problem. The models and approaches that have been used are varied, and this review paper brings them together in order for the readers to see them in a somewhat linear fashion.

Outline. The remainder of this article is organized as follows. Section 1.2 gives account of various mathematical programming-based static traffic assignment models that have been used. This section presents the user-equilibrium and system optimal formulations of the assignment problems, followed by the numerical schemes that have been used to solve those problems. Section 1.3 presents the fundamentals of the variational inequality framework which subsumes the mathematical programming methodology. Dynamic extension of the variational inequality framework is presented in Sect. 1.4. Section 1.5 presents the dynamic traffic assignment in continuous time. The discrete time and continuous time versions of this are presented. Section 1.7 presents the macroscopic DTA model including the new formulation and a new travel time partial differential equation. Section 1.8 presents a brief summary of the main features of simulation-based DTA. Finally, Sect. 1.10 gives the conclusions.

1.2 Mathematical Programming-based Static Traffic Assignment Model

To build the mathematical framework for our paper, we will start with terminology and framework used in Sheffi (1985). We illustrate a sample network that is also taken from Sheffi (1985) and is shown in Fig. 1.1. The digraph shows four nodes and four arcs. Nodes 1 and 2 are origin nodes and node 4 is the destination node. Hence there are two O–D pairs: $1-4$ and $2-4$.

There are two main classical traffic assignment optimization problems considered. Those two are user-equilibrium and system optimum.

1.2.1 User-Equilibrium

User-equilibrium problem is based on Wardrop's principle (Wardrop 1952) which is stated as:

> The journey times on all the routes actually used are equal, and less than those which would be experienced by a single vehicle on any unused route.

This equilibrium condition can be obtained as a solution of a mathematical programming problem presented below (Sheffi 1985).

1.2.1.1 Mathematical Programming Formulation

The user-equilibrium problem is stated as the mathematical programming problem [see Sheffi (1985), Dafermos and Sparrow (1969b)] shown in Eq. (1.1).

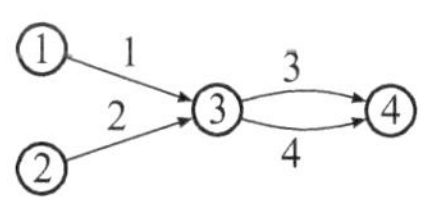

Fig. 1.1 Sample network

Table 1.1 Network notation

$\mathfrak{N}$	Set of nodes
$\mathfrak{A}$	Set of arcs
$\mathfrak{R}$	Set of origin nodes
$\mathfrak{S}$	Set of destination nodes
$\mathfrak{K}$	Set of paths connecting O–D pair $r-s$, $r \in \mathfrak{R}$, $s \in \mathfrak{S}$
x_a	Flow on arc $a \in \mathfrak{A}$
t_a	Travel time on arc $a \in \mathfrak{A}$
f_k^{rs}	Flow on path $k \in \mathfrak{K}$ between O–D pair $r-s$
c_k^{rs}	Travel time on path $k \in \mathfrak{K}$ between O–D pair $r-s$
q_{rs}	O–D Trip rate between O–D pair $r-s$
$\delta_{a,k}^{rs}$	$\delta_{a,k}^{rs} = 1$, if a is in path k between r and s, otherwise 0

Fig. 1.2 BPR link performance function

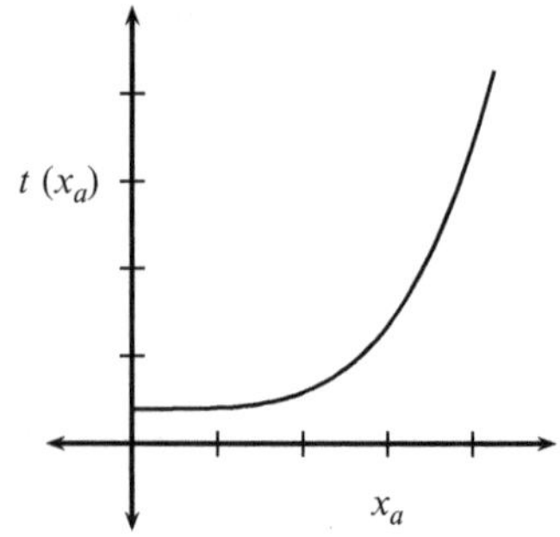

$$\min z(x) = \sum_a \int_0^{x_a} t_a(\omega)\mathrm{d}\omega \tag{1.1}$$

with the equality constraints:

$$\sum_k f_k^{rs} = q_{rs}\,\forall r,s \tag{1.2}$$

$$x_a = \sum_r \sum_s \sum_k f_k^{rs}\delta_{a,k}^{rs} \tag{1.3}$$

and the inequality constraint

$$f_k^{rs} \geq 0\,\forall r,s \tag{1.4}$$

The formulation given in Eq. (1.1) is the Beckmann transformation (Beckmann et al. 1955). The link performance function $t_a(x_a)$ is a function of traffic flow on the link and the link capacity c_a. According to the Bureau of Public Roads (BPR), it is given by Eq. (1.5)

$$t_a(x_a) = v_f\left(1 + 0.15\left(\frac{x_a}{c_a}\right)^4\right) \tag{1.5}$$

The plot of a BPR function is shown in Fig. 1.2

Wellposedness. The objective function is a smooth convex function ($\nabla^2(x)$ is positive definite), and the feasible region is convex, hence a unique solution exists.

1.2.1.2 Equivalence with Wardrop User-Equilibrium Condition

The Kuhn–Tucker conditions for the mathematical programming problem given by Eq. (1.1) can be obtained in terms of the Lagrangian given in Eq. (1.6).

$$\mathfrak{L}(f,\lambda) = z[x(f)] + \sum_{rs}\lambda_{rs}\left(q_{rs} - \sum_k f_k^{rs}\right) \tag{1.6}$$

Here, λ_{rs} is the Lagrangian multiplier. The Kuhn–Tucker conditions $\forall k, r, s$ are:

$$
\begin{aligned}
f_k^{rs} \frac{\partial \mathfrak{L}(f,\lambda)}{\partial f_k^{rs}} &= 0 \\
\frac{\partial \mathfrak{L}(f,\lambda)}{\partial f_k^{rs}} &\geq 0 \\
\frac{\partial \mathfrak{L}(f,\lambda)}{\partial \lambda^{rs}} &= 0
\end{aligned}
\tag{1.7}
$$

Applying these necessary conditions (1.7) to the mathematical program (1.1) we obtain the Wardrop conditions $\forall k, r, s$ as:

$$
\begin{aligned}
f_k^{rs}(c_k^{rs} - u_{rs}) &= 0 \\
c_k^{rs} - u_{rs} &\geq 0 \\
\sum_k f_k^{rs} &= q_{rs} \\
\sum_k f_k^{rs} &\geq 0
\end{aligned}
\tag{1.8}
$$

1.2.2 System Optimal Solution

System optimal solution is a solution that provides the total minimum time for the entire network. This condition can be obtained as a solution of a mathematical programming problem presented below (Sheffi 1985).

1.2.2.1 Mathematical Programming Formulation

The system optimal problem is stated as the mathematical programming problem [see Sheffi (1985), Dafermos and Sparrow (1969b)] shown in Eq. (1.9).

$$\min z(x) = \sum_a x_a t_a(x_a) \tag{1.9}$$

with the equality constraints:

$$\sum_k f_k^{rs} = q_{rs} \, \forall r, s \tag{1.10}$$

$$x_a = \sum_r \sum_s \sum_k f_k^{rs} \delta_{a,k}^{rs} \tag{1.11}$$

and the inequality constraint

$$f_k^{rs} \geq 0 \forall r, s \tag{1.12}$$

Wellposedness. The objective function is a smooth convex function ($\nabla^2(x)$ is positive definite), and the feasible region is convex, hence a unique solution exists.

1.2.2.2 Equivalence with Marginal User-Equilibrium Condition

Applying Kuhn–Tucker conditions in this case, we get $\forall k, r, s$:

$$\begin{aligned} f_k^{rs}(\tilde{c}_k^{rs} - \tilde{u}_{rs}) &= 0 \\ \tilde{c}_k^{rs} - \tilde{u}_{rs} &\geq 0 \\ \sum_k f_k^{rs} &= q_{rs} \\ \sum_k f_k^{rs} &\geq 0 \end{aligned} \tag{1.13}$$

Here, we have

$$\tilde{c}_k^{rs} = \sum_a \delta_{a,k}^{rs} \tilde{t}_a \tag{1.14}$$

where

$$\tilde{t}_a(x_a) = t_a(x_a) + x_a \frac{\mathrm{d}t_a(x_a)}{x_a} \tag{1.15}$$

1.2.3 Numerical Schemes

The numerical scheme for solving user-equilibrium is based on the Frank–Wolfe algorithm that obtains the feasible direction and the maximum step-size for each iteration in one step. In fact, for the static traffic assignment problem, this amounts to simply applying all or nothing assignment to the shortest path for each O–D pair. The next step for each iteration involves finding the step size in the direction of the link flow solution of the all-or-nothing assignment step. Appropriate stopping criterion can be applied using some convergence principle. Details of this are provided in Sect. 5.2, pages 116–122 of Sheffi (1985).

There are heuristic numerical methods available to perform the assignment to achieve user-equilibrium. Two of the common heuristic techniques are:

FHWA (modified capacity restraint) method In this method at each iteration an all-or-nothing assignment of the entire OD flow is performed on a single path. Travel times are updated by performing a weighted average of the travel time obtained by the latest assignment and the previous one. A convergence criterion is used to stop the iteration steps (e.g., when the maximum difference between two iterative steps of link flows is less than some ε). The final link flows assigned to the network are obtained by averaging the values from the last four iterative steps.

Incremental Assignment In incremental assignment, the OD values are divided into n parts, and then each part is assigned to the network using all or nothing assignment based on the previous travel time values.

Dafermos and Sparrow (1969a) applied the Frank–Wolfe method to traffic assignment problem. This method also results in an all-or-nothing assignment, followed by a line search step in each iteration. The details can be obtained from Sheffi (1985).

1.3 Variational Inequality-based Static Traffic Assignment Model

Variational inequality formulation for traffic equilibrium has been used as it generalizes the framework of mathematical programming even when the travel time function on one link depends on the conditions on other links as well (Dafermos 1980). Once the variational inequality model has been formulated, it can be solved using some appropriate numerical scheme, such as the ones based on projection method, linear approximation, relaxation method, or the more general iterative scheme of Dafermos (1983).

The variational inequality problem is stated as:

VI Problem. Given a continuous function $f : \mathbb{K} \to \mathbb{R}^n$, where $\mathbb{K}$ is a given closed and convex subset of $\mathbb{R}^n$, $\langle \cdot, \cdot \rangle$ denotes the inner product, find $x \in \mathbb{K}$, such that

$$\langle f(x), y - x \rangle \geq 0, \forall y \in \mathbb{K} \tag{1.16}$$

Figure 1.3 shows a convex set and the variational inequality condition at a corner. The relationship between variational inequalities and optimization problems is given by the following two theorems (Kinderlehrer and Stampacchia 2000).

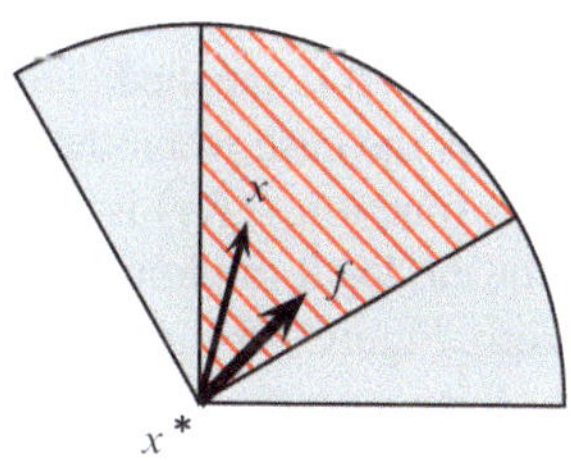

Fig. 1.3 Variational inequality

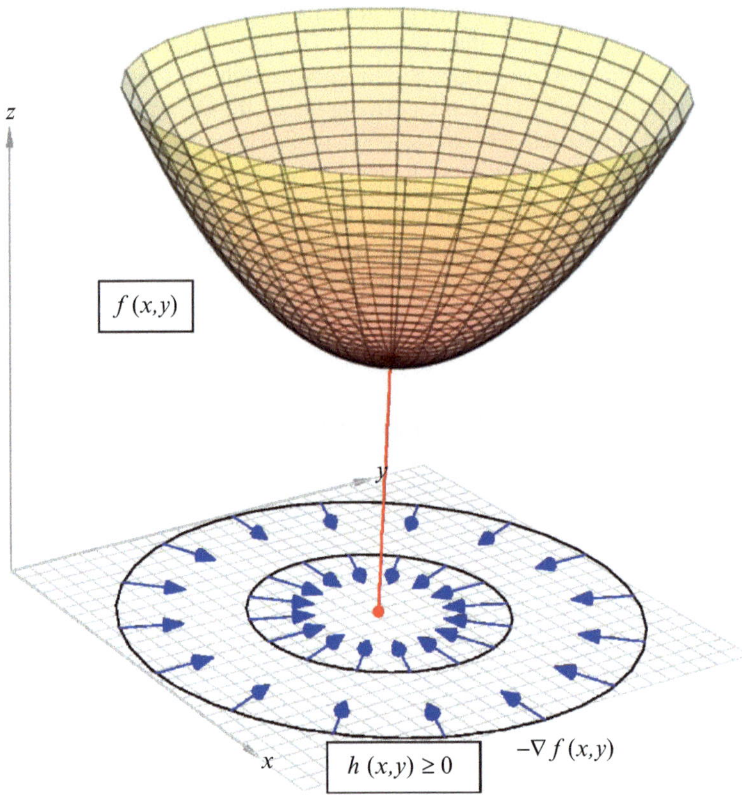

Fig. 1.4 Minimizer in the interior

Theorem 1. *$x \in \mathbb{K}$ s.t. $f(x) = \min_{y \in \mathbb{K}} f(y) \implies \langle \nabla f(x), y - x \rangle \geq 0, \forall y \in \mathbb{K}$.*

Theorem 2. *Convex f s.t. $\langle \nabla f(x), y - x \rangle \geq 0, \forall y \in \mathbb{K} \implies f(x) = \min_{y \in \mathbb{K}} f(y)$.*

To understand the constrained optimization problem and its interplay with variational inequalities, we present two figures (Figs. 1.4 and 1.5). The first quadrant in the $x - y$ plane is the constrained region of search where we have assumed that $h(x, y) \geq 0$ is satisfied. The function to be minimized is given by $f(x, y)$. Figure 1.4 shows the case when the minimizing point (on the $x - y$ plane) for a given smooth cost function $f(x, y)$ is contained in the interior of the region $\mathbb{K}$ given by $h(x, y) \geq 0$. For the local minimum to exist, it is necessary that the gradient of the function is zero. Figure 1.5 shows the case when the minimizing point (on the $x - y$ plane) for a given smooth cost function $f(x, y)$ is contained at the boundary of the region $\mathbb{K}$ given by $h(x, y) = 0$. For the given point to be the minimizer, any movement from this point in any feasible direction, i.e. in the direction of increasing $h(x, y)$, should increase the value of $f(x, y)$. This is the variational inequality statement. Moreover,

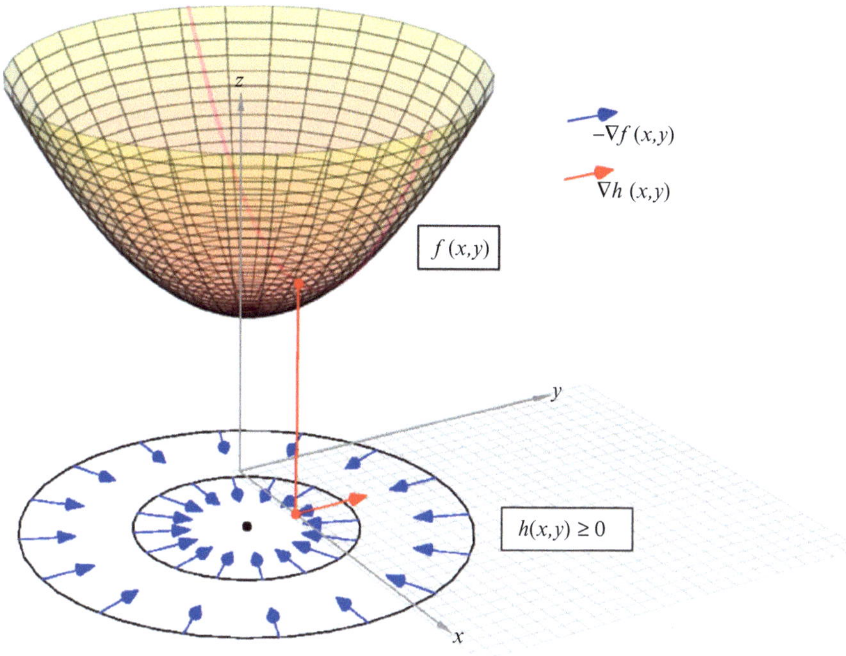

Fig. 1.5 Minimizer on the boundary

in this case (when certain regularity conditions are satisfied (Avriel 2003)), since, the boundary is given by $h(x,y) = 0$, the directional derivative of $f(x,y)$ in the direction of the tangent to the boundary should be zero. Moreover, the gradient of $h(x,y)$ as well as that of $f(x,y)$ should be pointing in the same direction. Kuhn–Tucker conditions (and Lagrangian method) state the condition on the relationship between the gradient of the cost function and that of the constraint functions. However, those are necessary conditions only if the problem satisfies certain regularity conditions [see Avriel (2003), Bazaraa et al. (2006), and Mangasarian (1994)].

Theorems 1 and 2 demonstrate that variational inequality framework is more general than the mathematical programming framework. The variational inequality formulations of the traffic equilibrium (user) problems are stated below.

Theorem 3. *$x \in \mathbb{K}$ is a solution to the user-equilibrium problem if and only*

$$\sum_{w \in W} \sum_{p \in P_w} C_p(x)(y-x) \geq 0, \forall x \in \mathbb{K}$$

Here, C_p is the travel time for the path p from the OD pair P_w from the set of OD pairs W. This variational inequality can also be written in terms of traffic flows instead of link flows (Nugurney 2000).

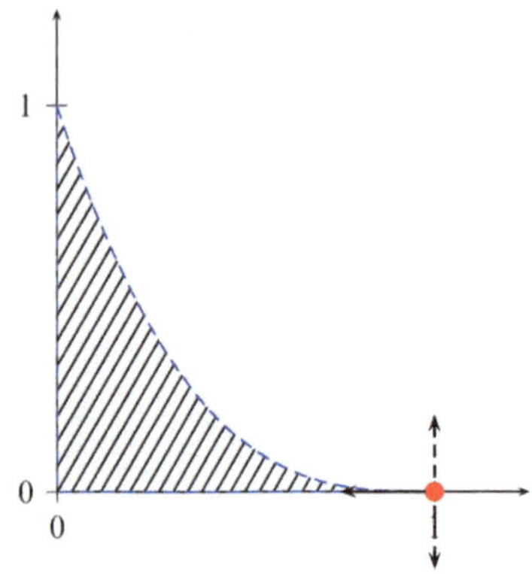

Fig. 1.6 Violation of Kuhn–Tucker condition

To understand how the variational inequality formulation is more general than the optimization problem, consider the variational inequality formulation again.

$$\langle f(x), y-x\rangle \geq 0, \forall y \in \mathbb{K} \tag{1.17}$$

Now, if $f(x) = \nabla\theta(x)$, then the condition

$$\langle \nabla\theta(x), y-x\rangle \geq 0, \forall y \in \mathbb{K} \tag{1.18}$$

is the necessary condition for the optimization problem

$$\text{minimize } \theta(x),\ x \in \mathbb{K} \tag{1.19}$$

The variational inequality has a corresponding gradient relationship based on the following theorem that is about the symmetry of second partial derivatives (Facchinei and Pang 2003).

Theorem 4. *Given $f : \mathbb{K} \to \mathbb{R}^n$, a continuously differentiable function on the open convex set $\mathbb{K} \subset \mathbb{R}^n$, then the following three conditions are equivalent.*

1. $\exists\theta$, *s.t.* $f(x) = \nabla\theta(x)$
2. $\nabla f(x) = [\nabla f(x)]^T\ \forall x \in \mathbb{K}$
3. f *is integrable on* $\mathbb{K}$

Theorem 4 shows that if the function f has a symmetric Jacobian then there is a corresponding optimization problem associated with it. However, if the Jacobian is asymmetric, for instance, when the user-equilibrium cost is asymmetric with respect to traffic flows, then the Wardrop solution (variational inequality) is the framework without a corresponding mathematical programming problem.

On a cautionary note, Kuhn–Tucker conditions (and Lagrangian method) state the condition on the relationship between the gradient of the cost function and that of the constraint functions. However, those are necessary conditions only if the problem satisfies certain regularity conditions [see Avriel (2003), Bazaraa et al. (2006), and Mangasarian (1994)]. For instance Fig. 1.6 shows a function

$f(x,y) = -x$ to be minimized which at the minimum point $(x,y) = (1,0)$ does not satisfy the Kuhn–Tucker conditions for the region constrained by the first quadrant and the curve $y = 1 - x^3$.

1.4 Projected Dynamical Systems: Dynamic Variational Equation Model

Dynamics of route switching has been analyzed using dynamic variational inequality by Nagurney and Zhang (1996; 1995; 1997; 1996; 1988). They developed the theory for projected dynamical systems in Zhang and Nagurney (1995) and applied the theory to traffic assignment in Zhang and Nagurney (1996) and Nagurney and Zhang (1997). The paper by Dupuis and Nagurney (1993) shows the main results in the theory and applications of projected dynamical systems including its relationship to the Skorokhod problem (Skorokhod 1961) for the study of its wellposedness.

Since variational inequality is related to the solution of a fixed point problem, we can relate the variational inequality solution to be the equilibrium point of a dynamic system. The stability of the equilibrium point can be studied within the framework of this dynamic system, and then those dynamics can be used to model a time-varying route assignment problem. This is precisely what Nagurney and Zhang do in their various papers. We summarize the technical results here.

1.4.1 Dynamic Route Choice

The dynamics of route choice adjustment are given by (Nagurney and Zhang 1996):

$$\dot{x} = \Pi_{\mathbb{K}}(x, -C(x)) \tag{1.20}$$

where

$$\Pi_{\mathbb{K}}(x,v) = \lim_{\varepsilon \to 0} \frac{P_{\mathbb{K}}(+\varepsilon v) - x}{\varepsilon} \tag{1.21}$$

and

$$P_{\mathbb{K}}(x) = \text{Argmin}_{x \in \mathbb{K}} \|x - z\| \tag{1.22}$$

Figure 1.7 shows the convex region inside which the vector field of the dynamics is shown. The equilibrium point as well as the solution of the variational inequality is at $(0,0)$.

The path flow vector $x^* \in \mathbb{K}$ is the solution of

$$0 = \Pi_{\mathbb{K}}(x^*, -C(x^*)) \tag{1.23}$$

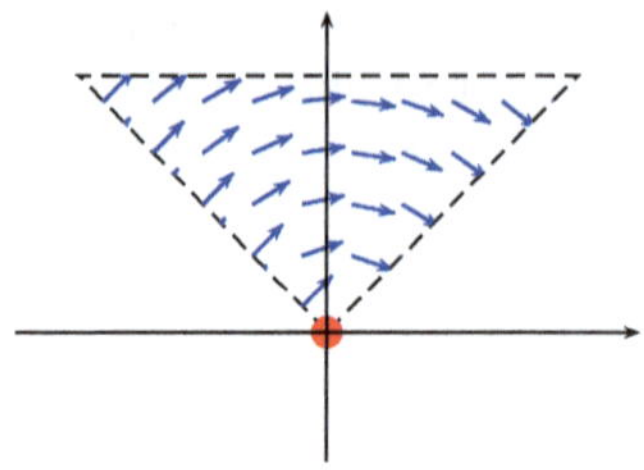

Fig. 1.7 The vector field

if and only if it satisfies

$$\langle C(x^*), x - x^* \rangle \geq 0, \forall y \in \mathbb{K} \tag{1.24}$$

The following theorem from Nagurney and Zhang (1996) gives the condition for asymptotic stability of the equilibrium point of the projected dynamics related to the route adjustment process.

Theorem 5. *If the link cost is a strictly monotonic continuous function of link flows, then the equilibrium point for dynamics shown in Eq.* (1.20) *is asymptotically stable.*

The major result from Nagurney and Zhang (1996) for applying the discrete algorithm for the dynamic route choice problem is the following.

Theorem 6. *The Euler method given by*

$$x^{\tau+1} = P_{\mathbb{K}}(x^{\tau} - a_{\tau}C(x^{\tau})) \tag{1.25}$$

when

$$\lim_{\tau \to \infty} a_{\tau} = 0 \tag{1.26}$$

and

$$\sum_{\tau=1}^{\infty} a_{\tau} = \infty \tag{1.27}$$

for $\mathbb{K}$ *being the positive orthant converges to some traffic network equilibrium path flow.*

1.5 Dynamic Traffic Assignment

There are some nice reviews that provide summary of the models and work that has been performed in the area of dynamic traffic assignment (DTA), such as Chiu et al. (2009), Peeta and Ziliaskopoulos (2001), Ran and Boyce (1996), and Friesz (2001). Our review will focus on the mathematical aspects of these developments.

1.5.1 Dynamic Traffic Assignment: Discrete Time

Merchant and Nemhauser (1978a; 1978b) were the first to present a dynamic traffic assignment problem where time-varying O–D flows are considered. Their formulation uses a state difference equation to represent the link dynamics, a conservation equation at the nodes of the digraph, and a cost function to minimize which leads to the following mathematical programming problem.

$$\min z(x) = \sum_{i=1}^{I} \sum_{j=1}^{a} t_{ij}(x_{ij}) \tag{1.28}$$

with the link discrete time dynamics as equality constraints:

$$x_j[i+1] = x_j[i] - g_j(x_j[i]) + d_j[i], i = 0, 1, \cdots I-1, \forall j \in \mathfrak{A} \tag{1.29}$$

the node conservation equation as

$$\sum_{j \in A(q)} d_j[i] = F_q[i] + \sum_{j \in B(q)} g_j(x_j[i]),\ i = 0, 1, \cdots I-1, \forall q \in \mathfrak{N} \tag{1.30}$$

and the inequality constraints

$$x_j[i] \geq 0\, i = 0, 1, \cdots I-1, \forall j \in \mathfrak{A} \tag{1.31}$$

$$d_j[i] \geq 0\, i = 0, 1, \cdots I-1, \forall j \in \mathfrak{A} \tag{1.32}$$

$$x_j[0] = x_0[j]\, \forall j \in \mathfrak{A} \tag{1.33}$$

Here, $x_j[i]$ is the number of vehicles at the beginning of time period i in link j, $g_j(x_j[i])$ is the number of vehicles exiting the link in the unit time as a function of $x_j[i]$, and $d_j[i]$ is the number of vehicles entering the link j. This problem formulation is a single destination network model. $F_q[i]$ show the inflow rates as the time-varying O–D flows. This can be extended to a multiorigin multidestination formulation.

1.5.2 Dynamic Traffic Assignment: Continuous Time

Now we present a continuous time formulation of the DTA problem (Boyce et al. 2001) where a dynamic variational inequality is used. The traffic dynamics utilize ordinary differential equations instead of finite difference equation as was the case for the discrete time formulation. There are other models that use dynamic continuous time models in optimal control or in variational setting such as Friesz et al. (1989; 1993), and Chen (1999).

The time-dependent Wardrop condition for the DTA are

$$
\begin{aligned}
&f_k^{rs}(t)(c_k^{rs}(t) - u_{rs}(t)) = 0 \\
&c_k^{rs}(t) - u_{rs}(t) \geq 0 \\
&\sum_k f_k^{rs}(t) = q_{rs}(t) \\
&\sum_k f_k^{rs}(t) \geq 0
\end{aligned} \tag{1.34}
$$

The traffic dynamics for this DTA problem are the continuous version of the difference equation for the Merchant Nemhauser model, and are given by the following conservation ordinary differential equation.

$$\dot{x}_{ak}^{rs}(t) = u_{ak}^{rs}(t) - g_{ak}^{rs}(x_a(t)) \tag{1.35}$$

Here, $u_{ak}^{rs}(t)$ is the time-varying inflow to link a on path k from origin r to destination s, and $g_{ak}^{rs}(x^{rs}(t))$ is the corresponding time-varying outflow which is the exit function which depends on the link density $x_a(t)$.

We have the following equality among matching constraints for various flows and links (Boyce et al. 2001).

$$\sum_r \sum_s \sum_k x_{ak}^{rs}(t)\delta_{a,k}^{rs} = x_a(t) \tag{1.36}$$

Numerical techniques are available to solve this variational inequality [see Boyce et al. (2001)]. Optimal control formulation for this problem can also be obtained which can be solved by calculus of variations or dynamic programming methods.

1.6 Travel Time and FIFO Issue

One major issue in dynamic traffic assignment problem is that of First In First Out (FIFO) constraint as discussed in Carey (1992). According to FIFO if $x_{t\tau a} > 0$ where $x_{t\tau a}$ is the traffic flow that enters link a at time t and exits at time τ, then any flow that enters before time t cannot exit after time τ at an average. This condition is shown to be nonconvex in Carey (1992) and is presented in Eq. (1.37).

$$(x_{t\tau a} > 0) \Rightarrow \left(\sum_{t'\tau' a} x_{t'\tau' a} | t' < t, \tau' > \tau \right) = 0 \tag{1.37}$$

A violation of this condition is shown in Fig. 1.8. The violation essentially occurs because of the nature of the exit function and also the time and space discretization

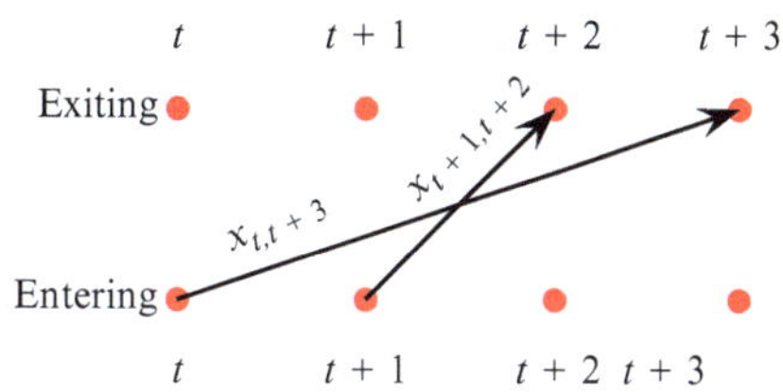

Fig. 1.8 FIFO violation

of the traffic link and dynamics. Both of these issues get resolved by a proper choice of space and time discretization that is chosen after the original modeling is performed in a hydrodynamic setting using the dynamic distributed parameter traffic flow theory. This theory allows for a proper development of a travel time function as well as a travel time vector field. This development is the main original technical contribution of this paper.

1.7 Macroscopic Model for DTA

We propose to use a hydrodynamic traffic model in the framework of the DTA problem. The Lighthill–Whitham–Richards (LWR) model, named after the authors in Lighthill and Whitham (1955) and Richards (1956), is a macroscopic one-dimensional traffic model. The conservation law for traffic in one dimension is given by

$$\frac{\partial}{\partial t}\rho(t,x)+\frac{\partial}{\partial x}f(\rho(t,x))=0 \tag{1.38}$$

In this equation ρ is the traffic density (vehicles or pedestrians) and f is the flux which is the product of traffic density and the traffic speed v, i.e. $f=\rho v$. There are many model researchers have proposed for how the flux should be dependent on traffic conditions. This relationship is given by the *fundamental diagram*.

1.7.1 Greenshield's Model

Greenshield's model [see Greenshields (1935)] uses a linear relationship between traffic density and traffic speed.

$$v(\rho)=v_f\left(1-\frac{\rho}{\rho_m}\right) \tag{1.39}$$

where v_f is the free flow speed and ρ_m is the maximum density. Free flow speed is the speed of traffic when the density is zero. This is the maximum speed. The maximum density is the density at which there is a traffic jam and the speed is equal

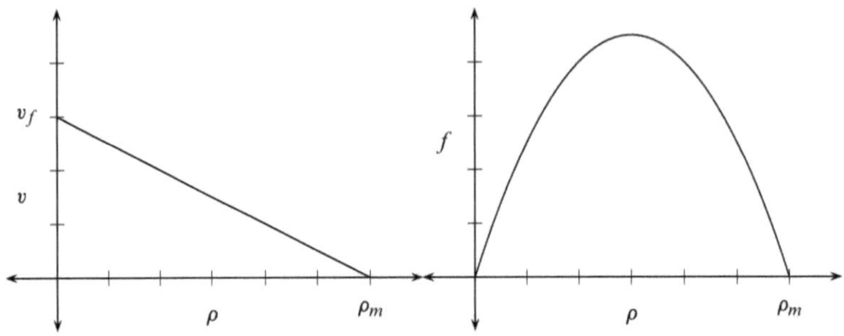

Fig. 1.9 Fundamental diagram using Greenshield model

to zero. The flux function is concave as can be confirmed by noting the negative sign of the second derivative of flow with respect to density, i.e. $\partial^2 f/\partial\rho^2 < 0$. The fundamental diagram refers to the relationship that the traffic density ρ, traffic speed v, and traffic flow f have with each other. These relationships are shown in Fig. 1.9.

1.7.2 Generalized/Weak Solution for the LWR Model

The hyperbolic partial differential equation (PDE) for the LWR model given by Eq. (1.38) can be solved by using the method of characteristics (LeVeque 1994). Figure 1.10 shows a $x-t$ plot for traffic density $\rho(t,x)$. Initially the traffic density is constant at ρ_0. At time $t=0$, there is a traffic light at $x=0$ that turns red. We see the shockwaves traveling backward so that there is a discontinuity between traffic density being ρ_0 to the left of the shock line and being ρ_m to the right of it. On the right there is another shockwave traveling to the right between zero traffic density and ρ_0. At time $t=t_c$, the light turns green and we see rarefaction of traffic starting at $x=0$. Corresponding to time $t=t_u$ we see the plot of traffic density $\rho(t_u,x)$ that shows to the two shock waves as well as rarefaction of the traffic density. This shows that the traffic solution has discontinuities and a weak solution of the LWR model is required that allows for these discontinuous solutions.

1.7.2.1 Generalized Solutions

For a conservation law

$$\rho_t + f(\rho)_x = 0 \tag{1.40}$$

with initial condition

$$\rho(x,0) = \rho_0(x), \tag{1.41}$$

where $u_0(x) \in L^1_{\text{loc}}(R;R^n)$, solution in the distributional sense is defined below for smooth vector field $f: R^n \to R^n$ [see Bressan (2005)].

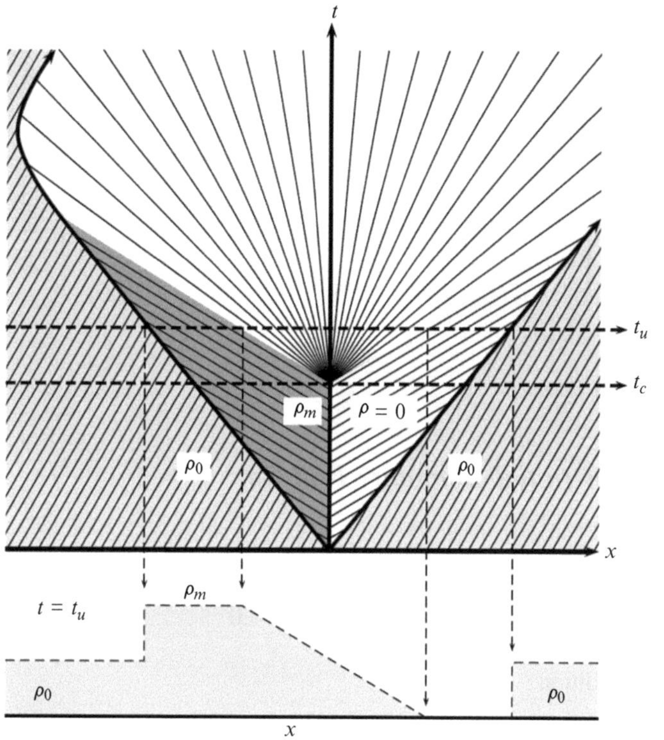

Fig. 1.10 Traffic characteristics

Definition 1.7.1. A measurable locally integrable function $\rho(t,x)$ is a solution in the distributional sense of the Cauchy problem (1.40) if for every $\phi \in C_0^\infty(R^+ \times R) \mapsto R^n$

$$\iint_{R^+\times R} [\rho(t,x)\,\phi_t(t,x) + f(\rho(t,x))\,\phi_x(t,x)]\,\mathrm{d}x\,\mathrm{d}t + \int_R u_0(x)\,\phi(x,0)\,\mathrm{d}x = 0 \quad (1.42)$$

1.7.2.2 Weak Solutions

A measurable locally integrable function $u(t,x)$ is a weak solution in the distributional sense of the Cauchy problem (1.40) if it is a distributional solution in the open strip $(0,T) \times \mathbb{R}$, satisfies the initial condition (1.41) and if u is continuous as a function from $[0,T]$ into L^1_{loc}. We require $u(t,x) = u(t,x^+)$ and

$$\lim_{t\to 0} \int_R |u(t,x) - u_0(x)|\,\mathrm{d}x = 0 \quad (1.43)$$

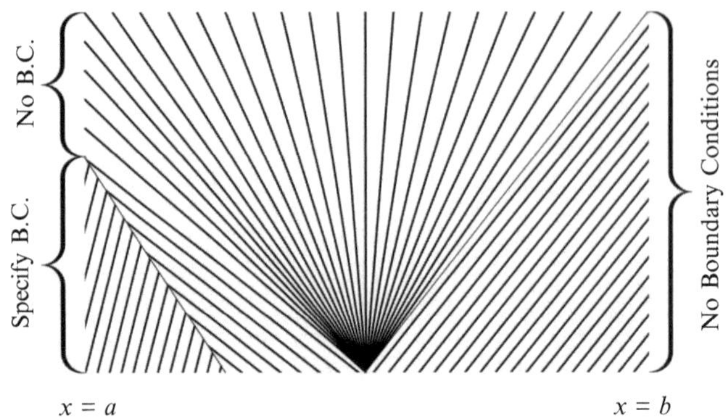

Fig. 1.11 Boundary data

1.7.3 Scalar Initial-Boundary Problem

Consider the scalar conservation law given here.

$$u_t + f(t,x,u)_x = 0 \tag{1.44}$$

with initial condition

$$u(0,x) = u_0(x), \tag{1.45}$$

and boundary conditions

$$u(t,a) = u_a(t) \text{ and } u(t,b) = u_b(t), \tag{1.46}$$

The boundary conditions cannot be prescribed point-wise since characteristics from inside the domain might be traveling outside of the boundary. If there is any data at the boundary for that time, that has to be discarded. Moreover, the data also must satisfy entropy condition at the boundary so as to render the problem wellposed. This is shown in Fig. 1.11 where for some time boundary data on the left can be prescribed when characteristics from the boundary can be *pushed in* [see Strub and Bayen (2006)]. However when the characteristics are coming from inside, the boundary data cannot be prescribed.

1.7.4 Macroscopic (PDE) Traffic Network

The network problem for traffic flow has been studied by researchers (Garavello and Piccoli 2006; Holden and Risebro 1995; Lebacque 1996; Coclite and Piccoli 2002). They consider a traffic node with incoming n junctions and outgoing m junctions as shown in Fig. 1.12.

Fig. 1.12 Traffic node with incoming and outgoing links

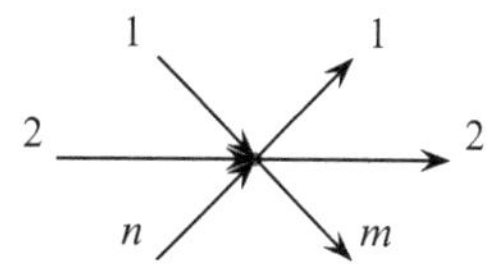

The traffic distribution at the junction is performed based on a traffic distribution matrix that must be provided for the node as well as using an entropy condition at the node that is equivalent to maximizing the flow at the node.

We present the summary of the Coclite/Piccoli model for the network (Coclite and Piccoli 2002; Garavello and Piccoli 2005, 2006). That summary is also used in Gugat et al. (2005). The formulation in terms of demand and supply is shown in the work by Lebacque (1996), Lebacque and Khoshyaran (2004), and Buisson et al. (1996). This formulation is equivalent to the Coclite/Piccoli formulation, and both then show numerical method using the Godunov scheme.

Each arc of the traffic network is an interval $[a_i, b_i]$. The model for the network is

$$\frac{\partial}{\partial t}\rho^i(t,x) + \frac{\partial}{\partial x} f(\rho^i(t,x)) = 0 \, \forall x \in [a_i, b_i], t \in [0,T] \tag{1.47}$$

$$\frac{\partial}{\partial t}\pi^i(t,x,k,r,s) + v^i(\rho^i(t,x))\frac{\partial}{\partial x}\pi^i(t,x,k,r,s) = 0 \, \forall x \in [a_i, b_i], t \in [0,T] \tag{1.48}$$

Here $\pi(t,x,k,r,s)$ is a function whose range is $[0,1]$ and gives the fraction of the traffic density on path k of the OD pair (r,s) on the arc i. Hence, we have

$$\rho^i(t,x,k,r,s) = \pi^i(t,x,k,r,s)\rho^i(t,x) \tag{1.49}$$

This ensures the FIFO condition automatically since vehicle speed is a function of traffic density, and hence vehicles don't cross each other in this model (unless we add lane modeling with lane change logic).

At any node the following flow conservation condition (Kirchoff's law) must be satisfied. This equation says that the total inflow to a node equals its outflow.

$$\sum_{i=1}^{n} f_i(\rho_i(b_i,t)) = \sum_{i=n+1}^{n+m} f_i(\rho_i(a_i,t)), \, \forall t \geq 0 \tag{1.50}$$

At the nodes, we have traffic splitting factors $\alpha_{j,i}$ that tell us what fraction of a given incoming arc i is going to an outgoing arc j of that node. The factors $\alpha_{j,i}$ have to be consistent with $\pi^i(t,x,k,r,s)$.

$$\alpha_{j,i} = \sum_r \sum_s \sum_k \pi^i(t, b_i-, k, r, s) \tag{1.51}$$

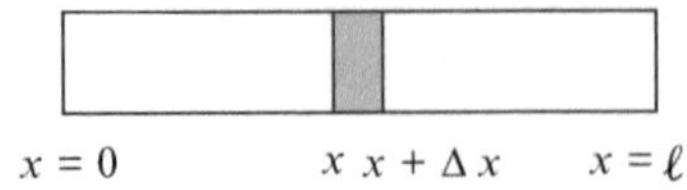

Fig. 1.13 Travel time on a link

The weak solution of the traffic density at a node is given by a collection of functions ρ_i such that the following is satisfied.

$$\sum_{i}^{n+m} \int_0^{\infty} \int_{a_i}^{b_i} \left(\rho_i \frac{\partial \phi_i}{\partial t} + f(\rho_i) \frac{\partial \phi_i}{\partial x} \right) \mathrm{d}x\mathrm{d}t = 0 \tag{1.52}$$

All the details of this model can be obtained from Garavello and Piccoli (2006). The Wardrop condition for this macroscopic DTA model become the following.

$$\begin{aligned}
&(\delta_{a,k}^{rs} \pi^i(t, a_i, k, r, s))(c_k^{rs}(t) - u_{rs}(t)) = 0 \\
&c_k^{rs}(t) - u_{rs}(t) \geq 0 \\
&\sum_k \delta_{a,k}^{rs} \pi^i(t, a_i, k, r, s) = q_{rs}(t) \\
&\sum_k \delta_{a,k}^{rs} \pi^i(t, a_i, k, r, s) \geq 0
\end{aligned} \tag{1.53}$$

Here, i in the expression $\pi^i(t, a_i, k, r, s)$ is the link connected to the source r for the particular k and s. The travel time $c_k^{rs}(t)$ is developed in the next section.

1.7.5 Travel Time Dynamics

This section provides a model for obtaining the experienced travel time function for the hydrodynamic model that can be used for the macroscopic DTA model.

Consider a link as shown in the Fig. 1.13. We want to develop a travel time function $T(t,x)$ that provides the travel time for a vehicle at position x and time t to reach $x = \ell$. It takes a vehicle time $\Delta x / v(t,x)$ to move from x to $x + \Delta x$. Hence, we have the following travel time condition.

$$T(t + \Delta t, x + \Delta x) = T(t,x) - \frac{\Delta x}{v(t,x)} \tag{1.54}$$

Taking the Taylor series first terms for $T(t,x)$ and simplifying, we obtain

$$\frac{\partial T(t,x)}{\partial t} \Delta t + \frac{\partial T(t,x)}{\partial x} \Delta x = -\frac{\Delta x}{v(t,x)} \tag{1.55}$$

Multiplying by $v(t,x)$, dividing by Δx, and then taking limits and simplifying we get the travel time partial differential equation.

$$\frac{\partial T(t,x)}{\partial t} + \frac{\partial T(t,x)}{\partial x} v(\rho(t,x)) + 1 = 0 \tag{1.56}$$

Hence, the one-way coupled PDE system for LWR and travel time for a link is given by

$$\frac{\partial}{\partial t}\rho(t,x) + \frac{\partial}{\partial x}[\rho(t,x)v(\rho(t,x))] = 0 \tag{1.57}$$

$$\frac{\partial T(t,x)}{\partial t} + \frac{\partial T(t,x)}{\partial x} v(\rho(t,x)) + 1 = 0 \tag{1.58}$$

$$v(\rho(t,x)) = v_f\left(1 - \frac{\rho}{\rho_m}\right) \tag{1.59}$$

1.8 Simulation-Based DTA

With the availability of faster processors and computers, using simulation-based DTA is becoming more and more popular (Peeta and Ziliaskopoulos 2001; Mahmassani et al. 1998; Ben-Akiva et al. 1998). Summary of simulation-based DTA and the methodology is presented in Chiu et al. (2009) and Peeta and Ziliaskopoulos (2001). In principle, the simulation of the network can be accomplished using microscopic, mesoscopic, or macroscopic simulations. Microscopic simulation is based on car-following models and they model the vehicle dynamics for each individual vehicle. Macroscopic simulations are based on discretization and numerical solutions of the macroscopic models, such as LWR-based models. Mesoscopic simulations use the fundamental relationship for obtaining vehicle speeds (macroscopic behavior) but also have individual vehicles (microscopic behavior) modeled with the tracking of their location and speeds. Since the mesoscopic modeling-based DTA is more prevalent, we will focus on that in this section.

There are two main steps to prepare the simulation-based DTA. A three-stage iterative process to obtain user-equilibrium behavior and a field data-based calibration process. Once these two processes have been successful, the software can be used for various studies.

1.8.1 Iterations for User-Equilibrium

This equilibration process is performed in three steps (Chiu et al. 2009). These three steps are iterated till the user-equilibrium condition is obtained within some tolerance limit.

Network Loading. This step is obtained by running the network simulation for a given time-varying OD and traffic assignment to various paths between each OD pairs. The result is the set of travel times for each path.

Path Set Update. The traffic loading obtained from the previous step is used to calculate the set of k-shortest paths between each OD pair.

Path Assignment Adjustment. In this step the OD flows are assigned to new updated paths from the previous step.

1.8.2 Calibration from Field Data

Data obtained from field surveys and sensors can be used to calibrate the simulation-based DTA models. Some parameters that can be tuned include the time-varying OD values, road capacities, and vehicle speed density parameters. The calibration can be performed in order to maximize the match between the simulated outputs and the observed data. Various numerical optimization methods have been used such as gradient-based methods and SPSA. The general scheme is to find the parameter vector that will minimize the least squared error of the observations, where the observations are y_i, and the output from simulation is dependent on the parameters as $h_i(\theta)$.

$$\theta^* = \text{Argmin}_\theta \sum_i (y_i - h_i(\theta))^2 \tag{1.60}$$

A typical iterative scheme if it is gradient based to find the optimal parameters can be

$$\theta^*[k+1] = \theta^*[k] - \eta \nabla_\theta \sum_i (y_i - h_i(\theta))^2 \tag{1.61}$$

OD estimation has been performed [see Ben-Akiva et al. (1998)] using an auto regressive model for OD variations from nominal values, and then applying Kalman filter techniques on it.

1.9 Traffic Operation Design and Feedback Control

Traffic assignment problem and its solutions have very strong roots in the transportation planning process, especially the four-stage process shown in Sect. 1.1. It is very important to keep this context in mind in order to ensure its proper use. DTA models can help in performing before and after studies for various transportation projects. They can also help in many other studies by enhancing its basic framework with additional features such as environment effects of congestion and costs.

For real-time traffic operations we must use and develop techniques specifically for real-time operations. For instance, if we have to design an isolated ramp control at one location, the entire OD matrix obtained and calibrated from field studies during some limited time is not relevant to that problem. Feedback control-based methods are extremely suited for design of traffic control and real-time operations. The details of many specific feedback control designs for traffic operations such as real-time traffic routing and ramp metering are available in multiple publications (Kachroo and Ozbay 1999; Kachroo and Özbay 2003, 1998, 2006, 2005).

1.10 Conclusions

This paper provided a mathematical survey of the static and dynamic traffic assignment problems. It presented the macroscopic DTA model using the LWR distributed parameter model as the basis. The paper presented a new partial differential equation for travel time function for a link. It also provided a brief summary of simulation-based DTA.

References

Avriel M. Nonlinear programming: analysis and methods. Mineola, NY: Dover Publications; 2003.

Bazaraa MS, Sherali HD, Shetty CM. Nonlinear programming: theory and algorithms. New York: Wiley; 2006.

Beckmann MJ, McGuire CB, Winsten CB. T Studies in the economics of transportation, Yale University Press, 1956.

Ben-Akiva M, Bierlaire M, Koutsopoulos H, Mishalani R. Dynamit: a simulation-based system for traffic prediction. In: DACCORS short term forecasting workshop, The Netherlands, Citeseer; 1998.

Ben-Akiva ME, Lerman SR. Discrete choice analysis: theory and application to travel demand. MIT Press series in transportation studies. Cambridge, MA: MIT Press; 1985.

Boyce D, Lee D, Ran B. Analytical models of the dynamic traffic assignment problem. Network Spatial Econ. 2001;1:377–90.

Bressan A. Hyperbolic systems of conservation laws: the one-dimensional cauchy problem. Oxford: Oxford University Press; 2005.

Buisson C, Lebacque JP, Lesort JB. Strada, a discretized macroscopic model of vehicular traffic flow in complex networks based on the godunov scheme. In: CESA'96 IMACS multiconference: computational engineering in systems applications; 1996. p. 976–81.

Carey M. Nonconvexity of the dynamic traffic assignment problem. Transport Res Part B 1992;26(2):127–33.

Chen HK. Dynamic travel choice models: a variational inequality approach. Berlin: Springer; 1999.

Chiu YC, Bottom J, Mahut M, Paz A, Balakrishna R, Waller T, et al. A primer for dynamic traffic assignment. Washington, DC: Transportation Research Board; http://www.nextrans.org/ADB30/index.php. Accessed 6 Jan 2013.

Coclite GM, Piccoli B. Traffic flow on a road network. Arxiv preprint math/0202146 (2002).

Dafermos S. Traffic equilibrium and variational inequalities. Transport Sci. 1980;14(1):42.

Dafermos S. An iterative scheme for variational inequalities. Math Program. 1983;26(1):40–7.

Dafermos S. Sensitivity analysis in variational inequalities. Math Oper Res. 1988;13:421–34.

Dafermos SC, Sparrow FT. The traffic assignment problem for a general network. J Res Natl Bur Stand Ser B 1969a;73(2):91–118.

Dafermos SC, Sparrow FT. The traffic assignment problem for a general network. J Res Natl Bur Stand. 1969b;73B:91–118.
Dupuis P, Nagurney A. Dynamical systems and variational inequalities. Ann Oper Res. 1993;44(1):7–42.
Facchinei F, Pang JS. Finite-dimensional variational inequalities and complementarity problems. vol. 1. New York: Springer; 2003.
Friesz TL. Special issue on dynamic traffic assignment, Part I: networks and spatial economics. vol. 1. Springer; New York 2001.
Friesz TL, Bernstein D, Smith TE, Tobin RL, Wie BW. A variational inequality formulation of the dynamic network user equilibrium problem. Oper Res. 1993;41(1):179–91.
Friesz TL, Luque J, Tobin RL, Wie BW. Dynamic network traffic assignment considered as a continuous time optimal control problem. Oper Res. 1989;37(6):893–901.
Garavello M, Piccoli B. Source-destination flow on a road network. Comm Math Sci. 2005;3(3): 261–83.
Garavello M,Piccoli B. Traffic flow on networks. Springfield, MO: American Institute of Mathematical Sciences; 2006.
Gazis DC. Traffic science. New York, NY: Wiley-Interscience; 1974.
Greenshields BD. A study in highway capacity. Highway Res Board 1935;14:458.
Gugat M, Herty M, Klar A, Leugering G. Optimal control for traffic flow networks. J Optim Theor Appl. 2005;126(3):589–616.
Hitchcock FL. The distribution of a product from several sources to numerous localities. J Math Phys. 1941;20:224–23.
Holden H, Risebro NH. A mathematical model of traffic flow on a network of unidirectional roads. SIAM J Math Anal. 1995;26:999.
Kachroo P, Özbay K. Solution to the user equilibrium dynamic traffic routing problem using feedback linearization. Transport Res Part B 1998;32(5):343–360.
Kachroo P, Ozbay K. Feedback control theory for dynamic traffic assignment. London: Springer; 1999.
Kachroo P, Özbay K. Feedback ramp metering in intelligent transportation systems. Springer; New York 2003.
Kachroo P, Özbay K. Feedback control solutions to network level user-equilibrium real-time dynamic traffic assignment problems. Network Spatial Econ. 2005;5(3):243–60.
Kachroo P, Özbay K. Modeling of network level system-optimal real-time dynamic traffic routing problem using nonlinear h-∞ feedback control theoretic approach. J Intell Transport Syst. 2006;10(4):159–71.
Kinderlehrer D, Stampacchia G. An introduction to variational inequalities and their applications. vol. 31. Philadelphia: Society for Industrial Mathematics; 2000.
Lebacque JP. The godunov scheme and what it means for first order traffic flow models. In: Internaional symposium on transportation and traffic theory; 1996. p. 647–77.
Lebacque JP, Khoshyaran MM. First order macroscopic traffic flow models for networks in the context of dynamic assignment. Transport Plann. 2004;64:119–40.
LeVeque RJ. Numerical methods for conservation laws. Basel: Birkhäuser; 1994.
Lighthill MJ, Whitham GB. On kinematic waves. i:flow movement in long rivers. ii:a theory of traffic on long crowded roods. Proc R Soc. 1955;A229:281–345.
Mahmassani HS, Hawas YE, Abdelghany K, Abdelfatah A, Chiu YC, Kang Y, et al. Dynasmart-x; volume ii: Analytical and algorithmic aspects. Technical Report ST067-85-Volume II, Center for Transportation Research, The University of Texas at Austin; 1998.
Mangasarian OL. Nonlinear programming. vol. 10. Philadelphia: Society for Industrial Mathematics; 1994.
Merchant DK, Nemhauser GL. A model and an algorithm for the dynamic traffic assignment problems. Transport Sci. 1978a;12(3):183–99.
Merchant DK, Nemhauser GL. Optimality conditions for a dynamic traffic assignment model. Transport Sci. 1978b;12(3):183–99.

Nagurney A, Zhang D. Projected dynamical systems and variational inequalities with applications. vol. 2. Boston: Kluwer Academic Publishers; 1996.

Nagurney A, Zhang D. Projected dynamical systems in the formulation, stability analysis, and computation of fixed-demand traffic network equilibria. Transport Sci. 1997;31(2):147–58.

Nugurney A. Sustainable transportation networks. Northampton, MA: Edward Elgar Publication; 2000.

Peeta S, Ziliaskopoulos AK. Foundations of dynamic traffic assignment: the past, the present and the future. Network Spatial Econ. 2001;1(3/4):233–65.

Potts RB, Oliver RM. Flows in transportation networks. Mathematics in science and engineering. New York: Academic Press; 1972.

Ran B, Boyce DE. Modeling dynamic transportation networks: an intelligent transportation system oriented approach: with 51 figures. New York: Springer; 1996.

Richards PI. Shockwaves on the highway. Oper Res. 1956;4:42–51.

Sheffi Y. Urban transportation networks: equilibrium analysis with mathematical programming methods. Englewood Cliffs, NJ: Prentice-Hall; 1985.

Skorokhod AV. Stochastic equations for diffusion processes in a bounded region. Theor Probab Appl. 1961;6:264.

Stouffer SA. Intervening opportunities: A theory relating mobility and distance. Am Socio Rev. 1940;5:845–67.

Strub I, Bayen A. Weak formulation of boundary conditions for scalar conservation laws: an application to highway modeling. Int J Robust Nonlinear Control 2006;16:733–48.

Voorhees AM. A general theory of traffic movement. In: Proceedings, Institute of Traffic Engineers; 1956.

Wardrop JG. Some theoretical aspects of road traffic research. In: Proceedings, Institute of Civil Engineers, PART II, 1; 1952. p. 325–78.

Wilson AG. A statistical theory of spatial distribution models. Transport Res. 1967;1:253–69.

Zhang D, Nagurney A. On the stability of projected dynamical systems. J Optim Theor Appl. 1995;85(1):97–124.

Zhang D, Nagurney A. On the local and global stability of a travel route choice adjustment process. Transport Res Part B 1996;30(4):245–62.

Chapter 2
The Max-Pressure Controller for Arbitrary Networks of Signalized Intersections

Pravin Varaiya

Abstract The control of an arbitrary network of signalized intersections is considered. At the beginning of each cycle, a controller selects the duration of every stage at each intersection as a function of all queues in the network. A stage is a set of permissible (non-conflicting) phases along which vehicles may move at pre-specified saturation rates. Demand is modeled by vehicles entering the network at a constant average rate with an arbitrary burst size and moving with pre-specified average turn ratios. The movement of vehicles is modeled as a "store and forward" queuing network. A controller is said to stabilize a demand if all queues remain bounded. The max-pressure controller is introduced. It differs from other network controllers analyzed in the literature in three respects. First, max-pressure requires only local information: the stage durations selected at any intersection depends only on queues adjacent to that intersection. Second, max-pressure is provably stable: it stabilizes a demand whenever there exists any stabilizing controller. Third, max-pressure requires no knowledge of the demand, although it needs turn ratios. The analysis is conducted within the framework of "network calculus," which, for fixed-time controllers, gives guaranteed bounds on queue size, delay, and queue clearance times.

2.1 Introduction

This chapter presents the *max-pressure* feedback policy for the control of an arbitrary network of signalized intersections. The cycle length at each intersection is fixed, although it may be different at different intersections. The policy determines at the beginning of each cycle, the fraction of the cycle that is allocated to each stage.

P. Varaiya (✉)
University of California, Berkeley, CA 94720, USA
e-mail: varaiya@eecs.berkeley.edu

S.V. Ukkusuri and K. Ozbay (eds.), *Advances in Dynamic Network Modeling in Complex Transportation Systems*, Complex Networks and Dynamic Systems 2,
DOI 10.1007/978-1-4614-6243-9_2,

(A stage is a set of permissible simultaneous movements.) The movement of traffic is modeled as a store and forward (SF) queuing system (Aboudolas et al. 2009a). Feedback policies based on queue measurements to control an SF model have been extensively studied, including in (Robertson and Bretherton 1991; Mirchandani and Head 2001; Heydecker 2004; Aboudolas et al. 2009b; Cai et al. 2009). There are two major differences between the max-pressure policy and those proposed in these studies.

First, the max-pressure policy is *decentralized*: the decision at any intersection depends only on the queues adjacent to that intersection; the policies in the other studies are *centralized*: the decision at each intersection depends on the queues at all intersections. This distinction may be practically important, since the communication infrastructure required to implement max pressure is much simpler to build. More significantly perhaps, max pressure may be implemented *incrementally*: remarkably, if a new intersection is added to the network, the max pressure policy for the original network does not change. In the centralized control of the cited studies, an expansion of the controlled network leads to changes in the policy of all intersections.

Second, the max-pressure policy is provably stable. That is, if external arrivals and turn ratios are stationary, and if there is *any* policy that keeps all queues bounded, then max-pressure also keeps all queues bounded. None of the cited studies provides such a stability guarantee. Also, max-pressure requires no knowledge of the external arrivals (but it does require knowledge of turn ratios, which can be estimated from the queue measurements), whereas the other studies require knowledge of the external arrivals. Consequently, max-pressure automatically adapts to slow changes in demand patterns.

Analysis of the SF queuing system is carried out using network calculus, which was developed to model, analyze, and control communication networks. (The fundamental reference is (Cruz 1991); we quote results from (Chang 2000); for a brief description see (Wikipedia 2009).) Network calculus is equally well suited for signal control studies, as the calculus describes traffic flows in terms of cumulative counts, commonly used in traffic studies. A flow is characterized by its average rate ρ and the maximum burst (or platoon) size σ. The service that an intersection provides to a flow is also characterized by two parameters: the saturation rate s and the maximum duration r (effective red) for which no service is provided. These parameters are easier to estimate empirically than parameters of stochastic queuing models. (The max-pressure policy for stochastic queueing models is studied in Varaiya (2009), which also proves stochastic stability.)

This chapter is organized as follows. The basic theory of the calculus for a single queue is recalled in Sect. 2.2 and used in Sect. 2.3 to study an isolated signalized intersection. The simplest case of the queue formed by a *single* constant flow of arriving vehicles and its dependence on the green duration is well known (Newell 1989, pp. 33–37). But even for this case, as seen in Sect. 2.3.1, network calculus offers a deeper analysis by providing bounds on queue length and busy period when vehicles arrive in bursts or platoons. This analysis readily extends to an intersection with multiple phases. The treatment in Sect. 2.3.2 of the fixed-cycle controller extends that of (Allsop 1972) and (Newell 1989, §2.2) by including the impact of traffic bursts.

A fixed-cycle controller is inevitably wasteful because sometimes it actuates phases with no queue even though other phases have queues. This waste is eliminated by the work-conserving controllers considered in Sect. 2.3.3. In the case that only one phase can be actuated at a time, work-conserving controllers are always superior to fixed-cycle controllers in the sense that the former minimize a weighted sum of queue lengths (Sect. 2.3.3.1, Theorem 3).

Two counter-examples in Sect. 2.3.3.2 show that obvious extensions of Theorem 3 are false. The first example presents an unstable work-conserving single-phase controller for a two-intersection network. The second example exhibits an unstable work-conserving controller for a single intersection in which each stage activates two phases. In both examples, there exist stabilizing fixed-time controllers.

These counterexamples motivates the problem: Construct a stable, adaptive, work-conserving controller. For an isolated intersection the "max-pressure" controller of Sect. 2.3.3.3 is one solution. It works as follows. At each time, the controller calculates the pressure exerted by each stage. The pressure is defined as the sum of the queue lengths multiplied by the saturation rates of all the phases actuated by the stage. The max-pressure controller selects the stage that exerts the maximum pressure. Theorem 5 states that the max-pressure controller is stabilizing whenever there exists a stabilizing fixed-cycle controller.

The problem for an arbitrary network of signalized intersections is treated in Sect. 2.4. The network calculus model is developed in Sect. 2.4.1. Once the model is laid out, Corollary 1 yields performance bounds for a fixed-cycle control scheme in Sect. 2.4.2. The max-pressure controller is described in Sect. 2.4.3. The pressure exerted by a stage is now different: it is the sum of the upstream queue lengths minus the downstream queue lengths (weighted by the turn ratio) multiplied by the saturation rates of all the phases actuated by the stage. Theorem 7 extends Theorem 5 to networks. This appears to be the first adaptive, provably stable controller in the literature.

The discussion in Sect. 2.5 provides the mathematical intuition underlying the max-pressure controller; compares it to other controllers presented in the literature; outlines the major limitations of the store and forward model; and poses some questions.

2.2 Network Calculus for a Single Queue

Time is discrete: $t = 0, 1, \cdots$. $\mathcal{F}$ ($\mathcal{F}_0$) denotes the set of all nonnegative, increasing functions f (with $f(0) = 0$). Consider a queuing system with cumulative arrivals $A \in \mathcal{F}_0$, cumulative departures $B \in \mathcal{F}_0$, and cumulative (virtual) service completions $C \in \mathcal{F}_0$. See Fig. 2.1. Let $q(t)$ denote the queue size at time t, with $q(0) = 0$. $q(t)$ satisfies Lindley's equation,

$$q(t+1) = [q(t) - c(t)]_+ + a(t+1), \quad t \geq 0. \tag{2.1}$$

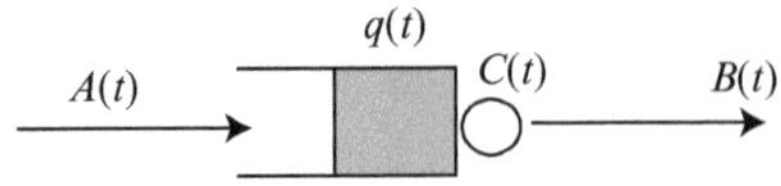

Fig. 2.1 A queuing system with arrivals A, departures B, service C

(Notation: $[x]_+ = \max\{0, x\}$.) In (2.1) $a(t) = A(t) - A(t-1)$ and $c(t) = C(t) - C(t-1)$ are respectively the numbers of arrivals and (virtual) service completions in period t. (Take $A(-1) = C(-1) = 0$.)

For $f \in \mathcal{F}$ and $s \le t$, let $f(t,s) = f(t) - f(s)$. Recall that $q(0) = 0$. Lemma 1 is proved in Appendix A.

Lemma 1 ((Chang 2000, Lemma 1.3.1)). *For all $t \ge 0$ the queue size is*

$$q(t) = \max_{0 \le s \le t} [A(t,s) - C(t-1,s)], \tag{2.2}$$

and the cumulative departures $B(t) = A(t) - q(t)$ are

$$B(t) = \min_{0 \le s \le t} [A(s) + C(t-1,s)]. \tag{2.3}$$

Definition 1. The arrival process $A \in \mathcal{F}_0$ is *upper-bounded* by $f_1 \in \mathcal{F}_0$ if $A(t,s) \le f_1(t-s)$ for all $t \ge s$. The service process $C \in \mathcal{F}_0$ *provides* service $f_2 \in \mathcal{F}_0$ if $C(t-1,s) \ge f_2(t-s)$ for all $t \ge s$.

Theorem 1 is proved in Appendix B.

Theorem 1 ((Chang 2000, Theorem 2.2.8)). *If A is upper-bounded by f_1 and C provides f_2,*

$$q(t) \le \max_{0 \le \tau \le t} [f_1(\tau) - f_2(\tau)], \tag{2.4}$$

$$B(t,s) \le A(t) - B(s) \le \max_{0 \le \tau} [f_1(t-s+\tau) - f_2(\tau)]. \tag{2.5}$$

The delay $d(t)$ of the last arrival before t is bounded by

$$d(t) \le \min\{d \ge 0 \mid f_1(\tau) \le f_2(\tau + d - 1), \tau = 1, \cdots, t\}. \tag{2.6}$$

The duration of any busy period is bounded by

$$BP = \max\{b \mid f_1(\tau) > f_2(\tau), \tau = 1, \cdots, b\}. \tag{2.7}$$

Remark. From (2.4) and (2.6) the queue size and delay are respectively bounded by the vertical and horizontal distances between f_1 and f_2, and from (2.7) the busy period is bounded by the first time the graphs of f_1, f_2 intersect, as in Fig. 2.2a. Thus the most important performance measures are captured by the curves f_1 and f_2.

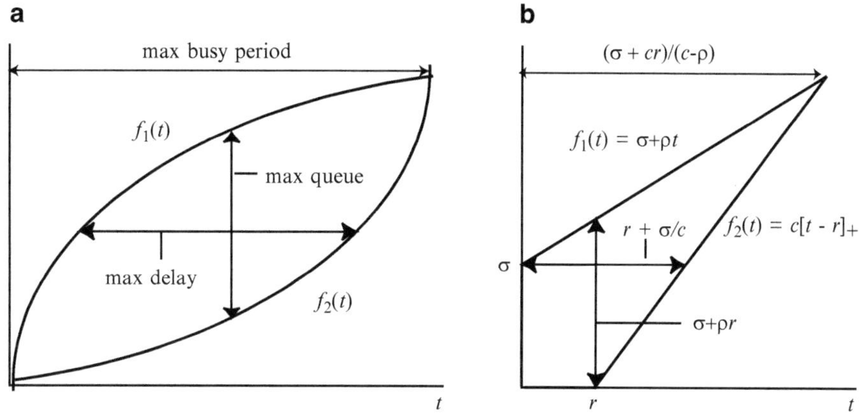

Fig. 2.2 Maximum queue size and delay: (**a**) general case, (**b**) Corollary 1

Definition 2. The arrival process A is (σ, ρ) upper-bounded if it is upper-bounded by $f_1(t) = \sigma + \rho t$. The service process C provides (c, r) service if it provides service $f_2(t) = c[t - r]_+$. One also says that A is bounded by rate ρ with burst size σ and C serves at rate c with delay r.

Theorem 1 is used in the paper in the simpler setting of Corollary 1.

Corollary 1. *Suppose A is (σ, ρ) upper-bounded, C provides service (c, r), and $c \geq \rho$. Then*

$$q(t) \leq \sigma + \rho \min\{t, r\} \leq (\sigma + \rho r), \tag{2.8}$$

$$B \text{ is } (\sigma + \rho r, \rho) \text{ bounded.} \tag{2.9}$$

The maximum queue size, delay, and busy period are bounded as

$$q(t) \leq \sigma + \rho r, \;\; d(t) \leq r + \sigma / c, \;\; BP \leq (\sigma + cr)/(c - \rho). \tag{2.10}$$

Proof. Using $f_1(t) = \sigma + \rho t$ and $f_2(t) = c[t - r]_+$ in (2.4), (2.5), (2.6), and (2.7) yield (2.8)–(2.10) as can be seen from Fig. 2.2b. □

2.3 Performance Bounds for a Single Intersection

Consider a signalized intersection with input links $l \in I$, and output links $m \in O$. A vehicle arriving on input link l can cross the intersection and move to one of several output links m. A *phase* is any movement, denoted by the associated input–output pair (l, m). Not every movement is permitted, e.g., U-turns or left turns may be prohibited. A set of phases may be simultaneously permitted. Such a set U is called a *stage*; $\mathcal{U}$ denotes the set of all stages.

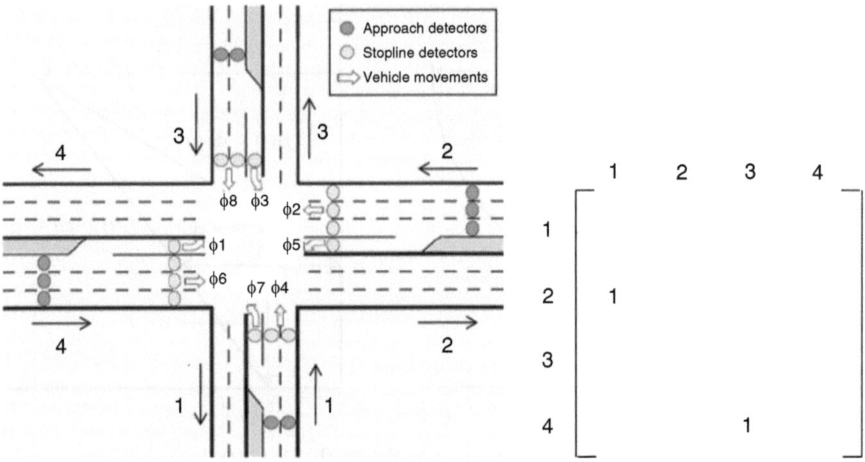

Fig. 2.3 The eight phases of a standard intersection (*left*) and the matrix representation for the stage $\{\phi_1, \phi_5\}$ (*right*)

For example, the standard intersection of Fig. 2.3 (left) has four input and four output links, both labeled $1,\ldots,4$; eight permitted phases, $\phi_1,\ldots,\phi_8$; and eight stages, each actuating two phases:

$$\{\phi_1,\phi_5\},\{\phi_1,\phi_6\},\{\phi_2,\phi_5\},\{\phi_2,\phi_6\},\{\phi_3,\phi_7\},\{\phi_3,\phi_8\},\{\phi_4,\phi_7\},\{\phi_4,\phi_8\}. \quad (2.11)$$

An intersection controller selects one stage $u(t) \in \mathcal{U}$ for each $t = 0, 1, \cdots$. If the sequence $u(t)$ is periodic, the controller is called *pre-timed* or *fixed-cycle* and the period T is the *cycle*. No vehicle movement is permitted for some portion of the cycle. This enforced idleness of duration L is required for pedestrian movement or for an amber light between successive transitions in $u(t)$. Thus within each cycle $T - L$ periods are available for vehicle movement.

A periodic control sequence selects stage $u(t) = U$ for duration $d_U \geq 0$ within each cycle. For performance analysis using network calculus, the parameters that matter are the durations $\{d_U, U \in \mathcal{U}\}$, whereas the order within a cycle in which $u(t)$ takes these values is not relevant. (The order is crucial in designing signal offsets.) Consequently, one may assume that each cycle is comprised of a fixed order of all phases; however, the duration of the phases may change from one cycle to the next.

The nonnegative durations must satisfy

$$\sum_{U \in \mathcal{U}} d_U \leq T - L. \quad (2.12)$$

A fixed-cycle controller is specified by the cycle T and durations $\{d_U\}$ satisfying (2.12).

We make two assumptions concerning the configuration of queues and service rates. First, vehicles entering input link l in order to make movement (l,m) join a *separate* queue dedicated to that movement. For example, the standard intersection of Fig. 2.3 has eight queues, one for each phase. A separate queue for each phase requires more space. For performance analysis, this assumption implies that vehicles intending different movements join different queues and do not block each other. Thus in Fig. 2.3, if the same queue was used by both phases ϕ_7 and ϕ_4, a vehicle intending to make a through movement ϕ_4 may be blocked by a vehicle in front of it intending to make a left turn ϕ_7. Such "head of line" blocking is precluded by this assumption. The loss of throughput due to head of line blocking could be evaluated as in the study of "input-buffered switches" (McKeown et al. 1993), but such an evaluation is not carried out here.

Second, it is assumed that whenever phase (l,m) is actuated, vehicles in queue (l,m) leave this queue at a known *saturation* rate of $s(l,m)$ vehicles per period, whereas if (l,m) is not actuated, no vehicle in this queue can leave. The saturation rate is associated with the phase and not with the stage. Thus in the standard intersection, in both stages $\{\phi_1,\phi_4\}$ and $\{\phi_1,\phi_6\}$, the queue associated with ϕ_1 is served at the same saturation rate.

2.3.1 *Analysis of a Single Movement*

The following notation is used.

$$
\begin{aligned}
(l,m) &= \text{ phase with input link } l \text{ and output link } m \\
s(l,m) &= \text{ saturation rate for phase } (l,m) \\
g(l,m)(r(l,m)) &= \text{ effective green (red) duration for phase } (l,m) \\
c(l,m)(t) &= \text{ service that phase } (l,m) \text{ receives in period } t \\
C(l,m)(t) &= \text{ cumulative service for phase } (l,m) \text{ up to } t
\end{aligned}
$$

Consider a fixed-cycle controller with cycle T and durations $\{d_U\}$ satisfying (2.12). Let $u(t), t \geq 0$, be the resulting periodic signal control sequence. The service $c(l,m)(t)$ that phase (l,m) receives depends on when $u(t)$ actuates the phase and its saturation rate:

$$
c(l,m)(t) = \begin{cases} s(l,m), & \text{if } (l,m) \in u(t) \\ 0, & \text{if } (l,m) \notin u(t) \end{cases} . \tag{2.13}
$$

The resulting cumulative service process is (with $C(l,m)(0) = 0$)

$$
C(l,m)(t) = \sum_{r=1}^{t} c(l,m)(r) = s(l,m) \sum_{r=1}^{t} \mathbf{1}[(l,m) \in u(r)],\ t \geq 1, \tag{2.14}
$$

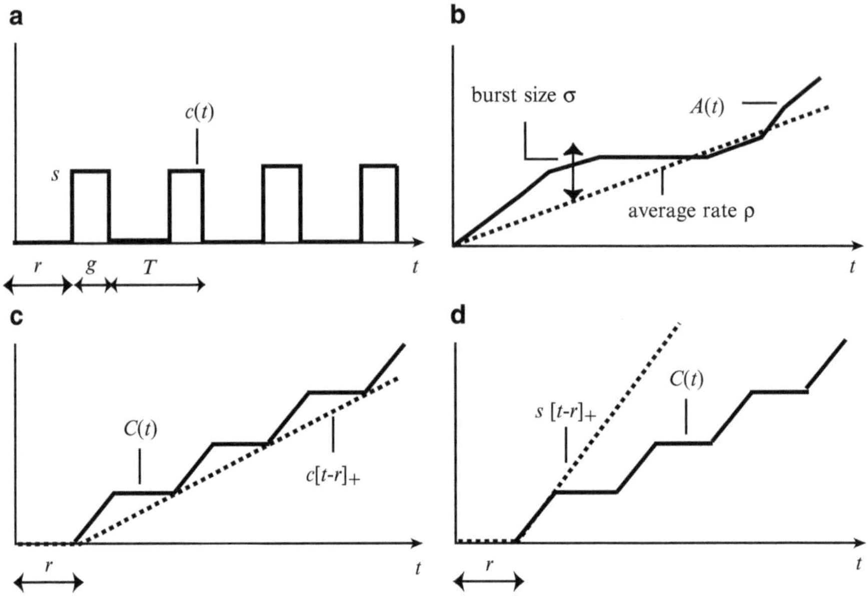

Fig. 2.4 (**a**) Service rate $c(t)$ for one phase: s is saturation rate, g, r, T are the durations of effective green, effective red, and cycle. (**b**) The cumulative arrival process $A(t)$ is (σ, ρ) upper-bounded. (**c**) The cumulative service process $C(t)$ provides service rate $c = sg/T$ with delay r. (**d**) $C(t,\tau) \geq s[t-\tau-r]_+$ for $t-\tau \leq T$

in which $\mathbf{1}[\cdot]$ is the indicator function. In each cycle phase (l,m) is actuated for a (green) duration $g(l,m)$, and it is not actuated for an effective (red) duration $r(l,m)$:

$$g(l,m) = \sum\{d_U \mid (l,m) \in U\}; \; r(l,m) = T - g(l,m). \tag{2.15}$$

So the average service rate for the queue at phase (l,m) is

$$\lim_{t\to\infty} C(l,m)(t)/t = s(l,m)g(l,m)/T.$$

Lemma 2. *The cumulative service process $C(l,m)(t)$ provides service rate $s(l,m)$ $g(l,m)/T$ with delay $r(l,m)$.*

Proof. Drop the phase index and write $C(t) = C(l,m)(t)$, $s = s(l,m)$, $g = g(l,m)$, $r = r(l,m)$, etc. Let $c = sg/T$ be the average service rate. Assisted by Fig. 2.4a and c one can see that if $0 \leq t-1-s = kT+\tau$ for some k and $\tau = (t-1-s)-kT$,

$$C(t-1,s) = C(t-1) - C(s) = ckT + \sum_{i=t-1-\tau}^{t-1} c(i) \geq ckT + s[\tau - r]_+ \geq c[t-s-r]_+, \tag{2.16}$$

so that $C(t)$ provides (c,r) service. □

Suppose arriving vehicles intending to move during phase (l,m) form the (σ,ρ) upper-bounded process $A(t)$:

$$A(t) - A(s) \leq \sigma + \rho(t-s),\ t \geq s. \tag{2.17}$$

Suppose

$$\rho \leq c = sg/T. \tag{2.18}$$

By Corollary 1 the queue size, the delay at the signal, and the busy period for this movement are then bounded by

$$q(t) \leq \sigma + \rho(T-g), \tag{2.19}$$

$$d(t) \leq (T-g) + \sigma/c, \tag{2.20}$$

$$BP \leq (\sigma + c(T-g))/(c-\rho). \tag{2.21}$$

The departure process $B(t)$ is $(\sigma+\rho(T-g),\rho)$ upper-bounded:

$$B(t) - B(s) \leq \sigma + \rho(T-g) + \rho(t-s). \tag{2.22}$$

Note that (2.16) and hence (2.19)–(2.22) hold even if the g duration is distributed anywhere within the cycle instead of contiguously as in Fig. 2.4a.

The model is simple. According to (2.17), vehicles arrive at average rate ρ and at most σ vehicles arrive in a "burst" or "platoon." If the average arrival rate is not more than the average service rate, the queue size is bounded by the maximum number of arrivals during an effective red, namely $\sigma+\rho(T-g)$; and the longest delay is faced by the last vehicle arriving in a burst of size σ just before red, namely $(T-g)+\sigma/c$. Lastly, the burst size of the departure process may exceed the arrival burst size σ by the number of vehicles $\rho(T-g)$ that can accumulate during red.

Suppose we know that the busy period never exceeds the cycle T, i.e., the queue clears in every cycle. In this case we can see from Fig. 2.4d or (2.16) that $C(t)$ provides the larger service $s[t-r]_+$, so in place of (2.21) we have the bound

$$BP \leq \frac{\sigma + sr}{s-\rho},$$

which is smaller than T if

$$\sigma + \rho T < gs. \tag{2.23}$$

For arrivals with no bursts, $\sigma = 0$, (2.23) reduces to $\rho < sg/T = c$, as in (Newell 1989, Eq. (2.1.6)). In reality, because of an upstream signal, the burst σ is likely to increase linearly with T. If $\sigma \approx \eta T$, (2.23) becomes

$$\eta + \rho < gs/T, \tag{2.24}$$

which requires a larger proportion of the cycle to be green than (2.18) in order to clear the queue in every cycle.

Remark. The parameters of the performance bounds (2.19)–(2.21) may be estimated if individual vehicle arrivals $a(t)$ are measured by a detector located sufficiently upstream of the signal so that the queue rarely reaches the location. Then:

Cumulative arrivals $A(t) = \sum_1^t a(\tau)$
Average arrival rate $\rho \approx A(t)/t$
Burst size $\sigma \approx \max_{s \le t}\{[\sum_s^t (a(\tau)] - \rho(t-s)]\}$
Service parameters g, r, T are known from the signal plan.

Estimating saturation rate s requires measurement of departures from the signal during green [see, e.g., (Kwong et al. 2009, §3.3)]. But note that the max pressure algorithm does *not* require knowledge of these parameters.

2.3.2 Analysis of All Movements at an Intersection

A stage U is henceforth represented by the binary $I \times O$ matrix U, with $U(l,m) = 1$ or 0 accordingly as U actuates phase (l,m) or not. (I (O) is the set of input (output) links at the intersection. See Fig. 2.3 (right).) $\mathcal{U}$ is the set of all stages or control matrices. Any signal controller is represented by a matrix sequence $u(t)$, $t \ge 0$, with values in $\mathcal{U}$.

Let $S = \{s(l,m),\ l \in I, m \in O\}$ denote the matrix of saturation rates of all phases. If phase (l,m) is not permitted, take $s(l,m) = 0$. The matrix $S \circ U$ defined by coordinate-wise multiplication, $(S \circ U)(l,m) = s(l,m)U(l,m)$, gives the service rates of all the phases simultaneously actuated by U.

Consider a fixed-cycle controller $u(t)$, $t \ge 0$, with cycle T. During each cycle $u(t)$ takes the value $U \in \mathcal{U}$ for duration d_U, with

$$\sum_U d_U \le T - L.$$

Expressed as proportions of the cycle, the durations

$$\lambda_U = d_U/T,\ U \in \mathcal{U},$$

satisfy

$$\sum_{U \in \mathcal{U}} \lambda_U \le 1 - L/T;\ \lambda_U \ge 0. \tag{2.25}$$

We identify a fixed-cycle controller with the array $[\lambda_U, U \in \mathcal{U}; T]$. In each cycle, this controller actuates phase (l,m) for an effective green duration

$$g(l,m) = T \sum_{U \in \mathcal{U}} \lambda_U U(l,m),$$

an effective red duration

$$r(l,m) = T - g(l,m) = T[1 - \sum_U \lambda_U U(l,m)],$$

and provides average service rate

$$c(l,m) = s(l,m)g(l,m)/T = (S \circ \sum_U \lambda_U U)(l,m).$$

By Lemma 2 the fixed-cycle controller $[\lambda_U, U \in \mathcal{U}; T]$ serves phase (l,m) at rate $c(l,m)$ with delay $r(l,m)$.

In a discrete-time setting, each duration $g(l,m)$ is an integer number of periods, so the proportions λ_U are multiples of $1/T$. If we allow the proportions to be arbitrary real numbers in $[0,1]$ the service that fixed-cycle controllers can provide is characterized by Theorem 2.

Theorem 2. *There is a fixed-cycle controller that serves each phase (l,m) at rate $c(l,m)$ with delay $r(l,m)$ if and only if there exist $\lambda_U \geq 0$, $\sum_U \lambda_U \leq 1 - L/T$ such that*

$$c(l,m) = \left(S \circ \sum_U \lambda_U U\right)(l,m), \; r(l,m) = T[1 - \left(\sum_U \lambda_U U\right)(l,m)]. \tag{2.26}$$

If vehicle arrivals for phase (l,m) are $(\sigma(l,m), \rho(l,m))$ upper-bounded and $\rho(l,m) \leq c(l,m)$, these vehicles will experience a queue size, delay, and busy period bounded by

$$q(l,m)(t) \leq \sigma(l,m) + \rho(l,m)r(l,m) \tag{2.27}$$

$$d(l,m) \leq r(l,m) + \sigma(l,m)/c(l,m) \tag{2.28}$$

$$BP(l,m) \leq [\sigma(l,m) + c(l,m)r(l,m)]/[c(l,m) - \rho(l,m)] \tag{2.29}$$

The departure process from phase (l,m) is bounded by rate $\rho(l,m)$ with burst size $\sigma(l,m) + \rho(l,m)r(l,m)$.

If the burst size $\sigma(l,m) = \eta(l,m)T$, the queue size bound is

$$q(l,m)(t) \leq T\left[\eta(l,m) + (1 - \left(\sum_U \lambda_U U\right)(l,m))\rho(l,m)\right], \tag{2.30}$$

and the queue is cleared in each cycle if

$$\eta(l,m) + \rho(l,m) \leq c(l,m) = \left(S \circ \sum_U \lambda_U U\right)(l,m). \tag{2.31}$$

Theorem 2 illustrates the use of network calculus. In a deterministic model that ignores bursts, $\sigma(l,m) = 0$, the stability condition is $\rho(l,m) \leq c(l,m)$ and so, by

(2.31), the queue must clear in every cycle; hence this deterministic model cannot explain why vehicles may wait at the intersection for one or more cycles, except by hypothesizing over-saturated traffic ($\rho(l,m) > c(l,m)$). By explicitly modeling bursts (which, in turn, may be due to a variety of conditions upstream of the intersection) (2.27) and (2.29) show how some vehicles may wait for a long time, even with undersaturated traffic ($\rho(l,m) < c(l,m)$). One way of explaining long delay with undersaturated traffic is to consider stochastic arrivals, whose variability creates bursts as in (Newell 1965). However, although there is no stochastic analysis of queues for a *network* of intersections, network calculus can be fruitfully used as will seen in Sect. 2.4.

A larger cycle T increases $(1 - L/T)$, so by (2.26) it increases the set of arrival rates $\rho(l,m)$ that can be accommodated, i.e., $\rho(l,m) \leq c(l,m)$. However, a larger T also increases the queue size bound (2.30), because it increases both the burst entering the queue (from upstream) and the red duration during which the queue grows (see (2.30)). Hence it is of interest to minimize T as in Corollary 2 which, for the no-burst case $\sigma(l,m) = 0$, is due to (Allsop 1972).

Corollary 2. *The shortest cycle needed by a fixed-cycle controller to accommodate all the arrivals bounded by rate* $\rho(l,m)$ *with burst size* $\sigma(l,m) = \eta(l,m)T$ *and clear all queues in every cycle is*

$$T = \frac{L}{1 - \sum \lambda_U^*}, \tag{2.32}$$

in which $\{\lambda_U^*\}$ *is the solution of the linear program:*

$$\begin{aligned} &\min \sum \lambda_U \\ &s.t.\ (S \circ \sum \lambda_U U)(l,m) \geq \eta(l,m) + \rho(l,m), \ \ all\ (l,m) \\ &\lambda_U \geq 0 \ \ all\ U \in \mathcal{U}. \end{aligned} \tag{2.33}$$

If $\sum \lambda_U^* > 1$, *no fixed-cycle controller can clear all queues in every cycle.*

Instead of minimizing the cycle, one can formulate a linear programming problem that minimizes (say) a linear combination of queue sizes, delays, or clearance times using (2.27)–(2.29), thereby extending the discussion in (Newell 1989, §2.2).

Remark. In the special case that each control value or stage U actuates only one phase (l,m), one may identify U with (l,m) and write $\lambda_U = \lambda_{(l,m)}$. The optimal solution to (2.33) is

$$\lambda_{(l,m)}^* = \frac{\rho(l,m) + \eta(l,m)}{s(l,m)},$$

and the shortest cycle is

$$T = \frac{L}{1 - \sum_{(l,m)}[(\rho(l,m) + \eta(l,m))/s(l,m)]},$$

which may be compared with Webster's rule.

2.3.3 Work-Conserving Controllers

A fixed-cycle controller $[\lambda_U, U \in \mathcal{U}; T]$ assigns the intersection to stage U for duration $\lambda_U T$ in each cycle. Consequently there will be time instants when stage $u(t) = U$ serves no queue even though there are nonempty queues at phases not served by U. To prevent this waste (which will lead to larger delays than necessary) the signal controller must select the control matrix as a function of the queue sizes, i.e., it must be traffic-responsive or in feedback form. Of special interest are work-conserving controllers, which are never idle when there is a nonempty queue. The controller still has a fixed cycle T, for L periods of which the intersection is not used by vehicles, but it need not be periodic.

We ignore the discrete-time restriction and allow $u(t)$ to take a value U for an arbitrary portion λ_U of a period. In effect at each t the controller selects $u(t)$ from the convex set $[\mathcal{U}]$:

$$[\mathcal{U}] = \left\{ \sum \lambda_U U \mid \lambda_U \geq 0, \sum \lambda_U \leq 1 - L/T \right\}. \tag{2.34}$$

Call $[\mathcal{U}]$ the set of *relaxed* controls. Let $u(t) = \sum \lambda_U(t) U$, $t \geq 0$, be a relaxed control sequence. Suppose vehicle arrivals $A_{(l,m)}(t)$ for phase (l,m) are $(\sigma(l,m), \rho(l,m))$ upper-bounded. These vehicles join queue (l,m), which therefore evolves as $(q_{(l,m)}(0) = 0)$

$$q_{(l,m)}(t+1) = [q_{(l,m)}(t) - \left(S \circ \sum \lambda_U(t) U\right)(l,m)]_+ + a_{(l,m)}(t+1),\ t \geq 0. \tag{2.35}$$

Here $a_{(l,m)}(t) = A_{(l,m)}(t) - A_{(l,m)}(t-1)$.

Definition 3. The controller $u(t) = \sum \lambda_U(t) U$, $t \geq 0$, is *work-conserving* if

$$\begin{aligned} &\exists U,\ \forall (l,m) \text{ with } U(l,m) = 1 : q_{(l,m)}(t) - (S \circ \sum \lambda_U(t) U)(l,m) < 0 \\ &\Rightarrow \forall (l,m) : q_{(l,m)}(t) - (S \circ \sum \lambda_U(t) U)(l,m) \leq 0. \end{aligned} \tag{2.36}$$

In words: control U may waste service in every phase that U actuates only if no phase has a nonzero queue.

Definition 4. The controller $u(t) = \sum_U \lambda_U(t) U$, $t \geq 0$, is *stabilizing* if all queues are bounded:

$$\max_{(l,m)} \sup_{t \geq 0} q_{(l,m)}(t) < \infty.$$

2.3.3.1 Actuating Single Phase Intersections

This section is devoted to single-phase intersections, in which each stage U actuates only one phase, say (l,m), $U = \delta_{(l,m)}$ is the $I \times O$ matrix whose (l,m)th entry is 1 and other entries are 0. A relaxed control matrix has the form $\sum \lambda_{(l,m)} \delta_{(l,m)}$. Let $u(t) = \sum \lambda_{(l,m)}(t)\delta_{(l,m)}$, $t \geq 0$, be the control sequence. Then (2.35) simplifies:

$$q_{(l,m)}(t+1) = [q_{(l,m)}(t) - s(l,m)\lambda_{(l,m)}(t)]_+ + a_{(l,m)}(t+1),\ t \geq 0. \tag{2.37}$$

Here $s(l,m)$ is the saturation rate for phase (l,m). Equation (2.36) also simplifies: $u(t) = \sum \lambda_{(l,m)}(t)\delta_{(l,m)}, t \geq 0$, is work-conserving if

$$\begin{aligned} &\exists (l,m):\ q_{(l,m)}(t) - s(l,m)\lambda_{(l,m)}(t) < 0 \\ \Rightarrow\ &\forall (l,m):\ q_{(l,m)}(t) - s(l,m)\lambda_{(l,m)}(t) \leq 0. \end{aligned} \tag{2.38}$$

Let $u(t)$, $t \geq 0$, be work-conserving and define the weighted total queue size

$$q(t) = \sum_{(l,m)} \frac{q_{(l,m)}(t)}{s(l,m)}.$$

From (2.37)

$$q(t+1) = \sum \left[\frac{q_{(l,m)}(t)}{s(l,m)} - \lambda_{(l,m)}(t)\right]_+ + \sum \frac{a_{(l,m)}(t+1)}{s(l,m)}. \tag{2.39}$$

Because of (2.38) terms within the square brackets [ ] all have the same sign, and so

$$q(t+1) = \left[\sum\left(\frac{q_{(l,m)}(t)}{s(l,m)} - \lambda_{(l,m)}(t)\right)\right]_+ + \sum \frac{a_{(l,m)}(t+1)}{s(l,m)} = [q(t) - c(t)]_+ + a(t+1),$$

in which $a(t) = \sum[a_{(l,m)}(t)/s(l,m)]$, and

$$c(t) = \sum \lambda_{(l,m)}(t) = 1 - L/T. \tag{2.40}$$

Theorem 3. *The weighted arrivals $A(t) = \sum[A_{(l,m)}(t)/s(l,m)]$ are upper-bounded by rate $\rho = \sum[\rho(l,m)/s(l,m)]$ with burst size $\sigma = \sum[\sigma(l,m)/s(l,m)]$. The cumulative service $C(t) = \sum_{s \leq t} c(s)$ serves at rate $1 - L/T$ with delay L. If*

$$\rho = \sum[\rho(l,m)/s(l,m)] \leq 1 - L/T, \tag{2.41}$$

the size, the delay at the signal, and the busy period of the weighted queue are bounded by

$$q(t) \leq \sigma + \rho L, \tag{2.42}$$

$$d(t) \leq L + \sigma/[1 - L/T], \tag{2.43}$$

$$BP \leq [\sigma + (1 - L/T)L][1 - L/T - \rho]\,. \tag{2.44}$$

If the bursts $\sigma(l,m) = \eta(l,m)T$ *and*

$$\rho + [\sum \eta(l,m)/s(l,m)] \leq 1 - L/T, \tag{2.45}$$

then every queue will be cleared in every cycle.

Consequently, if any fixed-cycle controller is stabilizing, then every work-conserving controller is also stabilizing.

Proof. First, for $s \leq t$,

$$A(t) - A(s) = \sum \frac{A_{(l,m)}(t) - A_{(l,m)}(s)}{s(l,m)} \leq \sum [\sigma(l,m) + \rho(l,m)(t-s)]/s(l,m) = \sigma + \rho(t-s).$$

Next, by (2.40) and Lemma 2, $C(t)$ serves at rate $[1 - L/T]$ with delay L. Then (2.18)–(2.21) translate into (2.41)–(2.44), and (2.24) into (2.45). The last assertion follows from Theorem 2. □

Consider the simplest example of an intersection with two phases, only one of which can be actuated at any time. The controller in (Mirchandani and Zou 2007) actuates one phase until its queue is empty, whereupon it switches to the other phase. The controller in (Lin and Lo 2008) switches from phase 1 to phase 2 accordingly as the ratio of the queues $q_1(t)/q_2(t)$ drops below or exceeds a desired ratio. In network calculus terms this ratio is analogous to $[(\rho_1 + \eta_1)/s_1]/[(\rho_2 + \eta_2)/s_2]$. One may consider a third controller that gives priority to say, phase 1, and actuates that phase whenever $q_1(t) > 0$; otherwise it actuates phase 2. (Priorities may be used for buses or emergency vehicles.) These three controllers are all work-conserving, and Theorem 2 gives the same bounds on the weighted queue size and delay. Of course, bounds on individual queue lengths will be different for each controller: for example, the queue at phase 1 will have the smallest bound for the third controller that gives priority to phase 1.

In (Mirchandani and Zou 2007) and (Lin and Lo 2008) arrivals are Poisson processes, and evaluating performance measures such as queue size and delay ultimately requires simulation, although (Mirchandani and Zou 2007) also provides an analytical approximation. The complexity of the analysis and simulations grows exponentially with the number of phases. By contrast, network calculus provides simple computable bounds for arbitrarily many phases. Furthermore, when we consider a *network* of intersections, arrivals are not Poisson and standard stochastic queuing approaches are inapplicable, even though network calculus bounds can be constructed as in Sect. 2.4.

Condition (2.45) to clear all queues is significantly weaker than its counterpart in (2.33) for fixed-cycle controllers, and shows the benefit of work-conserving controllers. In fact the following result proved in Appendix C.

Theorem 4. *Let* $Q^w(t) = \{q^w_{(l,m)}(t)\}$ *be the queues for any work-conserving controller and let* $Q(t) = \{q_{(l,m)}(t)\}$ *be the queues for any controller, with* $Q^w(0) = Q(0) = 0$. *Then for all* t,

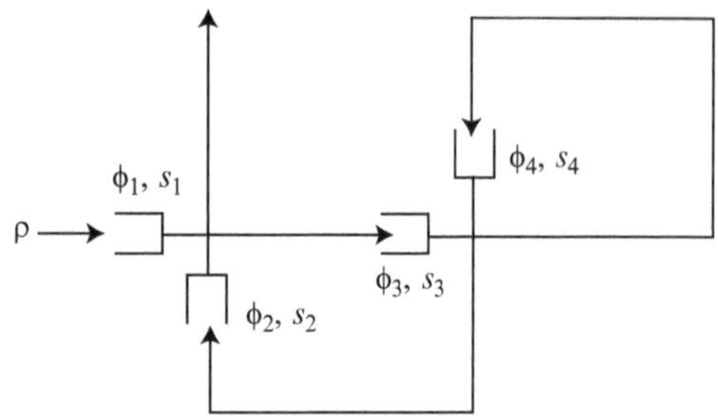

Fig. 2.5 One phase can be actuated at a time in each intersection

$$\sum_{(l,m)} \frac{q^w_{(l,m)}(t)}{s(l,m)} \leq \sum_{(l,m)} \frac{q_{(l,m)}(t)}{s(l,m)}. \tag{2.46}$$

2.3.3.2 Two Counter-Examples

The first example shows that Theorem 3 does not extend to a network of two intersections in which only one phase is actuated in each stage. In the network of Fig. 2.5 phases ϕ_1 and ϕ_4 are fast, with saturation rates $s_1 = s_4 = \infty$; ϕ_2 and ϕ_3 are slow, with $s_2 = s_3 = 1.5$. The arrival rate is $\rho = 1$, with no bursts. $T = 1, L = 0$. Clearly there is a stabilizing fixed-time controller for this network. Now consider work-conserving controllers that give priority to the slow phases, ϕ_2, ϕ_3, i.e., these phases are served immediately if they have a nonempty queue. Consider the initial condition: $q_1(0) = 1$, $q_2(0) = q_3(0) = q_4(0) = 0$. One can check that

$$q_1(4) = 2,\ q_1(8) = 4, \cdots, q_1(4n) = 2n,\ n \geq 1,$$

so that these controllers are unstable. This example is from (Lu and Kumar 1991). There are also examples that do not require infinite service rates, but these are more complex to describe, see, e.g., (Dai 1995).

The second example shows that Theorem 3 does not extend to the case of the isolated intersection of Fig. 2.3 in which multiple phases may be actuated simultaneously. The intersection depicted in Fig. 2.6 only includes part of the standard intersection. (The example obviously extends to the standard intersection.)

There are four phases and three stages, each actuating one phase pair (cf (2.11)):

$$\{\phi_1, \phi_2\}, \{\phi_3, \phi_4\}, \{\phi_2, \phi_4\}.$$

The cumulative arrivals at each phase have the same rate ρ with burst size σ. The saturation rate at all phases is the same, $s = 1$. Let $\alpha = [1 - L/T]$. Consider the fixed-cycle controller that actuates phase pairs $\{\phi_1, \phi_2\}$ and $\{\phi_3, \phi_4\}$ each for half the time, i.e., for duration $0.5[T - L] = 0.5\alpha T$ in each cycle. By Theorem 2, this controller serves every phase at rate 0.5α with delay $0.5[T + L]$ and if

$$\rho \leq 0.5\alpha, \tag{2.47}$$

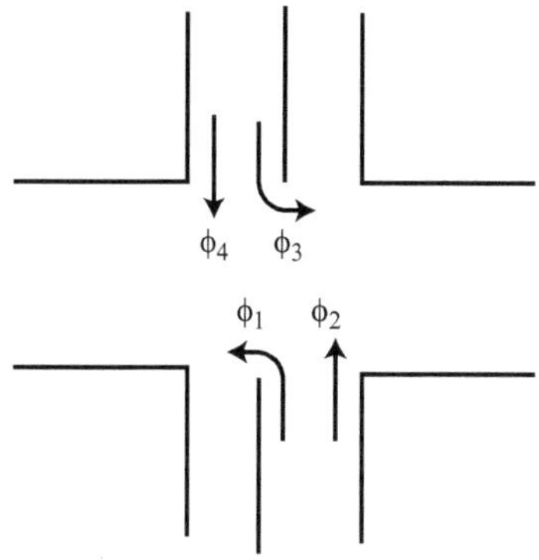

Fig. 2.6 The intersection permits four of eight phases of the standard intersection

the controller is stabilizing, and the queue in every phase is bounded by

$$q(t) \leq \sigma + \rho \times 0.5[T + L].$$

There will be instants when vehicles simultaneously arrive for phases ϕ_2 and ϕ_4. Suppose this occurs at rate $\delta > 0$. Formally:

$$\sum_{s<i\leq t} \mathbf{1}[a_2(i) > 0 \text{ and } a_4(i) > 0] \geq \delta(t - s). \tag{2.48}$$

Now consider *any* controller $u(t)$, $t \geq 0$, that selects stage $\{\phi_2, \phi_4\}$ whenever both $q_2(t) > 0$ and $q_4(t) > 0$, i.e., $\{\phi_2, \phi_4\}$ gets priority in the event that vehicles are queued up at both phases. Because of the priority and (2.48), $\{\phi_2, \phi_4\}$ receives service for duration at least δT in each cycle; hence the two remaining pairs $\{\phi_1, \phi_2\}, \{\phi_3, \phi_4\}$ together will receive service for duration at most $T - L - \delta T$. Consequently one of these two pairs, say $\{\phi_1, \phi_2\}$, will receive service for duration at most $0.5(T - L - \delta T)$ in each cycle that is at rate at most $0.5(1 - L/T - \delta) = 0.5(\alpha - \delta)$. Comparison with (2.47) shows that if

$$0.5(\alpha - \delta) < \rho \leq 0.5\alpha, \tag{2.49}$$

every controller with this priority is *unstable* and the queue length at phase ϕ_1 must become unbounded! Note that any controller that always serves a nonempty queue while keeping this priority is work-conserving.

A controller that gives priority to $\{\phi_2, \phi_4\}$ if either $q_2(t) > 0$ or $q_4(t) > 0$ will have worse performance, since the instability condition (2.49) is replaced by the weaker inequality,

$$0.5(\alpha - \rho) < \rho \leq 0.5\alpha.$$

Such a controller is commonly used to give priority to buses (Chada and Newland 2002).

It is easy to construct a stable work-conserving controller by modifying any stable fixed-cycle controller so that it actuates a phase with a nonempty queue whenever the controller becomes idle. (This recalls the common practice of terminating the green phase on a "cross street" when there is no queue.) However, this controller is *not* adaptive, since constructing a stable fixed-cycle controller requires knowledge of the demands. This suggests the following problem: *Construct a stable, adaptive, work-conserving controller*. The problem is solved in the next section for an isolated intersection.

2.3.3.3 The Adaptive Controller Problem

Here is the precise problem. For a relaxed control sequence $u(t) = \sum \lambda_U(t)U, t \geq 0$, the evolution of the intersection's queues is given by ($q_{(l,m)}(0) = 0$)

$$q_{(l,m)}(t+1) = \left[q_{(l,m)}(t) - \left(S \circ \sum \lambda_U(t)U\right)(l,m)\right]_+ + a_{(l,m)}(t+1), \ t \geq 0. \quad (2.50)$$

Let q denote the array $\{q_{(l,m)}\}$ of all the queues. The problem is to find a function $\lambda_U^*(q)$ of q such that the feedback control sequence $u(t) = \sum \lambda_U^*(q(t))U$ stabilizes the queues for any set of demands for which a stabilizing fixed-cycle controller exists. We exhibit such a feedback control.

Define the *pressure exerted by stage U* at q by

$$w(q,U) = \sum_{(l,m)} q_{(l,m)} S \circ U(l,m) = \sum_{(l,m)} q_{(l,m)} s(l,m) U(l,m), \quad (2.51)$$

i.e., it is the sum of the queue lengths multiplied by the saturation rates of the phases that U actuates. Extend linearly the pressure to any relaxed control $[U] = \sum \lambda_U U$,

$$w(q,[U]) = \sum \lambda_U w(q,U) = \sum q_{(l,m)} s(l,m)[U](l,m).$$

Define the *max-pressure stage* by

$$U^*(q) = \arg\max\{w(q,U) \mid U \in \mathcal{U}\}. \quad (2.52)$$

In (2.52) ties are broken arbitrarily. The name "max-pressure policy" was apparently first introduced in (Dai and Lin 2005), although similar policies were studied earlier; Tassiulas and Ephremides (1992) was the first study to investigate its stability properties in the context of wireless networks.

Definition 5. The *max-pressure controller* $u^*(t)$ selects the max-pressure stage at $q(t)$:

$$u^*(t) = (1 - L/T)U^*(q(t)).$$

Lemma 3. *$u^*(t)$ maximizes $w(q,[U])$ over the set $[\mathcal{U}]$ of relaxed controls.*

Proof. $w(q,[U])$ is linear in $[U]$ and $[\mathcal{U}]$ is the convex hull of its vertices $\{(1-L/T)U,\ U \in \mathcal{U}\}$. Hence the maximum of $w(q,[U])$ is achieved at $(1-L/T)U^*(q)$. □

Theorem 5. *Let $q(t)$ be the queues resulting from the max-pressure controller:*

$$q_{(l,m)}(t+1) = [q_{(l,m)}(t) - (S \circ (1-L/T)U^*(q(t)))(l,m)]_+ + a_{(l,m)}(t+1),\ \ t \geq 0. \tag{2.53}$$

Suppose that in (2.53) the arrivals $A_{(l,m)}$ are $(\sigma(l,m),\rho(l,m))$ upper-bounded and there exists a (fixed-cycle) relaxed control $[U]$ such that

$$c(l,m) = S \circ [U](l,m) > \rho(l,m),\ \text{all}\ (l,m). \tag{2.54}$$

Then $\{q(t),\ t \geq 0\}$ is a bounded sequence, i.e., the max-pressure controller is stabilizing.

Proof. Write $c^*(l,m)(t) = (S \circ (1-L/T)U^*(q(t)))(l,m)$, so under the max-pressure controller

$$q_{(l,m)}(t+1) = [q_{(l,m)}(t) - c^*(l,m)(t)]_+ + a_{(l,m)}(t+1). \tag{2.55}$$

For any q let $|q|^2 = \sum q^2_{(l,m)}$. It is shown in Appendix D that there exist $k < \infty$, $\varepsilon > 0$, and $\sigma(t) \geq 0$ with $\sum_t \sigma(t) < \infty$, so that

$$|q(t+1)|^2 - |q(t)|^2 \leq k - (2\varepsilon - \sigma(t))|q(t)|, \tag{2.56}$$

Suppose (2.56) holds. With T such that $\sigma(t) < \varepsilon$, $t \geq T$, (2.56) gives

$$|q(t+1)|^2 - |q(t)|^2 \leq k - \varepsilon|q(t)|,\ t > T,$$

and so

$$|q(t+1)|^2 - |q(t)|^2 < 0,\ |q(t)| > k/\varepsilon,\ t > T,$$

which implies that $|q(t)|$, $t \geq 0$, is bounded. □

The max-pressure controller is adaptive since it requires no knowledge of the parameters $(\sigma(l,m),\rho(l,m))$ of the arrival processes. It is robust in the sense that if any controller can keep queues bounded, so can the max-pressure controller. From the proof of Theorem 5 one gains the intuition that the max-pressure controller attempts at each t to minimize $|q(t+1)|^2$ given $q(t)$.

2.4 Performance Bounds for a Network of Intersections

The model of a network of signalized intersections is formulated in Sect. 2.4.1. The performance bounds of Corollary 1 are applied to the network with fixed-cycle controllers in Sect. 2.4.2. The extension of the max-pressure controller to an arbitrary network is carried out in Sect. 2.4.3.

2.4.1 Network Model

This section is based on (Chang 2000, §1.7). The concept of *router* is needed to extend the discussion of Sect. 2.2 to a network of intersections. A router $P \in \mathcal{F}$ is a network element with cumulative arrivals $A \in \mathcal{F}$ and departures $B \in \mathcal{F}$ given by $B(t) = P(A(t))$ for all t. The interpretation is that the router selects or samples $P(n)$ among its first n arrivals so that $B(t) = P(A(t))$ is the cumulative number of selections by time t. Routers are used to model turn movements.

Suppose A is (σ, ρ) upper-bounded, and P is (δ, γ) upper-bounded. Since

$$B(t) - B(s) = P(A(t)) - P(A(s)) \leq \delta + \gamma(A(t) - A(s)) \leq (\delta + \gamma\sigma) + \gamma\rho(t-s),$$

it follows that B is $(\gamma\sigma + \delta, \gamma\rho)$ upper-bounded.

Figure 2.7 will help explain the notation and the model.

$$\begin{aligned}
\mathcal{L} = \{1, \cdots, L\} &= \text{ set of all links, elements } l, m, k \\
\mathcal{N} &= \text{ set of nodes or intersections, elements } n \\
I_n &\subset \mathcal{L}, \text{ set of input links to } n \in \mathcal{N} \\
O_n &\subset \mathcal{L}, \text{ set of output links from } n \in \mathcal{N}
\end{aligned}$$

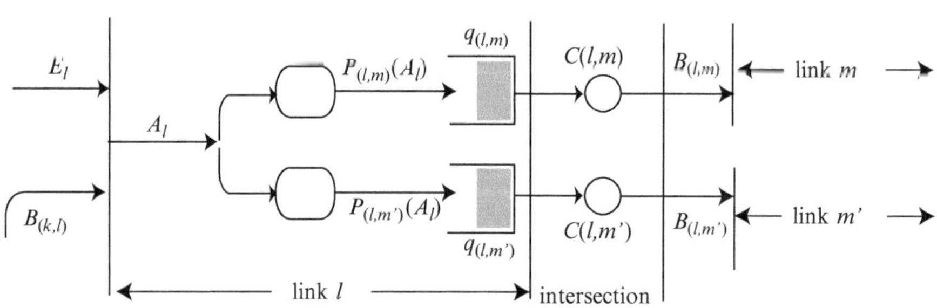

Fig. 2.7 E_l are external arrivals into link l, $B_{(k,l)}$ are internal arrivals routed from link k to l, and $C_{(l,m)}$ is the service process for phase (l,m)

$$
\begin{aligned}
E_l, (\alpha_l, \beta_l) &= \text{ external arrivals into link } l, \ (\alpha_l, \beta_l) \text{ upper-bounded} \\
B_{(l,m)} &= \text{ departures from } l \text{ to } m \\
C(l,m) &= \text{ service for phase } (l,m) \text{ at rate } c(l,m) \text{ with delay } r(l,m) \\
A_l &= E_l + \sum_k B_{(k,l)} \text{ total arrivals into link } l \\
P_{(l,m)}, (\delta_{(l,m)}, \gamma_{(l,m)}) &= \text{ router from link } l \text{ to link } m, \ (\delta_{(l,m)}, \gamma_{(l,m)}) \text{ upper-bounded} \\
q_{(l,m)} &= \text{ queue (in link } l\text{) for phase } (l,m) \\
s(l,m) &= \text{ saturation rate of phase } (l,m)
\end{aligned}
$$

Although $P_{(l,m)}, B_{(l,m)}, C(l,m), q_{(l,m)}$, etc. are only defined for permissible phases, it will be convenient to define them for all $(l,m) \in \mathcal{L} \times \mathcal{L}$ by setting their values to 0 for phases that are not permitted.

Suppose A_l is (σ_l, ρ_l) upper-bounded (σ_l, ρ_l are determined below). Then $P_{(l,m)}(A_l)$ is $(\delta_{(l,m)} + \gamma_{(l,m)}\sigma_l, \gamma_{(l,m)}\rho_l)$ upper-bounded. Suppose $C(l,m)$ provides service $(c(l,m), r(l,m))$ with $c(l,m) \geq \gamma_{(l,m)}\rho_l$. By Corollary 1

$$q_{(l,m)}(t) \leq \delta_{(l,m)} + \gamma_{(l,m)}\sigma_l + \gamma_{(l,m)}\rho_l r(l,m),$$

$$B_{(l,m)} \text{ is } (\delta_{(l,m)} + \gamma_{(l,m)}\sigma_l + \gamma_{(l,m)}\rho_l r(l,m), \gamma_{(l,m)}\rho_l) \text{ upper-bounded},$$

$$d_{(l,m)}(t) \leq r(l,m) + [\delta_{(l,m)} + \gamma_{(l,m)}\sigma_l]/c(l,m),$$

$$BP_{(l,m)} \leq [\delta_{(l,m)} + \gamma_{(l,m)}\sigma_l + c(l,m)r(l,m)]/[c(l,m) - \gamma_{(l,m)}\rho_l]. \tag{2.57}$$

So $A_l = E_l + \sum_k B_{(k,l)}$ is (σ_l, ρ_l) upper-bounded with

$$\sigma_l = \alpha_l + \textstyle\sum_k [\delta_{(k,l)} + \gamma_{(k,l)}\sigma_k + \gamma_{(k,l)}\rho_k r(k,l)], \tag{2.58}$$

$$\rho_l = \beta_l + \textstyle\sum_k \gamma_{(k,l)}\rho_k. \tag{2.59}$$

It is convenient to use vector–matrix notation. All vectors below are *row* vectors of dimension L and all matrices are of dimensions $L \times L$.

Let $\alpha = (\alpha_1, \cdots, \alpha_L)$, $\beta = (\beta_1, \cdots, \beta_L)$, $\sigma = (\sigma_1, \cdots, \sigma_L)$, $\rho = (\rho_1, \cdots, \rho_L)$. Define matrices $\Gamma = \{\gamma_{(l,m)}\}$, $\Delta = \{\delta_{(l,m)}\}$, $R = \{r(l,m)\}$, $\Gamma \circ R = \{\gamma_{(l,m)} r(l,m)\}$. Let $e = (1, \cdots, 1)$ be the row vector with all entries 1, and $\delta = e\Delta$. Write (2.58)–(2.59) as

$$
\begin{aligned}
\sigma &= \alpha + \delta + \sigma\Gamma + \rho\Gamma \circ R \\
\rho &= \beta + \rho\Gamma
\end{aligned}
$$

Assume that for all l, $\sum_m \gamma_{(l,m)} \leq 1$, and the spectral radius of Γ, which equals its maximum eigenvalue, is strictly less than 1. (This is equivalent to the condition that every vehicle eventually leaves the network.) Then

$$[I-\Gamma]^{-1} = I+\Gamma+\Gamma^2+\cdots,$$

$$\rho = \beta+\rho\Gamma = \beta[I-\Gamma]^{-1}, \tag{2.60}$$

$$\sigma = \alpha+\delta+\rho\Gamma\circ R+\sigma\Gamma = (\alpha+\delta+\rho\Gamma\circ R)[I-\Gamma]^{-1}. \tag{2.61}$$

Let $q = \{q_{(l,m)}\}$, $C = \{c(l,m)\}$, $B = \{B_{(l,m)}\}$, and let $[\sigma],[\rho]$ denote diagonal matrices with entries σ_l, ρ_l.

Lemma 4. *Suppose the external arrivals E_l are (α_l, β_l) upper-bounded, the spectral radius of the routing matrix Γ is strictly less than 1, and $C(l,m)$ provides service $(c(l,m), r(l,m))$ with $c(l,m) \geq \gamma_{(l,m)}\rho_l$. Then, with $A = (A_1,\cdots,A_L)$, $B = \{B_{(l,m)}\}$, $q = \{q_{(l,m)}\}$, the following bounds hold:*

$$A(t) \text{ is } (\sigma,\rho) \text{ upper-bounded}, \tag{2.62}$$

$$B(t) \text{ is } (\Delta+[\sigma]\Gamma+[\rho]\Gamma\circ R, [\rho]\Gamma) \text{ upper bounded}, \tag{2.63}$$

$$q(t) \leq \Delta+[\sigma]\Gamma+[\rho]\Gamma\circ R, \tag{2.64}$$

$$d_{(l,m)}(t) \leq r(l,m)+[\delta_{(l,m)}+\sigma_l\gamma_{(l,m)}]/c(l,m), \tag{2.65}$$

$$BP_{(l,m)} \leq [\delta_{(l,m)}+\gamma_{(l,m)}\sigma_l+c(l,m)r(l,m)]/[c(l,m)-\gamma_{(l,m)}\rho_l]. \tag{2.66}$$

Above ρ and σ are given by (2.60) and (2.61).

2.4.2 Performance of Fixed-Cycle Controller

We extend the notation of Sect. 2.3 for a single controller to that for a network, and use Lemma 4 to design fixed-cycle controllers for the network.

A node or intersection $n \in \mathcal{N}$ is specified by input links $l \in I_n$, output links $m \in O_n$, and a set of $I_n \times O_n$ binary control matrices $U_n \in \mathcal{U}_n$ representing all permissible stages at n. Let $S_n = \{s(l,m),\ l \in I_n, m \in O_n\}$ be the matrix of saturation rates of all phases at intersection n, with $s(l,m) = 0$ if (l,m) is not permitted. If U_n is the stage selected at intersection n, the matrix $S_n \circ U_n$ is the matrix of service rates at t provided by U_n to the phases at n.

We can take the "direct product" of the control matrices U_n at each intersection n to obtain a *network stage* matrix $U = \prod_n U_n$ of dimension $L \times L$ for the entire network,

$$U(l,m) = \begin{cases} U_n(l,m), & \text{if } (l,m) \in I_n \times O_n \\ 0, & \text{otherwise} \end{cases}.$$

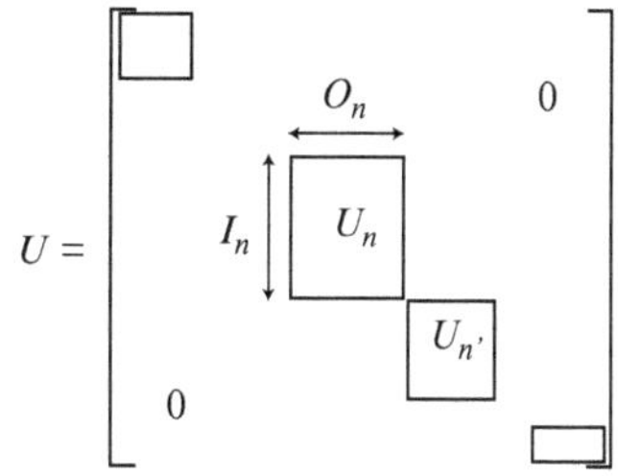

Fig. 2.8 A network stage matrix U is a block-diagonal matrix with intersection stage matrices $U_n, U_{n'}$ along the diagonal

One may picture U in block-diagonal form as in Fig. 2.8. Analogously, let $S = \prod_n S_n$ be the $L \times L$ matrix of the saturation rates of all phases (l,m). Let $\mathcal{U} = \prod \mathcal{U}_n$ be the set of all network stage matrices. We now proceed as in Sect. 2.3.2. For simplicity, assume that all intersections have the same cycle T and the same lost time L. (Having different cycles at different intersections only complicates the notation.) Let

$$[\mathcal{U}] = \left\{ \sum_U \lambda_U U \mid \lambda_U \geq 0, \sum_U \lambda_U \leq 1 - L/T \right\}, \tag{2.67}$$

be the set of all relaxed controls. Theorem 6 is the network counterpart of Theorem 2.

Theorem 6. *There is a fixed-cycle network controller that serves each phase (l,m) at rate $c(l,m)$ with delay $r(l,m)$ if and only if there exist $\lambda_U \geq 0$, $\sum_U \lambda_U \leq 1 - L/T$ such that*

$$c(l,m) = \left(S \circ \sum_U \lambda_U U \right)(l,m); \quad r(l,m) = T \left[1 - \left(\sum_U \lambda_U U \right)(l,m) \right]. \tag{2.68}$$

Suppose the external arrivals E_l are (α_l, β_l) upper-bounded, and $c(l,m) \geq \rho_l \gamma_{(l,m)}$ for every (l,m), or in matrix notation

$$S \circ \sum_U \lambda_U U \geq [\rho]\Gamma = [\beta[I - \Gamma]^{-1}]\Gamma. \tag{2.69}$$

Then the performance of the controller satisfies the bounds (2.62)–(2.66).

The external arrival rates β are "inflated" by the routing matrix multiplier $[I - \Gamma]^{-1}$ to give the aggregate arrival rates $\rho = \beta[I - \Gamma]^{-1}$, as is to be expected. More interesting is the impact of routing on transforming the bursts α in the external arrivals into the bursts σ in (2.61). Both impacts affect the maximum queue size, delay, and clearance times (2.64)–(2.66). Corollary 3 is the network counterpart of Corollary 2.

Corollary 3. *The shortest cycle needed by a stabilizing fixed-cycle controller for all external arrivals A_l bounded by rate β_l with burst size α_l is*

$$T = \frac{L}{1 - \sum \lambda_U^*}, \tag{2.70}$$

in which $\{\lambda_U^*\}$ is the solution of the linear program:

$$\begin{aligned} &\min \sum \lambda_U \\ &s.t. (S \circ \sum \lambda_U U)(l,m) \geq \left[\beta[I - \Gamma]^{-1}\Gamma\right](l,m), \ all\ (l,m) \\ &\lambda_U \geq 0 \ \ all\ U \in \mathcal{U}. \end{aligned} \tag{2.71}$$

If $\sum \lambda_U^* > 1$, there is no stabilizing fixed-cycle controller.

Because U and S are block-diagonal, the linear program decomposes into a set of independent linear programs, one per intersection.

One of the $L \times L$ inequalities in (2.71), corresponding to say (l^*, m^*), will hold as an *equality* in the solution. One could call (l^*, m^*) the *critical phase* and the intersection n^* for which $l^* \in I_{n^*}, m^* \in O_{n^*}$ as the *critical intersection*. Following (Allsop 1972) one may also define the capacity of this intersection.

2.4.3 Max-Pressure Controller

The max-pressure controller of Sect. 2.3.3.3 is extended to a network in this section. Define the *pressure exerted by network stage U* at $q = \{q_{(l,m)}\}$ by

$$w(q,U) = \sum_{(l,m)} \left[q_{(l,m)} - \sum_p \gamma_{(m,p)} q_{(m,p)} \right] S \circ U(l,m), \tag{2.72}$$

and linearly extend the definition to $[U] = \sum \lambda_U U$,

$$w(q,[U]) = \sum \lambda_U w(q,U) = \sum_{(l,m)} \left[q_{(l,m)} - \sum_p \gamma_{(m,p)} q_{(m,p)} \right] S \circ [U](l,m).$$

This definition of pressure differs from (2.51) in that for each phase (l,m) we take the product of its queue length $q_{(l,m)}$ and saturation rate $s(l,m)$ and *subtract* the corresponding amount from the downstream queue $q_{(m,p)}$ weighted by the average turn ratio $\gamma_{(m,p)}$. For the isolated intersection considered in Sect. 2.3.3.3 with no downstream queue, (2.72) reduces to (2.51).

Note that to calculate the pressure (2.72) exerted by a network stage one needs to know the turn ratios $\{\gamma_{(l,m)}\}$ in addition to the queue lengths. (It is of course easy to estimate turn ratios.) However, no knowledge of the parameters (α_l, β_l) of the external demands E_l is needed.

Define the *max-pressure stage* as

$$U^*(q) = \arg\max\{w(q,U) \mid U \in \mathcal{U}\}. \tag{2.73}$$

In (2.73) ties are broken arbitrarily. Let $q(t) = \{q_{(l,m)}(t)\}$ be the queue length array.

Definition 6. The *max-pressure network controller* $u^*(t)$ selects the max-pressure stage at $q(t)$:

$$u^*(t) = (1 - L/T)U^*(q(t)).$$

The pressure (2.72) of a network stage is the sum of the pressures exerted at each intersection stage, so the max-pressure network stage (2.73) is simply the collection of the max-pressure stages at all the intersections. Hence, the max-pressure controller is *decentralized*. If the network of intersections is expanded, the max-pressure controller for the original network is unchanged, so the max-pressure controller can be introduced *incrementally*.

The proof of Lemma 5 is identical to that of Lemma 3.

Lemma 5. *$u^*(t)$ maximizes $w(q,[U])$ over the set $[\mathcal{U}]$ of relaxed controls.*

Referring to Fig. 2.7, let $\tilde{A}_{(l,m)}(t) = P_{(l,m)}(A_l(t))$ be the cumulative number of vehicles routed from link l to m. Under the max-pressure controller $\tilde{A}_{(l,m)}$ receives service

$$c^*(l,m)(t) = S \circ (1 - L/T)U^*(q(t))(l,m),$$

so the evolution of the array $q(t)$ is governed by these equations:

$$q_{(l,m)}(t+1) = [q_{(l,m)}(t) - c^*(l,m)(t)]_+ + \tilde{a}_{(l,m)}(t+1), \tag{2.74}$$

$$\tilde{a}_{(l,m)}(t+1) = \gamma_{(l,m)} a_l(t+1) + \delta_{(l,m)}(t), \tag{2.75}$$

$$a_l(t+1) = e_l(t+1) + \textstyle\sum_k b_{(k,l)}(t), \tag{2.76}$$

$$b_{(k,l)}(t) = \min\{q_{(k,l)}(t), c^*(k,l)(t)\}. \tag{2.77}$$

Above as elsewhere, $\tilde{a}_{(l,m)}(t+1) = \tilde{A}_{(l,m)}(t+1) - \tilde{A}_{(l,m)}(t)$, $a_l(t+1) = A_l(t+1) - A_l(t+1)$, etc. Since $P_{(l,m)}$ is $(\delta_{(l,m)}, \gamma_{(l,m)})$ upper-bounded,

$$\sum_t \delta_{(l,m)}(t) \leq \delta_{(l,m)}, \quad \text{where } \delta_{(l,m)}(t) - a_{(l,m)}(t) - \gamma_{(l,m)} a_l(t). \tag{2.78}$$

Substitution into (2.74) gives the evolution of $q(t)$ directly in terms of the external arrivals:

$$q_{(l,m)}(t+1) = \left[q_{(l,m)}(t) - c^*(l,m)(t)\right]_+ + \gamma_{(l,m)}\left[e_l(t+1) + \sum_k \min\{q_{(k,l)}(t), c^*(k,l)(t)\}\right] + \delta_{(l,m)}(t+1). \tag{2.79}$$

Theorem 7 extends Theorem 5 to the network case.

Theorem 7. *Let $q(t)$ be the queues resulting from the max-pressure controller. Suppose that the external arrivals E_l are (α_l, β_l) upper-bounded, the routers $P_{(l,m)}$ are $(\delta_{(l,m)}, \gamma_{(l,m)})$ upper-bounded and there exists a (fixed-cycle) relaxed network control matrix $[U]$ such that*

$$c(l,m) = S \circ [U](l,m) > \rho_l \gamma_{(l,m)}, \text{ all } (l,m), \tag{2.80}$$

in which $\rho = \beta[I - \Gamma]^{-1}$. Then $\{q(t),\, t \geq 0\}$ is a bounded sequence and the max-pressure controller is stabilizing.

Proof. Under the max-pressure controller the queues evolve according to (2.79). Let $|q|^2 = \sum q^2_{(l,m)}$. It is shown in Appendix E that there exist $k < \infty$, $\varepsilon > 0$, and $\sigma(t) \geq 0$ with $\sum_t \sigma(t) < \infty$, so that

$$|q(t+1)|^2 - |q(t)|^2 \leq k - (2\varepsilon - \sigma(t))|q(t)|, \tag{2.81}$$

Suppose (2.81) holds. With T such that $\sigma(t) < \varepsilon$, $t \geq T$, (2.81) gives

$$|q(t+1)|^2 - |q(t)|^2 \leq k - \varepsilon|q(t)|,\ t > T,$$

and so

$$|q(t+1)|^2 - |q(t)|^2 < 0,\ |q(t)| > k/\varepsilon,\ t > T, \tag{2.82}$$

which implies that $|q(t)|$, $t \geq 0$ is bounded. □

2.4.4 Two Extensions of Max-Pressure Controller

The pressure $w(q,U)$ defined in (2.72) treats all queues equally. It may be desirable to treat them differently by giving them weights. Let $\kappa_{(l,m)} > 0$ be pre-specified weights and define the weighted pressure exerted by stage U as

$$w_\kappa(q,U) = \sum_{(l,m)} \left[\kappa_{(l,m)} q_{(l,m)} - \sum_p \gamma_{(m,p)} \kappa_{(m,p)} q_{(m,p)} \right] S \circ U(l,m), \tag{2.83}$$

simply by replacing $q_{(l,m)}$ in (2.72) by $\kappa_{(l,m)} q_{(l,m)}$. Define the max-pressure stage as

$$U^*_\kappa(q) = \arg\max\{w_\kappa(q,U) \mid U \in \mathcal{U}\},$$

and the max-pressure controller at $q(t)$ by

$$u^*_\kappa(t) = (1 - L/T) U^*_\kappa(q(t)).$$

Theorem 7 remains true with this definition of the max-pressure controller. The proof of Theorem 7 applies with appropriate changes.

The weighted pressure (2.83) can be used to give preference to the clearance of certain queues. For example, (Aboudolas et al. 2009b, Eq. (11)) suggests using

$$\kappa_{(l,m)} = [Q_{(l,m)}]^{-1},$$

where $Q_{(l,m)}$ is the maximum permissible queue length for phase (l,m). Another possibility is to give more weight to phases that are restricted to buses, giving them greater priority.

The second extension might be termed *max-pressure-lite*. Suppose the intersection controllers already have in place several timing plans, scheduled depending on time of day. In our notation a timing plan is just a relaxed control. Suppose K timing plans are in place, denoted as in (2.67) by

$$[U^i] = \sum_U \lambda_U^i U, \; \sum \lambda_U^i \leq 1 - L/T, \quad i = 1, \cdots, K. \tag{2.84}$$

Depending on the time of day, the controller selects one of the $[U^i]$ without regard to traffic conditions. If queue measurements are available, one can select the timing plan that exerts the maximum pressure:

$$[U^*](q) = \arg\max\{w(q, [U^i]) \mid i = 1, \cdots, K\}.$$

The max-pressure-lite controller is given by

$$u^*_{lite}(t) = [U^*](q(t)).$$

The following result can be proved in a way similar to Theorem 7.

Theorem 8. *Suppose there exists a convex combination of the fixed timing plans* $[U] = \sum_{i=1}^K \mu_i [U^i]$, $\mu_i \geq 0$, $\sum \mu_i = 1$ *such that*

$$S \circ [U](l,m) > \rho_l \gamma_{(l,m)}, \text{ all } (l,m).$$

Then the max-pressure-lite controller is stabilizing.

2.5 Discussion

We present the intuition underlying the max-pressure controller. This is followed by a comparison with other controller designs. Lastly, we discuss model limitations, followed by some open problems.

2.5.1 Intuition

For any relaxed network control sequence $[U(t)]$ let $c(l,m)(t) = S \circ [U(t)](l,m)$ be the resulting service rates. The evolution of queues in response to the control and the external arrivals is given by (2.79):

$$q_{(l,m)}(t+1) = \left[q_{(l,m)}(t) - c(l,m)(t)\right]_+ + \gamma_{(l,m)} \left[e_l(t+1) + \sum_k \min\{q_{(k,l)}(t), c(k,l)(t)\}\right] + \delta_{(l,m)}(t+1).$$

If the queues are sufficiently large (saturated case, $q_{(l,m)}(t) > c(l,m)(t)$) this simplifies to

$$q_{(l,m)}(t+1) - q_{(l,m)}(t) = -c(l,m)(t) + \gamma_{(l,m)} \sum_k c(k,l)(t) + \gamma_{(l,m)} [\beta_l + \alpha_l(t+1)] + \delta_{(l,m)}(t+1). \tag{2.85}$$

Regard (2.85) as a discrete-time approximation of the differential equation

$$\frac{\mathrm{d}}{\mathrm{d}t} q_{(l,m)}(t) = -c(l,m)(t) + \gamma_{(l,m)} \sum_k c(k,l)(t) + \gamma_{(l,m)} \beta_l + \psi_{(l,m)}(t), \tag{2.86}$$

in which $\psi_{(l,m)}(t) = \gamma_{(l,m)} \alpha_l(t) + \delta_{(l,m)}(t)$ is a “disturbance” input. Then

$$\begin{aligned}
\frac{1}{2}\frac{\mathrm{d}}{\mathrm{d}t}|q(t)|^2 &= \langle q(t), \dot{q}(t) \rangle \\
&= -\sum q_{(l,m)}(t) \left[c(l,m)(t) - \gamma_{(l,m)} \sum_k c(k,l)(t)\right] + \sum q_{(l,m)}(t) \gamma_{(l,m)} \beta_l \\
&\quad + \sum q_{(l,m)}(t) \psi_{(l,m)}(t) \\
&= -w(q(t), [U(t)]) + \sum q_{(l,m)}(t) \gamma_{(l,m)} \beta_l + \sum q_{(l,m)}(t) \psi_{(l,m)}(t).
\end{aligned} \tag{2.87}$$

The third term on the right is evanescent and may be ignored since $\sum \psi_{(l,m)}(t) < \infty$. The max-pressure controller selects stage $[U(t)] \in [\mathcal{U}]$ that makes the first term as negative as possible. The second constant forcing term is due to the average rate of external arrivals. Theorem 7 says that if there is a stabilizing fixed-cycle controller the first term will dominate the second term. In fact, (2.108) says that

$$-w(q(t), [U(t)]) + \sum q_{(l,m)}(t) \gamma_{(l,m)} \beta_l < -\varepsilon |q(t)|,$$

which is why the max-pressure controller is stabilizing. The fact that the inequality above does *not* require knowledge of the arrival rates $\{\beta_l\}$ explains why the max-pressure controller is adaptive.

2.5.2 Comparison with Other Designs

Previous designs of traffic-responsive controllers (Robertson and Bretherton 1991; Mirchandani and Head 2001; Heydecker 2004; Aboudolas et al. 2009b; Cai et al. 2009) require (or make) an estimate of the arrivals over a finite or infinite horizon and select the control that minimizes queues or delay over that horizon. The presumption is that the longer is the horizon, the better will be the controller performance. By contrast, the max-pressure controller is "myopic" and does not make any estimate of the arrivals. In addition, previous designs require estimates of the queues to be communicated to a central controller. By contrast, the max-pressure controller requires only local communication, since the pressure of a stage at any intersection depends only on the queues adjacent to the intersection. Lastly, none of the cited references proves that their controller design is stabilizing as is the case with the max-pressure controller.

We compare in some detail the max-pressure controller with that of (Aboudolas et al. 2009b), which uses the same model as (2.85), expresses $c(l,m)(t) = S \circ [\sum_U \lambda_U(t)](l,m)$, and also takes the proportions $\{\lambda_U(t),\ U \in \mathcal{U}\}$ of the available time $(T-L)$ as the control vector. The control vector is decomposed as

$$\lambda_U(t) = \lambda_U^F + \Delta\lambda_U(t),$$

in which $\{\lambda_U^F\}$ is, by assumption, a *known* stabilizing fixed-cycle controller for the external arrivals $\{\beta_l\}$. With this assumption, (2.85) simplifies to

$$q_{(l,m)}(t+1) - q_{(l,m)}(t) = -S \circ \left[\sum_U \Delta\lambda_U(t)\right](l,m) + \gamma_{(l,m)} \sum_k S \circ \left[\sum_U \Delta\lambda_U(t)\right](k,l) + \psi_{(l,m)}(t). \tag{2.88}$$

The control deviations $\{\Delta\lambda_U(t), U \in \mathcal{U}\}$ are selected to minimize the quadratic cost

$$\sum_t |q_{(l,m)}(t)|^2 + p \sum_t \sum_{U \in \mathcal{U}} |\Delta\lambda_U(t)|^2.$$

Neglecting the evanescent disturbances $\{\psi_{(l,m)}(t)\}$, this cost is minimized by easily calculated linear feedback rules:

$$\Delta\lambda_U(t) - \sum_{l,m} G_U(l,m) q_{(l,m)}(t),\ U \in \mathcal{U}.$$

Since the resulting proportions will not satisfy the constraint

$$\sum_U [\lambda_U^F + \Delta\lambda_U(t)] \le T - L, \tag{2.89}$$

the "gain" matrices G_U are changed to $\tilde{G}_U$ so that the modified deviations $\{\Delta\lambda_U(t)\} = \sum_{l,m} \tilde{G}_U(l,m) q_{(l,m)}(t)$ do meet this constraint.

Note that if the true arrivals are different from the assumed arrivals, a steady-state bias in the queue lengths must be present to compensate for the error in the assumed arrivals. Of course, the linear model (2.85) will be quite inaccurate when the queues are small. Two more complex optimization methods are considered in (Aboudolas et al. 2009b) but not discussed here.

2.5.3 Model Limitations

We discuss four limitations. In a "store and forward" (SF) model, there is no limit to how much a queue can grow, so the condition in which a downstream queue blocks upstream vehicles is not modeled. It is straightforward to modify (2.79) to model blocking. The first term on the right of (2.79), namely $[q_{(l,m)}(t) - c(l,m)(t)]_+$ indicates that the queue $q_{(l,m)}(t)$ is decremented by the saturation rate $s(l,m)$ whenever the phase (l,m) is actuated, regardless of the congestion in the downstream link m. If link m has a queue capacity of $Q(m)$ and its average queue size is $q(m)(t) = \sum \gamma_{(m,p)} q_{(m,p)}(t)$, one could replace $[q_{(l,m)}(t) - c(l,m)(t)]_+$ by

$$\mathbf{1}[q(m)(t) < Q(m)] \times [q_{(l,m)}(t) - c(l,m)(t)]_+,$$

so that the movement of vehicles from link l to m is blocked when $q(m)(t)$ exceeds $Q(m)$. Unfortunately, with this model of blocking, it is easy to construct examples that create "gridlock" in such a way that there is *no* stabilizing controller. On the other hand, (2.82) implies that the max-pressure controller is stabilizing if the $Q(m)$ are large enough.

Second, the model does not take into account that it takes time for vehicles to traverse a link. If this time is constant (so called free flow travel time), it can be modeled by a constant delay network element as in (Chang 2000, Lemma 2.3.9). It is not difficult to see that the max-pressure controller is stabilizing in this case as well.

The third limitation is related to the second. The store and forward model leaves no room for signal offset. A signal offset design can be grafted on to the max-pressure controller in the same manner as in Diakaki et al. (2003).

Fourth, the model assumes turn ratios as opposed to O–D patterns. If O–D patterns are fixed, i.e., each O–D demand is distributed in fixed proportions over a set of routes, the demand can be equivalently described by turn ratios. But if drivers respond to delays by changing their route, the turn ratios will also change, and the max-pressure controller needs to adapt to the changes. The resulting system could be studied in a two-level control framework similarly to Wong and Yang (1997).

2.5.4 Future Work

Several questions seem worth investigation. The first concerns priorities: Which priorities in a single-phase network or in a multiphase isolated intersection permit stabilizing work-conserving controllers?

The second question concerns an evacuation scenario in which there is an initial set of queues $q(0)$ and no external inputs, and one wishes to design a fixed-time controller and a feedback controller that minimize $\sum_t q(t)$. How should one restrict the phases actuated by each stage U to a subset $[U']$ (i.e., $U'(l,m) = 1 \rightarrow U(l,m)=1$) so as to minimize $\sum_t q(t)$? The idea here is that whereas U may permit left-turns (say) it may be more efficient to prevent such turns.

The third question concerns over-saturated networks in which the average rates $\{\beta_l\}$ of external arrivals $\{e_l\}$ in (2.79) are such that there is no stabilizing controller. How should one design an adaptive scheme to "meter" these arrivals so that the network can be stabilized? For a macroscopic discussion see (Daganzo 2007).

2.6 Conclusion

The max-pressure controller appears to offer advantages over other adaptive controllers. The controller at each intersection only needs to know the queues on adjoining links and the computation required to select the max-pressure stage is trivial. No knowledge of demand (or even the network topology) is needed, although each intersection controller does need to know local turn ratios. Max-pressure is provably stable whenever there exists any stabilizing controller. Lastly, max-pressure is attractive from an implementation viewpoint: it requires less communication and computational infrastructure than other adaptive controllers; and it can be incrementally deployed since addition of new intersections entails no change in the control of existing intersections.

Acknowledgements This research was supported by National Science Foundation Award CMMI-0941326 and the California Department of Transportation. The author is very grateful to François Baccelli for guiding his study of network calculus and for discussions about the proofs, and to Andy Chow, Nik Geroliminis, Gabriel Gomes, Werner Kraus, Pitu Mirchandani, Markos Papageorgiou, and Bart De Schutter for important critical comments concerning signal control.

Appendices

A Proof of Lemma 1

By induction. Since $q(0) = 0$, (2.2) holds for $t = 0$. Suppose (2.2) holds for t. Then

$$\begin{aligned} q(t+1) &= \max\left\{0, \max_{0 \le s \le t}[A(t,s) - C(t-1,s)] - c(t)\right\} + a(t+1) \\ &= \max\left\{a(t+1), \max_{0 \le s \le t}[A(t+1,s) - C(t,s)]\right\} \\ &= \max_{0 \le s \le t+1}[A(t+1,s) - C(t,s)], \end{aligned}$$

so (2.2) holds for $t+1$. Since the queue size is the difference between arrivals and departures,

$$\begin{aligned} B(t) &= A(t) - [q(t) - q(0)] \\ &= A(t) - \max_{0 \le s \le t} [A(t,s) - C(t-1,s)] \\ &= \min_{0 \le s \le t} [A(s) + C(t-1,s)], \end{aligned}$$

which proves (2.3). □

B Proof of Theorem 1

Equation (2.4) follows from

$$q(t) = \max_{0 \le s \le t} [A(t,s) - C(t-1,s)] \le \max_{0 \le s \le t} [f_1(t-s) - f_2(t-s)] = \max_{0 \le \tau \le t} [f_1(\tau) - f_2(\tau)].$$

Since always $B(t) \le A(t)$,

$$\begin{aligned} B(t,s) &\le A(t) - B(s) \\ &= A(t) - \min_{0 \le r \le s} [A(r) + C(s-1,r)] \\ &= \max_{0 \le r \le s} [A(t,r) - C(s-1,r)] \\ &\le \max_{0 \le r \le s} [f_1(t-r) - f_2(s-r)] \\ &= \max_{0 \le \tau \le s} [f_1(t-s+\tau) - f_2(\tau)] \\ &\le \max_{0 \le \tau} [f_1(t-s+\tau) - f_2(\tau)], \end{aligned}$$

which proves (2.5). Next $t+d(t)$ is the least time by which there are $A(t)$ cumulative departures, so

$$d(t) = \min\{d \mid B(t+d) \ge A(t)\}.$$

From (2.3),

$$\begin{aligned} B(t+d) - A(t) &= \min_{s \le t+d} \{A(s) + C(t+d-1,s)\} - A(t) \\ &\ge \min\{0, \min_{s \le t} \{-A(t,s) + C(t+d-1,s)\}\}, \text{ as } A(s) - A(t) \\ &\quad + C(t+d-1,s) \ge 0, s \ge t \\ &\ge \min\{0, \min_{s \le t} \{-f_1(t-s) + f_2(t-s+d)\}\} \\ &= \min\{0, \min_{0 \le \tau \le t} \{-f_1(\tau) + f_2(\tau+d)\}\}. \end{aligned}$$

Hence $B(t+d) \geq A(t)$ if $f_1(\tau) \leq f_2(\tau+d)$ for $\tau = 1,\cdots t$, which implies (2.6). Lastly, a busy period starting at s lasts until t if

$$A(s) = B(s),\ A(t+1) = B(t+1),\ \text{ and } A(s+\tau) > B(s+\tau),\ \tau = 1,\cdots,t-s,$$

and so

$$0 < A(s+\tau,s) - B(s+\tau,s) \leq f_1(\tau) - f_2(\tau),\ \tau = 1,\cdots,t-s,$$

from which (2.7) follows. □

C Proof of Theorem 4

Proof. According to (2.39) and (2.40) $\Sigma_{(l,m)}[q^w_{(l,m)}(t)/s(l,m)]$ is the same for all work-conserving controllers. So it is enough to exhibit one work-conserving controller for which (2.46) holds. For any controller $\sum \lambda_{(l,m)}(t)\delta_{(l,m)}$ write (2.37) in vector form ($Q(t) = \{q_{(l,m)}(t)\}$)

$$Q(t+1) = f(Q(t),t).$$

Because of (2.38) one can construct a work-conserving feedback controller $\sum \lambda^w_{(l,m)}(Q,t)\delta_{(l,m)}$ such that

$$[q^w_{(l,m)} - s(l,m)\lambda^w_{(l,m)}(Q'^w,t)]_+ \leq [q_{(l,m)}(t) - s(l,m)\lambda_{(l,m)}(t)]_+ \tag{2.90}$$

for all t, (l,m) and $Q^w \leq Q$ (the vector $\leq$ is interpreted component-wise). Write (2.37) for this work-conserving controller as

$$Q^w(t+1) = g(Q^w(t),t).$$

It is not difficult to see that the functions $f(Q,t)$ and $g(Q,t)$ are both monotonic in Q, i.e.,

$$Q^w \leq Q \Rightarrow f(Q^w,t) \leq f(Q,t) \text{ and } g(Q^w,t) \leq g(Q,t).$$

We claim that if $Q^w(0) = Q^w(0)$ then

$$Q^w(t) \leq Q(t),\ \ t \geq 0. \tag{2.91}$$

Equation (2.91) is clear for $t = 0$. Suppose it is true for t. Then

$$Q^w(t+1) = g(Q^w(t),t) \leq g(Q(t),t) \leq f(Q(t),t) \leq Q(t+1),$$

in which the first inequality is due to monotonicity of g and the second follows from (2.90). Thus this, and hence all, work-conserving controllers satisfy (2.46). □

D Proof of (2.56)

We prove (2.56) in a few steps. For arrays $x = \{x_{(l,m)}\}$ and $y = \{y_{(l,m)}\}$ write $\langle x,y\rangle = \sum x_{(l,m)}y_{(l,m)}$, $|x|^2 = \langle x,x\rangle$, $\min\{x,y\} = \{\min(x_{(l,m)},y_{(l,m)})\}$, $\max\{x,y\} = \{\max(x_{(l,m)},y_{(l,m)})\}$. Then (2.55) can be written as

$$q(t+1) = [q(t) - c^*(t)]_+ + a(t+1) = \max\{q(t) - c^*(t), 0\} + a(t+1),$$

so

$$\delta = q(t+1) - q(t) = \max\{-c^*(t), -q(t)\} + a(t+1) = -\min\{c^*(t), q(t)\} + a(t+1).$$

Next,

$$|q(t+1)|^2 - |q(t)|^2 = 2\langle\delta, q(t)\rangle + |\delta|^2 = 2\alpha + \beta, \text{say.} \tag{2.92}$$

We separately upper-bound α and β.

Bound on α

$$\alpha = \langle\delta, q(t)\rangle = \sum q_{(l,m)}(t)[a_{(l,m)}(t+1) - \min\{c(l,m)^*(t), q_{(l,m)}(t)\}] \tag{2.93}$$

$$= \sum q_{(l,m)}(t)[a_{(l,m)}(t+1) - c(l,m)^*(t) + \max\{c(l,m)^*(t) - q_{(l,m)}(t), 0\}] \tag{2.94}$$

$$= \sum q_{(l,m)}(t)[a_{(l,m)}(t+1) - c(l,m)^*(t)]$$
$$+ \sum q_{(l,m)}(t)\max\{c(l,m)^*(t) - q_{(l,m)}(t), 0\} \tag{2.95}$$

$$= \alpha_1 + \alpha_2, \text{say.} \tag{2.96}$$

Let $K = \max\{a_{(l,m)}(t+1), c^*(l,m)(t)\}$, the maximum taken over all $(l,m), t$. Then

$$\alpha_2 \leq \sum q_{(l,m)}(t)c^*(l,m)(t+1)\mathbf{1}[q_{(l,m)}(t) < c^*(l,m)(t)] \leq NK^2, \tag{2.97}$$

in which N is the number of (l,m) pairs. Next

$$\alpha_1 = \sum q_{(l,m)}(t)[a_{(l,m)}(t+1) - c^*(l,m)(t)]$$

$$\begin{aligned}
&= \sum q_{(l,m)}(t)[a_{(l,m)}(t+1) - \rho(l,m)] + \sum q_{(l,m)}(t)[\rho(l,m) - c(l,m)] \\
&\quad + \sum q_{(l,m)}(t)[c(l,m) - c^*(l,m)(t)] \\
&= \alpha_{11} + \alpha_{12} + \alpha_{13}, \text{ say.}
\end{aligned}$$

Let $\sigma_{(l,m)}(t+1) = a_{(l,m)}(t+1) - \rho(l,m)$. Since $A_{(l,m)}(t)$ is $(\sigma(l,m), \rho(l,m))$ upper-bounded,

$$\alpha_{11} = \sum q_{(l,m)}(t)\sigma_{(l,m)}(t+1), \text{ with } \sum_t \sigma_{(l,m)}(t) \le \sigma(l,m).$$

By (2.54) $\rho(l,m) - c(l,m) < 0$ for all (l,m), so there exists $\eta > 0$ such that

$$\alpha_{12} \le -\eta \sum q_{(l,m)}(t).$$

Lastly, since $u^*(t)$ maximizes the pressure $w(q(t),[U])$, it follows that

$$\alpha_{13} = \sum q_{(l,m)}(t)[c(l,m) - c^*(l,m)(t)] = w(q(t),[U]) - w(q(t),u^*(t)) \le 0.$$

Combining these three estimates gives

$$\alpha_1 \le \sum(-\eta + \sigma_{(l,m)}(t))q_{(l,m)}(t), \text{ with } \sum_t \sigma_{(l,m)}(t) \le \sigma(l,m). \tag{2.98}$$

Bound on β

$$\begin{aligned}
\delta_{(l,m)} &= a_{(l,m)}(t+1) - \min\{c(l,m)^*(t), q_{(l,m)}(t)\} \\
&= a_{(l,m)}(t+1) - c^*(l,m)(t)\mathbf{1}[q_{(l,m)}(t) > c^*(l,m)(t)] - q_{(l,m)}(t)\mathbf{1}[q_{(l,m)}(t) \\
&\le c^*(l,m)(t)] \\
|\delta_{(l,m)}| &\le |a_{(l,m)}(t+1) - c^*(l,m)(t)| + q^*_{(l,m)}(t)\mathbf{1}[q_{(l,m)}(t) \\
&\le c^*(l,m)(t)] \\
&\le |a_{(l,m)}(t+1) - c^*(l,m)(t)| + c^*(l,m)(t) \le 2K
\end{aligned}$$

So

$$|\delta|^2 = \sum \delta^2_{(l,m)} \le 4NK^2. \tag{2.99}$$

Equation (2.56) follows from (2.92) to (2.99). □

E Proof of (2.81)

The proof follows the same lines as in Appendix D. Write (2.74) in vector–matrix form as

$$q(t+1) = [q(t) - c^*(t)]_+ + \tilde{a}(t+1).$$

Let

$$x = q(t+1) - q(t) = -\min\{c^*(t), q(t)\} + \tilde{a}(t+1).$$

Then

$$|x|^2 = 2\langle x, q(t)\rangle + |x|^2 = 2\mu + \nu, \text{ say.} \tag{2.100}$$

We separately bound μ, ν.

Bound on μ

$$\begin{aligned}
\mu &= \langle x, q(t)\rangle = \sum q_{(l,m)}(t)[\tilde{a}_{(l,m)}(t+1) - \min\{c^*(l,m)(t), q_{(l,m)}(t)\}] \\
&= \sum q_{(l,m)}(t)[\tilde{a}_{(l,m)}(t+1) - c^*(l,m)(t) + \max\{c^*(l,m)(t) - q_{(l,m)}(t), 0\}] \\
&= \sum q_{(l,m)}(t)[\tilde{a}_{(l,m)}(t+1) - c^*(l,m)(t)] + \sum q_{(l,m)}(t)\max\{c^*(l,m)(t) - q_{(l,m)}(t), 0\} \\
&= \mu_1 + \mu_2, \text{ say.}
\end{aligned}$$

Let $K = \max\{c^*(l,m)(t)\}$ be the maximum over all t, (l,m). Then

$$\begin{aligned}
\mu_2 &= \sum q_{(l,m)}(t)\max\{c^*(l,m)(t) - q_{(l,m)}(t), 0\} \\
&\le \sum q_{(l,m)}(t)c^*(l,m)(t)\mathbf{1}[c^*(l,m)(t) \ge q_{(l,m)}(t)] \\
&\le NK^2,
\end{aligned} \tag{2.101}$$

in which N is the number of (l,m) pairs in the network.

Using (2.74)–(2.77),

$$\begin{aligned}
\mu_1 &= \sum_{l,m} q_{(l,m)}(t)[\tilde{a}_{(l,m)}(t+1) - c^*(l,m)(t)] \\
&= \sum_{l,m} q_{(l,m)}(t)\left[e_l(t+1)\gamma_{(l,m)} + \sum_k b_{(k,l)}(t)\gamma_{(l,m)} + \delta_{(l,m)}(t) - c^*(l,m)(t)\right]
\end{aligned}$$

$$
\begin{aligned}
&= \sum_{l,m} q_{(l,m)}(t) \Bigg[e_l(t+1)\gamma_{(l,m)} + \sum_k \min\{q_{(k,l)}(t), c^*(k,l)(t)\}\gamma_{(l,m)} \\
&\quad + \delta_{(l,m)}(t) - c^*(l,m)(t) \Bigg] \\
&\le \sum_{l,m} q_{(l,m)}(t) \left[e_l(t+1)\gamma_{(l,m)} + \sum_k c^*(k,l)(t)\gamma_{(l,m)} - c^*(l,m)(t) \right] \\
&\quad + \sum_{l,m} q_{(l,m)}(t)\delta_{(l,m)}(t) \\
&= \sum_{l,m} q_{(l,m)}(t) \left[\beta_l \gamma_{(l,m)} + \sum_k c^*(k,l)(t)\gamma_{(l,m)} - c^*(l,m)(t) \right] \\
&\quad + \sum_{l,m} q_{(l,m)}(t) \left[\alpha_l(t+1)\gamma_{(l,m)} + \delta_{(l,m)}(t) \right] \\
&= \mu_{11} + \mu_{12} + \mu_{13}.
\end{aligned}
$$

Above, $\alpha_l(t+1) = e_l(t+1) - \beta_l$, so $\sum_t \alpha_l(t) \le \alpha_l$, since E_l is (α_l, β_l) upper-bounded;

$$
\mu_{11} = \sum_{l,m} q_{(l,m)}(t)\beta_l \gamma_{(l,m)}, \tag{2.102}
$$

$$
\begin{aligned}
\mu_{12} &= \sum_{l,m} q_{(l,m)}(t) \left[\sum_k c^*(k,l)(t)\gamma_{(l,m)} - c^*(l,m)(t) \right] \\
&= \sum_{l,m} \left[\sum_p q_{(m,p)}(t)\gamma_{(m,p)} - q_{(l,m)}(t) \right] c^*(l,m)(t) \\
&= -w(q(t), u^*(t)),
\end{aligned} \tag{2.103}
$$

$$
\mu_{13} = \sum_{l,m} q_{(l,m)}(t) \left[\alpha_l(t+1)\gamma_{(l,m)} + \delta_{(l,m)}(t) \right]. \tag{2.104}
$$

Substituting $\beta_l = \rho_l - \sum_k \rho_k \gamma_{(k,l)}$ from (2.59) into (2.102) gives

$$
\begin{aligned}
\mu_{11} &= \sum_{l,m} q_{(l,m)}(t) \left[\rho_l - \sum_k \rho_k \gamma_{(k,l)} \right] \gamma_{(l,m)} \\
&= \sum_{l,m} \rho_l \gamma_{(l,m)} q_{(l,m)}(t) - \sum_{l,m} q_{(l,m)}(t) \sum_k \rho_k \gamma_{(k,l)} \gamma_{(l,m)} \\
&= \sum_{l,m} \rho_l \gamma_{(l,m)} q_{(l,m)}(t) - \sum_m \left[\sum_l \rho_l \gamma_{(l,m)} \right] \sum_p q_{(m,p)}(t)\gamma_{(m,p)} \\
&= \sum_{l,m} \rho_l \gamma_{(l,m)} \left[q_{(l,m)}(t) - \sum_p q_{(m,p)}(t)\gamma_{(m,p)} \right].
\end{aligned} \tag{2.105}
$$

By (2.80) there exists $[U] \in [\mathcal{U}]$ such that $S \circ [U] > [\rho]\Gamma$. Since $0 \in [\mathcal{U}]$, this implies that $[\rho]\Gamma$ is in the interior of $S \circ [\mathcal{U}]$. Hence there exist (possibly different) $[U]$ and $\eta > 0$ such that

$$S \circ [U](l,m) = \begin{cases} \rho_l \gamma_{(l,m)} + \eta, & \text{if } q_{(l,m)}(t) - \sum_p q_{(m,p)}(t)\gamma_{(m,p)} > 0 \\ \rho_l \gamma_{(l,m)} - \eta, & \text{if } q_{(l,m)}(t) - \sum_p q_{(m,p)}(t)\gamma_{(m,p)} \leq 0 \end{cases},$$

and so

$$w(q,[U]) \geq \mu_{11} + \eta \sum_{l,m} |q_{(l,m)}(t) - \sum_p q_{(m,p)}(t)\gamma_{(m,p)}|. \tag{2.106}$$

The linear transformation $\{q_{(l,m)}\} \mapsto \{q_{(l,m)} - \sum_p q_{(m,p)}\gamma_{(m,p)}\}$ is 1:1 from the conditions imposed on Γ. Hence (2.106) implies that there exists $\varepsilon > 0$ so that

$$w(q(t),[U]) \geq \mu_{11} + \varepsilon |q(t)|,$$

which together with (2.103) gives

$$\mu_{11} + \mu_{12} \leq w(q(t),[U]) - w(q(t),u^*(t)) - \varepsilon|q(t)| \leq -\varepsilon|q(t)|, \tag{2.107}$$

since the pressure $w(q,[U]$ is maximized at $u^*(t)$. Together with (2.104) we get the bound

$$\mu_1 \leq -\varepsilon|q(t)| + \sigma(t)|q(t)|, \tag{2.108}$$

for some $\sigma(t) \geq 0$, $\sum \sigma(t) < \infty$.

Bound on v

From (2.100), $v = \sum_{l,m} |x_{(l,m)}|^2$, and

$$\begin{aligned} x_{(l,m)} &= \tilde{a}_{(l,m)}(t+1) - \min\{c^*(l,m)(t), q_{(l,m)}(t)\} \\ &= \tilde{a}_{(l,m)}(t+1) - c^*(l,m)(t) - \min\{q_{(l,m)}(t) - c^*(l,m)(t), 0\}, \end{aligned}$$

so

$$|x_{(l,m)}| \leq |\tilde{a}_{(l,m)}(t+1) - c^*(l,m)(t)| + |c^*(l,m)(t)| \leq |\tilde{a}_{(l,m)}(t+1)| + 2|c^*(l,m)(t)|.$$

From (2.75) to (2.77) it follows that $|\tilde{a}_{(l,m)}(t+1)|$ is bounded. Hence there is $k < \infty$ such that $v \leq k$, which together with (2.108) and (2.101) yield (2.81) as required.

□

References

Aboudolas K, Papageorgiou M, Kosmatopoulos E. Store-and-forward based methods for the signal control problem in large-scale congested urban road networks. Transport Res C 2009a;17:163–174.
Aboudolas K, Papageorgiou M, Kouvelas A, Kosmatopoulos E. A rolling-horizon quadratic-programming approach to the signal control problem in large-scale congested urban road networks. Transport Res C 2009b;17:680–694.
Allsop RE. Estimating the traffic capacity of a signalized road junction. Transport Res 1972;6:245–255.
Cai C, Wong CK, Heydecker BG. Adaptive traffic signal control using approximate dynamic programming. Transport Res C 2009;17(5):456–474.
Chada S, Newland R. Effectiveness of bus signal priority: final report. Technical Report NCTR-416-04, National Center For Transit Research (NCTR), University of South Florida, 2002.
Chang C-S. Performance guarantees in communication networks. New York: Springer; 2000.
Cruz RL. A calculus for network delay. Part I: Network elements in isolation. Part II: Network analysis. IEEE Trans Inform Theor. 1991;37(1):114–141.
Daganzo CF. Urban gridlock: macroscopic modeling and mitigation approaches. Transport Res B 2007;41:49–62.
Dai JG. On positive Harris recurrence of multiclass queueing networks: A unified approach via fluid limit models. Ann Appl Probab. 1995;5(1):49–77.
Dai JG, Lin W. Maximum pressure policies in stochastic processing networks. Oper Res. 2005;53(2):197–218.
Diakaki C, Dinopoulou V, Aboudolas K, Papageorgiou M, Ben-Shabat E, Seider E, Leibov A. Extensions and new applications of the traffic-responsive urban control strategy. Transport Res Rec. 2003;1856:202–211.
Heydecker BG. Objectives, stimulus and feedback in signal control of road traffic. J Intell Transport Syst. 2004;8:63–76.
Kwong K, Kavaler R, Rajagopal R, Varaiya P. Arterial travel time estimation based on vehicle re-identification using wireless sensors. Transport Res C 2009;17(6):586–606.
Lin WH, Lo HK. A robust quasi-dynamic traffic signal control scheme for queue management. In: Proceedings of the 13th International Conference of Hong Kong Society for Transportation Studies, pages 37–46, Hong Kong, 2008.
Lu SH, Kumar PR. Distributed scheduling based on due dates and buffer priorities. IEEE Trans Automat Contr. 1991;36(12):1406–1416.
McKeown N, Varaiya P, Walrand J. Scheduling cells in an input-buffered switch. Electron Lett. 1993;29(25):2174–2175.
Mirchandani P, Head L. A real-time traffic signal control system: architecture, algorithms, and analysis. Transport Res C 2001;9:415–432.
Mirchandani PB, Zou N. Queuing models for analysis of traffic adaptive signal control. IEEE Trans Intell Transport Syst. 2007;8(1):50–59.
Newell GF. Approximation methods for queues with application to the fixed-cycle traffic light. SIAM Rev. 1965;7(2):223–240.
Newell GF. Theory of highway traffic signals. Course Notes UCB-ITS-CN-89-1, Institute of Transportation Studies, University of California, Berkeley, CA 94720, 1989.
Robertson DI, Bretherton RD. Optimizing networks of traffic signals in real time-the SCOOT method. IEEE Trans Veh Tech. 1991;40(1):11–15.
Tassiulas L, Ephremides A. Stability properties of constrained queuing systems and scheduling policies for maximum throughput in multihop radio networks. IEEE Trans Auto Contr. 1992;37(12):1936–1948.

Varaiya P. A universal feedback control policy for arbitrary networks of signalized intersections, September 2009. http://paleale.eecs.berkeley.edu/~varaiya/papers_ps.dir/090801-Intersectionsv5.pdf.

Wikipedia. Network calculus. http://en.wikipedia.org/wiki/Network_calculus, accessed December 23, 2009.

Wong SC, Yang H. Reserve capacity of a signal-controlled road network. Transport Res B 1997;31(5):397–402.

Chapter 3
Coordinated Feedback-Based Freeway Ramp Metering Control Strategies "C-MIXCROS and D-MIXCROS" that Take Ramp Queues into Account

Ilgin Gokasar, Kaan Ozbay, and Pushkin Kachroo

Abstract In this paper, C-MIXCROS and D-MIXCROS, two feedback-based coordinated ramp metering strategies that explicitly consider ramp queues, are proposed. They are evaluated using both macroscopic (Rutgers Macroscopic Simulation Environment) and microscopic (PARAMICS) simulation models (on an 11-mile-long corridor of I-295 in South Jersey) under different demand conditions. In addition to the newly proposed coordinated ramp metering strategies, a well-known coordinated strategy (METALINE [Papageorgiou et al. Transport Res. 1990;24A:361–370]) and three other local strategies (ALINEA [Papageorgiou et al. Transportation research record, No. 1320, Washington, D.C.: TRB, National Research Council; 1991. p. 58–64], New Control [Kachroo and Ozbay. Feedback ramp metering in intelligent transportation systems. New York: Kluwer Academics; 2003], and MIXCROS [Kachroo and Ozbay. Feedback ramp metering in intelligent transportation systems. New York: Kluwer Academics; 2003) are also implemented using the same network and results are compared. The proportional-derivative state feedback control logic and direct regulation of on-ramp queues are employed in the derivation of this new proposed coordinated ramp metering strategy. The simulation results are consistent with the macroscopic simulation results, where D-MIXCROS and C-MIXCROS both perform more competently than all other control strategies tested for every demand scenario. The deteriorating effect of enormous on-ramp queues on the total travel time is observed especially in METALINE results; the

I. Gokasar
Department of Civil Engineering, Bogazici University, Istanbul, Turkey
e-mail: ilgin.gokasar@boun.edu.tr

K. Ozbay (✉)
Department of Civil & Environmental Engineering, Rutgers University, NJ 08854, USA
e-mail: kaan@rci.rutgers.edu

P. Kachroo
Department of Electrical Engineering, University of Nevada, Las Vegas, USA
e-mail: pushkin@unlv.edu

S.V. Ukkusuri and K. Ozbay (eds.), *Advances in Dynamic Network Modeling in Complex Transportation Systems*, Complex Networks and Dynamic Systems 2,
DOI 10.1007/978-1-4614-6243-9_3, © Springer Science+Business Media New York 2013

total travel time for METALINE is approximately 22% greater compared with C-MIXCROS. MIXCROS also successfully maintains the on-ramp queues at a reasonable level for each ramp. However, because it is a local ramp metering strategy, coordinated versions of MIXCROS are observed to be more beneficial both for the ramp system and at the network level.

3.1 Introduction

The continuous rise in traffic demand has led to increasingly severe congestion, both recurrent (occurring daily during rush hours) and nonrecurrent (resulting from incidents). One of the most efficient and direct control measures typically employed in freeway networks is ramp metering. Ramp metering provides improvement on freeway flow by dispersing platoons of vehicles and accommodating more efficient merging, as well as reducing accidents and fuel consumption. Freeway control can be categorized as open-loop (in general, time-of-day–dependent) or closed-loop (traffic-responsive) control. In the first case, controls are derived from a priori known traffic data such as demands and occupancies (e.g., Demand Capacity [Masher et al. 1975]), whereas closed-loop controls directly react to existing traffic conditions (e.g., ALINEA and MIXCROS).

There are basically two types of ramp metering, namely, local and coordinated. Local ramp metering considers an isolated section of the network consisting of a freeway segment with one on-ramp, and the controller responds only to changes in the local conditions. Coordinated ramp metering is applied to a series of entrance ramps with the goal of coordinating the response of all the ramps in the system.

Coordinated traffic-responsive ramp metering was first implemented in the 1970s and has been gradually adopted by the USA and countries around the world for many freeway control systems. It is used to control a series of ramps in order to optimize the performance of a freeway facility at the network level. The coordination of the controls allows the metering rate at any ramp to be influenced by conditions at other locations within the network. Circumstances such as multiple bottlenecks on the freeway, nonrecurrent congestion problems (e.g., incidents and environmental conditions), the urgency for optimization of throughput on freeway corridors, and the need for flexibility in addressing changing conditions over time more rapidly lead to the selection of coordinated ramp metering strategies versus isolated ramp metering strategies.

A number of coordinated traffic-responsive control strategies have been proposed but few have been implemented. Some of the implemented coordinated ramp metering strategies include the Zone algorithm (Stephanedes 1994) along I-35 East in Minneapolis/St. Paul, Minnesota, in 1970; Helper ramp algorithm (Lipp et al. 1991) along the I-25 freeway in Denver, Colorado, in March 1981; Bottleneck algorithm (Jacobsen et al. 1989) on I-5, north of the Seattle central business district in Seattle, Washington; Sperry ramp metering algorithm (VDOT) along I-395 and I-66 in northern Virginia during 1985; Fuzzy logic algorithm (Meldrum and Taylor 1995) along I-405 in Seattle, Washington, in 1999; Linear programming algorithm

(Yoshino et al. 1995) in Kobe, Japan; FLOW (Jacobsen et al. 1989) in Seattle, Washington, on September 30, 1981; METALINE (Papageorgiou et al. 1990) in Paris (France), Milwaukee (Wisconsin), and Amsterdam (Netherlands); and SWARM (Paesani et al. 1997) in Orange County, California. There are also more sophisticated new ramp metering algorithms that combine feedback-based control with online learning such as the "iterative learning approach" proposed by Hou et al. (2008).

The Zone algorithm divides the freeway facility into zones with a variable length of 3–6 miles. These zones may contain several metered or non-metered on-ramps. The upstream end of a zone is a free-flow area, and the downstream end of a zone is usually a critical bottleneck. The system-level metering rate is determined by volume control of each zone. The basic concept of the algorithm is to balance the volume of traffic leaving the zone. Comprehensive evaluations of the Zone algorithm show increased freeway speeds, as well as reduced freeway accidents and air pollution, after 10 years of operation. A recent enhancement of the system is the stratified zone algorithm, which is capable of considering maximum queue constraints while the ramp metering rates are calculated. However, the dynamic nature of the traffic flow process is not appraised in the Zone algorithm; therefore, it may not perform efficiently under incident conditions when fast changes of traffic flow occur. Also, the parameters for the algorithm have to be tuned carefully to reflect the local traffic and freeway characteristics; this is laborious to accomplish because the relation between the control parameters and the control objective is not clear in the Zone algorithm (Bogenberger and May 1999; Zhang et al. 2001; Kotsialos et al. 2004).

Helper ramp algorithm is the local traffic-responsive metering algorithm combined with a centralized coordinated operational override feature. The local ramp metering strategies belong to the class of demand capacity. The coordination is achieved through a heuristic site-specific logic. The on-ramps being controlled are divided into six location groups containing 1–7 on-ramps. Based on the localized conditions, each meter selects 1 of 6 available metering rates. Predetermined queue thresholds are used, and, when activated, upstream on-ramps are assigned restrictive metering rates. However, the algorithm requires experience with local traffic patterns to ascertain the superior performance measures. According to the field implementations, HELPER ramp algorithm is found to be very effective in reducing congestion when speeds are less than 90 km/h. However, when the local traffic-responsive control can maintain a speed of 90 km/h, centralized control offer limited or no benefit (Zhang et al. 2001; Kotsialos et al. 2004).

Bottleneck algorithm uses both local traffic-responsive upstream occupancy data and bottleneck data to determine a local metering rate (from historical data) and a bottleneck metering rate. The more restrictive of the two rates is then implemented at each ramp. Queue override is used to prevent spillback onto the arterial street network. A 6-year Bottleneck algorithm (Jacobsen et al. 1989) evaluation study is performed, consisting of 17 southbound ramps during the AM peak and five northbound during the PM peak along a 6.9-mile test corridor in Seattle, Washington. Over the study period, it is reported that travel time drops from 22 to 11.5 min after metering despite higher volumes (mainline volumes increased by more than

86% northbound and 62% southbound). The accident rate drops by about 39%, and average metering delays at each ramp remain at or below 3 min (Zhang et al. 2001; Bogenberger and May 1999; Kotsialos et al. 2004).

Sperry ramp metering algorithm is based on the demand capacity strategy. It attempts to maintain centralized demand below the capacity at each detector station to maximize freeway vehicle miles of travel. It uses only flow measurements and estimated capacities as input. The strategy operates at two distinct modes, the restrictive and the nonrestrictive when ramp spillbacks occur. Algorithm implementation begins with the entrance ramp furthest downstream in the section and then continues upstream, one ramp at a time (Bogenberger and May 1999; Kotsialos et al. 2004).

Fuzzy logic algorithm is based on rules that incorporate human expertise; in this way, it can balance several performance objectives simultaneously and consider many types of information. Fuzzy logic control is especially suitable when an accurate system model is unavailable. Evaluation of the Fuzzy logic algorithm in 9 m locations (1989–1995) on the A12 freeway between Utrecht and Hague (Zoetermeer, Netherlands) reveals higher speeds during congested periods and shorter travel times within the 11-km study area. However, ramp delays are increased. According to Zhang et al. (2001), this algorithm is theoretically very attractive but too complicated to configure, requiring concentrated effort to calibrate the tuning rules and membership functions. It performs poorly when not configured properly, which limits the practical value of this algorithm in the field (Bogenberger and May 1999; Zhang et al. 2001).

Linear programming algorithm identifies an objective function that needs to be minimized/maximized and a series of constraint equations to work within while optimizing this function. Although mathematically more complex than most algorithms, linear programming algorithm can be solved very efficiently using off-the-shelf software programs. One drawback of this algorithm is that its performance relies heavily on accurate origin–destination data. Furthermore, it is static; that is, it neglects the variation of travel time in its computation of ramp metering rates (Zhang et al. 2001).

FLOW is an integrated, traffic-responsive metering algorithm in which metering rates are calculated in real time based on system and local capacity conditions. In addition, queuing conditions on the ramps are also incorporated in the final calculation of metering rates. Hence, the metering algorithm has three components: local metering rate, bottleneck metering rate, and metering rate based on on-ramp queue lengths. The bottleneck metering rate is computed to achieve coordination among ramp meters, and it accounts for the interdependencies among entrance ramp operations. A brief description of FLOW, including its limitations and advantages, is presented in Jacobsen et al. (1989), and Hasan et al. (2002).

METALINE is the coordinated version of the local ramp metering strategy ALINEA. The control logic of METALINE is proportional-integral state feedback. The main challenge to the successful operation of METALINE is the proper choice of the control matrices and the target occupancy vector. There is no direct consideration of on-ramp queues in METALINE (Papageorgiou et al. 1990). According to Zhang et al. (2001), this algorithm is theoretically sound and potentially robust; however, it is difficult to calibrate and operate. Based on the field implementations, it is reported that METALINE performs slightly more effectively for nonrecurrent congestion compared with ALINEA (Papageorgiou et al. 1991).

SWARM (System-wide adaptive ramp metering) seeks to maximize the overall flow of cars on the freeway. The algorithm consists of two levels. The local control determines ramp metering rates based on the local density. The global control determines the overall volume reduction from the ramps upstream a critical bottleneck and then distributes it to upstream ramps according to a set of predetermined fractions to obtain a new set of ramp metering rates. The more restrictive of the two is selected for each ramp. SWARM uses predicted volumes, rather than solely measured conditions, to locate bottlenecks. Therefore, its performance is very sensitive to the accuracy of the predictions (Zhang et al. 2001). SWARM is tested off-line extensively in the Caltrans District 12 Traffic Management Center but is never implemented due to the operational and functional problems, as well as lack of an operator's manual. Caltrans did not detect any evidence that SWARM had affected any ramps (Bogenberger and May 1999; Zhang et al. 2001; Kotsialos et al. 2004).

Kwon et al. (2001) performed a macroscopic simulation evaluation of the Zone algorithm, Fuzzy logic algorithm, and a coordinated algorithm used in Colorado. Because the Zone algorithm did not use queue control, it resulted in the most restrictive metering rates, the least mainline congestion, but the longest ramp queues. In contrast, analysis of Fuzzy logic demonstrated that queue control could reduce mainline efficiency. However, this test also indicated that the Fuzzy logic algorithm is very sensitive to the weights used for each rule.

Chu et al. (2001) evaluated three adaptive ramp metering algorithms, namely, ALINEA, Bottleneck, and Zone, over a stretch of freeway I-405, California, using PARAMICS. It was reported that the two coordinated ramp metering algorithms Bottleneck and Zone performed more satisfactorily than the current fixed-time control and ALINEA algorithm under both morning and afternoon scenarios.

Another simulation study evaluated two ramp control algorithms, a local control algorithm (ALINEA) and a coordinated algorithm (FLOW), using the MITSIM microscopic traffic simulator on a network including part of the Central Artery/Tunnel (CA/T) Project in Boston (Hasan 2002). The performance of ALINEA was satisfactory when there was not a bottleneck downstream of the metered ramps. FLOW outperformed ALINEA under a downstream bottleneck scenario. The improvements of total travel time in FLOW were greater compared with ALINEA when demand was elevated. The study indicated the superiority of system-wide optimization of ramp meter control.

Some of the proposed coordinated ramp metering strategies, which have not been implemented, include Ball Aerospace/FHWA, ARMS (Advanced Real-time Metering System; Liu et al. 1993) and coordinated metering using artificial neural networks (Wei and Wu 1996).

On the other hand, there have been some new papers reporting results from field evaluation of various ramp metering algorithms. For example, Bhouri et al. (2011)

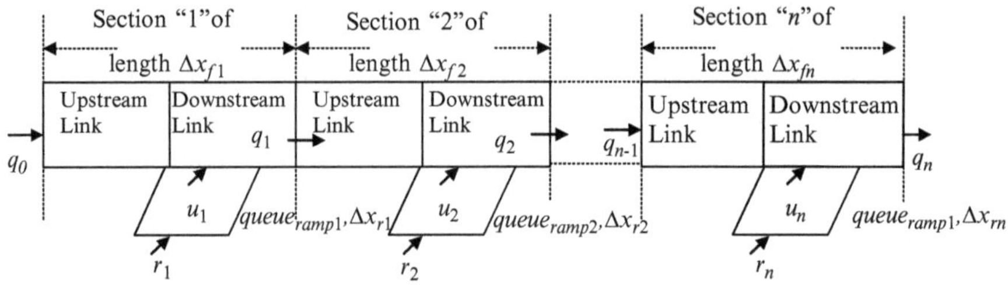

Fig. 3.1 "*n*"Freeway sections with 1 on-ramp

compare isolated and coordinated ramp metering strategies based on the field data. They report major travel time around 24–37% for both isolated and coordinatedramp metering. While coordinated ramp metering is observed to improve travel time reliability, isolated ramp metering has not been found to have an impact on travel time reliability. The ramp metering controls that are evaluated in this paper, namely, ALINEA, New Control, METALINE, MIXCROS, and the coordinated version of MIXCROS, are briefly described below.

ALINEA, a closed-loop local ramp metering strategy recommended by Papageorgiou et al. (1991), to be applied at the time instants $kT, k = 0, 1, 2, \ldots,$ for any sample time interval T (e.g., $T = 60$s) is

$$r(k) = r(k-1) + K\left[\hat{o} - o_{\text{out}}(k)\right],$$

where $K > 0$ is a regulator parameter, $\hat{o}$ is a set (desired) value for the downstream occupancy (typically, but not necessarily, $\hat{o} = o_{\text{cr}}$ may be set, in which case the downstream freeway flow approaches the value of q_{cap}; refer to Fig. 3.1), $r(k-1)$ is the last on-ramp volume, and $r(k)$ is the current ramp volume.

Most of the ramp control strategies proposed so far, such as ALINEA and a new section-based control law (New Control) introduced in Kachroo and Ozbay (2003) shown below, do not directly consider on-ramp queues. They are therefore handled through overriding restrictive metering rates where the metering rate is set to maximum when the on-ramp queue reaches a predetermined level:

$$r(k) = -K\left[o(k) - o_{\text{cr}}\right] + \left[q_{\text{out}}(k) - q_{\text{in}}(k)\right],$$

where $o(k)$ is the current downstream occupancy at time step k, o_{cr} is the set occupancy value, $q_{\text{in}}(k)$ is the flow entering the freeway section at time step k, and $q_{\text{out}}(k)$ is the flow leaving the freeway section at time step k. This control law guarantees that $\lim_{k\to\infty}(\rho - \rho_{\text{cr}})^2 \to 0$, which is the objective of the controller. In fact, it guarantees that the rate of convergence of $o - o_{\text{cr}}$ is geometric at a rate dictated by the control gain K.

METALINE (Papageorgiou et al. 1990) control law is

$$\vec{u}(k) = \vec{u}(k-1) - K_1\left[\vec{o}(k) - \vec{o}(k-1)\right] - K_2\left[\vec{O}(k) - \vec{O}_{\text{cr}}\right],$$

where $\vec{u}(k) \in \Re^m$ is the vector of metering rates for the m controlled ramps at time step k; $\vec{o}(k) \in \Re^n$ is the vector of n measured occupancies within the directional freeway segment at time step k; and $\vec{O}, \vec{O}_{\text{cr}} \in \Re^n$ are, respectively, the measured and desired occupancy downstream of m controlled ramps. K_1 and K_2 are two gain matrices. The main challenge to the successful operation of METALINE is the proper choice of the control matrices K_1 and K_2 and the target occupancy vector $\vec{O}_{\text{cr}}$. There is no direct consideration of queue overflow, which could be handled by override tactics as in ALINEA.

MIXCROS, a traffic-responsive local ramp metering control law proposed by Ozbay and Kachroo (2003), is developed to maximize the throughput on the freeway without creating long queues on the ramp via the use of carefully calibrated weight parameters for the freeway and ramp, namely, w_1 and w_2. The control logic of MIXCROS is proportional-derivative state feedback. MIXCROS proved to be very effective in reducing the congestion on the ramp system while keeping the on-ramp queue at an acceptable level. However, because it is a local feedback-based ramp metering strategy, it produced meager improvement at the network level (Ozbay et al. 2004).

The control objective is defined as

$$e(k) = w_1\left|\rho(k) - \rho_{\text{cr}}\right| + w_2 queue_{\text{ramp}},$$

where ρ is the density of the freeway section (veh/mi), ρ_{cr} is the critical density of the freeway section (veh/mi), and $queue_{\text{ramp}}$ is the queue length on the ramp (veh/mi). The error function, which takes these two objectives (term 1 and term 2) into account, determines how much significance should be attributed to freeway density and queue length on the ramp with the aid of weights w_1 and w_2. Appropriate values for w_1 and w_2 are determined by carefully scrutinizing the control objective of the system. The system can be categorized in two regions. In one region, the traffic density is greater than the critical density and in the other region, the traffic density is less than or equal to the critical density.

These two regions can be combined to devise an integrated control law (i.e., one that is applicable in both regions). The overall control law is therefore given by

$$u(k) = G^{-1}\left[-F - Ke(k)\right],$$

where

$$F = sign(\rho(k) - \rho_{\text{cr}})w_1\left[\rho(k) - \rho_{\text{cr}} + \frac{T}{L_f}(f_1(k) - q_{\text{out}}(k))\right] + w_2\left[queue_{\text{ramp}}(k) + \frac{T}{L_r}f_2(k)\right],$$

and

$$G = sign(\rho(k) - \rho_{cr})w_1\frac{T}{L_f} - w_2\frac{T}{L_r}.$$

The variable f_1 (veh/hr) is the flow entering the freeway section, f_2 is the flow entering the ramp (veh/hr), and L_f and L_r are the length of the freeway and ramp section (mi), respectively; moreover, T is the time step duration (hr), and K is the control gain $(0 < K < 1), k$ is the time step $(k = 0, 1, \ldots)$ The variable $sign(\rho(k) - \rho_{cr})$ equals 1 when $\rho(k)$ is greater than ρ_{cr}. Otherwise, $sign(\rho(k) - \rho_{cr})$ equals -1. The complete derivation of the above control law, which is beyond the scope of this paper, is provided by Kachroo and Ozbay (2003).

3.2 Motivation

3.3 Description of the Coordinated Version of MIXCROS

The coordinated ramp metering problem refers to a freeway system that has ramps on it at various points. The challenge lies in how the ramp metering should be designed while taking into account the interactions among the different ramps. The coordinated ramp problem is illustrated in Fig. 3.1 below. In designing this new feedback-based ramp metering strategy, the coordinated ramp metering problem is expressed as the problem of controlling the traffic density on the mainline while minimizing the on-ramp queues through the use of assiduously calibrated weight parameters for the freeway and each on-ramp, namely, $w_{1(i)}$ and $w_{2(i)}$. Decoupled or coupled approaches (D-MIXCROS and C-MIXCROS, respectively) can be used as a solution.

3.4 Freeway Traffic Model

The basic model used for the design of the coordinated MIXCROS control law is shown in Fig. 3.1.

In this figure, n is the number of freeway sections, f_i (veh/hr) is the flow entering the freeway at the first section, q_i is the flow leaving the freeway section i (veh/hr), r_i is the flow entering the ramp (veh/hr), u_i is the metered flow (veh/hr), ρ_i is the freeway density (veh/mi), $\rho_{c(i)}$ is the critical density (veh/mi), T is the time step duration (hr), $w_{1(i)}$ and $w_{2(i)}$ are the weight factors $(w_{1(i)} + w_{2(i)} = 1)$, $queue_{\text{ramp}i}$ is the queue length on the ramp (veh/mi), K_i is the control gain $(0 < K < 1)$, and L_{fi} and L_{ri} are the length of the freeway and ramp section i(mi), respectively.

In time Δt, the traffic density of the section of length Δx_{fi} changes from $\rho_i(t)$ to $\rho_i(t + \Delta t)$. This change is caused by the effective inflow at the section. The effective inflow is given by the sum of the freeway and ramp inflows after removing the outflow from the sum. This relationship in an equation form is

$$\rho_i(t+\Delta t)-\rho_i(t)=\Delta t\frac{(-q_{i+1}(t)+u_i(t)+q_i(t))}{\Delta x_{fi}}.$$

By placing the function for infinitesimal time on the left-hand side of the equation and taking the limits, the following equation is created

$$\lim_{\Delta t\to\infty}\frac{\rho_i(t+\Delta t)-\rho_i(t)}{\Delta t}=\frac{(-q_{i+1}(t)+u_i(t)+q_i(t))}{\Delta x_{fi}}.$$

This can be expressed as the following ordinary differential equation:

$$\dot{\rho}_i=\frac{\mathrm{d}\rho_i(t)}{\mathrm{d}t}=\frac{1}{\Delta x_{fi}}(-q_{i+1}(t)+u_i(t)+q_i(t)).$$

The ramp dynamics is derived using conservation equation on the ramp. In time Δt, the "amount" of vehicles that have entered the ramp is expressed by $queue_{\text{ramp}i}(t+\Delta t)-queue_{\text{ramp}i}(t)$. Due to the conservation law, this should equal the change caused by the inflow and outflow during the same time, given by $\frac{(r_i(t)-u_i(t))}{\Delta x_{ri}}\Delta t$. By equating both as

$$queue_{\text{ramp}i}(t+\Delta t)-queue_{\text{ramp}i}(t)=\frac{(r_i(t)-u_i(t))}{\Delta x_{ri}}\Delta t$$

and observing limits, the ramp dynamics are obtained:

$$\lim_{\Delta t\to 0}\frac{queue_{\text{ramp}i}(t+\Delta t)-queue_{\text{ramp}i}(t)}{\Delta t}=\lim_{\Delta t\to 0}\frac{(r_i(t)-u_i(t))}{\Lambda x_{ri}},$$

$$(queue_{\text{ramp}})'=\frac{(r_i(t)-u_i(t))}{\Delta x_{ri}}.$$

Here $(queue_{\text{ramp}i})'$ indicates differentiation with respect to time. Combining the above equations, the overall system dynamics can be written as

$$\text{System Dynamics}:\begin{cases}\dot{\rho}_i=\frac{\mathrm{d}\rho_i(t)}{\mathrm{d}t}=\frac{1}{\Delta x_{fi}}(-q_{i+1}(t)+u_i(t)+q_i(t))\\(queue_{\text{ramp}i})'=\frac{1}{\Delta x_{ri}}(-r_i(t)-u_i(t))\end{cases},\quad(3.1)$$

$$\text{Initial conditions}:\begin{cases}\rho_i(0)=\rho_{0(i)}\\queue_{\text{ramp}i}(0)=queue_{\text{ramp}0(i)}\end{cases}.$$

The flow relationship is

$$q_i(t)=\rho_i(t)v_i(t),$$

The speed relationship is taken as

$$v_i(t) = v_f\left(1 - \frac{\rho_i(t)}{\rho_{\text{jam}(i)}}\right)$$

3.5 The Control Objective

The control objective for this new control law is to make the error term approaches the value 0 asymptotically. That is,

$$\operatorname*{Lim}_{t\to\infty} e(t) = 0.$$

This can be achieved by designing a control law that makes the system follow the below closed-loop dynamics:

$$\dot{e}(t) + Ke(t) = 0\,(0 < K < 1). \tag{3.2}$$

The coordinated version of MIXCROS is applied to several on-ramps to provide network-wide improvements. It aims to maximize the throughput on all the freeway sections without creating extensive queues on all the metered ramps. Therefore, the error variable that accomplishes this is designed as

$$e(t) = \sum_{i=1}^{n}\left(\left|w_{1(i)}x_{1(i)}(t)\right| + \left|w_{2(i)}x_{2(i)}(t)\right|\right)_i, \tag{3.3}$$

where $x_{1(i)}(t) = \rho_i(t) - \rho_{\text{cr}(i)}$, $x_{2(i)}(t) = queue_{\text{ramp}i}(t)$, and $i = 1, 2, \ldots, n$, ("i"=section index).

The proportional-derivative state feedback control logic (3.2) and direct regulation of on-ramp queues (3.3) are employed in the derivation of this newly proposed coordinated ramp metering strategy.

The critical value for the density of the freeway section "i" is $\rho_{\text{cr}(i)}$ (Fig. 3.1), in which case the freeway flow approaches the value of q_{capacity} (Fig. 3.2).

This system can be represented in a vector form as

$$x(t) = [x_1(t)x_2(t)]^{\text{T}}.$$

Hence, substituting $x_{1(i)}$ and $x_{2(i)}$ into the control objective (3.3) yields

$$\begin{aligned} e(t) &= w_{1(1)}|\rho_1(t) - \rho_{\text{cr}(1)}| + w_{2(1)}queue_{\text{ramp1}}(t) + w_{1(2)}|\rho_2(t) - \rho_{\text{cr}(2)}| \\ &\quad + w_{2(2)}queue_{\text{ramp2}}(t) + \ldots + w_{1(n)}|\rho_n(t) - \rho_{\text{cr}(n)}| + w_{2(n)}queue_{\text{ramp}n}(t). \end{aligned} \tag{3.4}$$

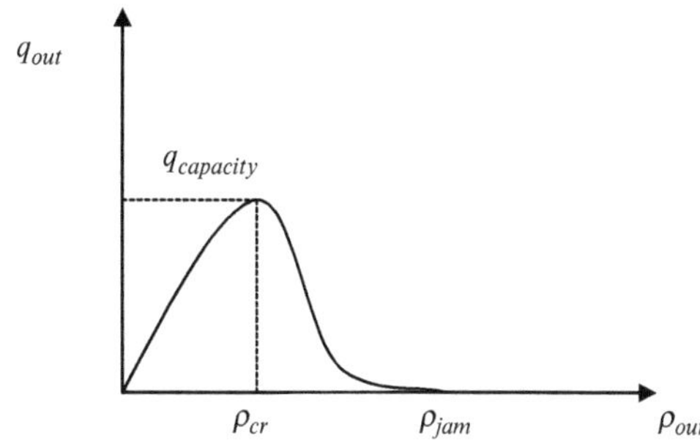

Fig. 3.2 The fundamental diagram (May 1990)

The error function is defined as the sum of the absolute values of the state variables, $x_{1(i)}$ and $x_{2(i)}$. The state variable $x_{1(i)}$ represents the freeway section of the network, and $x_{2(i)}$ represents the on-ramp queues. This function takes these state variables into account and determines how much importance should be given to freeway density and queue length on the ramp with the help of weights, $w_{1(i)}$ and $w_{2(i)}$.

Appropriate values of the parameters $w_{1(i)}$ and $w_{2(i)}$ are determined by taking the two objectives of the system into consideration simultaneously. That is, these parameters are selected in such a way that they ensure maximization of the throughput on the freeway ($w_{1(i)}$) without creating long queues ($w_{2(i)}$) on the ramp.

When $w_{1(i)}$ is greater than $w_{2(i)}$ (i.e., $0.5 < w_{1(i)} \leq 1$), in order to minimize the error function, the amount of variance from the critical density on the freeway segment is restricted by decreasing the amount of vehicles released from the on-ramp. In other words, choosing $w_{1(i)}$ as greater than $w_{2(i)}$ improves the freeway throughput. However, it can eventually lead to increased on-ramp delays.

If $w_{2(i)}$ is excluded by setting it to zero, then the term that considers the queue length on the ramp disappears from the error equation. This way, the queue length on the ramp is no longer included in the control law.

3.6 Coordinated MIXCROS Control Law

Coordinated MIXCROS control law can be best described with the assistance of a block representation of the algorithm (Fig. 3.3). The process under control is the traffic flow on n freeway sections with one on-ramp (Fig. 3.1).

Control systems are affected by certain process inputs called disturbances, which cannot be manipulated. In this system, represented by the block diagram in Fig. 3.3, q_i and r_i are the measurable disturbances, or freeway and ramp demands, respectively. The values for q_i and r_i are real-time data gathered from the detectors located on the upstream portion of the freeway and on the ramp, respectively. The states of the system, $x_{1(i)}$ and $x_{2(i)}$, are functions of these disturbances (i.e., q_i and r_i). Traffic density of the section and queue length on the ramp, both of which constitute the system outputs, are obtained using sensors located on the upstream

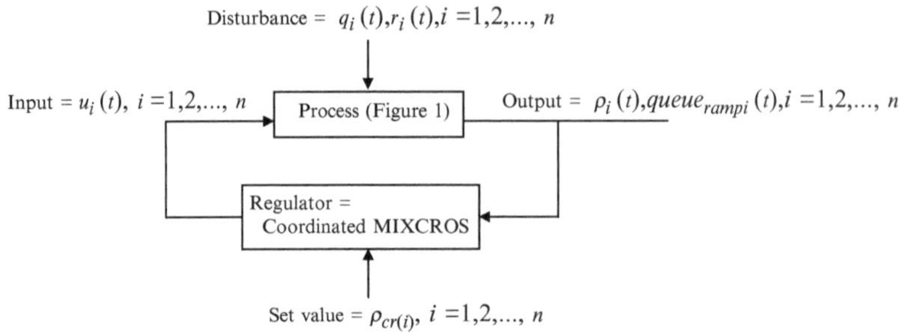

Fig. 3.3 The proposed traffic-responsive ramp metering control system

and downstream of the freeway and ramp sections. Then, these output data are fed back into the controller to obtain a new value for the system input u_i, which is the metered ramp flow. In the following section, the derivation of the discrete version of the feedback-based coordinated ramp metering strategies, namely, D-MIXCROS and C-MIXCROS, is introduced.

3.7 Derivation of the Discrete Version of Coordinated MIXCROS Control Law

The derivation of the discrete version of coordinated MIXCROS control law is obtained by time-discretizing the model. By time-discretizing Eq. (3.1), the following difference equation dynamics is obtained, which is still valid after replacing the time t variable by time instant k variable:

$$\text{System Dynamics}: \begin{cases} \frac{\rho_i(k+1)-\rho_i(k)}{T} = \frac{1}{\Delta x_{fi}}(-q_{i+1}(t)+u_i(t)+q_i(t)) \\ \frac{queue_{\text{rampi}}(k+1)-queue_{\text{rampi}}(k)}{T} = \frac{1}{\Delta x_{ri}}(r_i(k)-u_i(k)) \end{cases}.$$

In the above equation, T is the sampling time. Therefore, the system dynamics becomes

$$\text{System Dynamics}: \begin{cases} \rho_i(k+1) = \rho_i(k) + \frac{T}{\Delta x_{fi}}(-q_{i+1}(t)+u_i(t)+q_i(t)) \\ queue_{\text{rampi}}(k+1) = queue_{\text{rampi}}(k) + \frac{T}{\Delta x_{ri}}(r_i(k)-u_i(k)). \end{cases} \tag{3.5}$$

The feedback control design commences by incrementing Eq. (3.4). The system can be in 2^n regions, where n is the number of freeway sections in the network based on the first term in the error function $\left(\left|w_{1(i)}(\rho_i(t)-\rho_{\text{cr}(i)})\right|\right)$. These 2^n regions can be combined to form a control law that is applicable to all regions with the help of a function *sign* such that

$$sign = \begin{cases} 1 & \text{if } \rho_i(t) > \rho_{\mathrm{cr}(i)} \\ -1 & \text{else.} \end{cases}$$

So, the error function Eq. (3.4) becomes

$$\begin{aligned} e(k) = {} & w_{1(1)} sign\left[\rho_1(k) - \rho_{\mathrm{cr}(1)}\right] + w_{2(1)} queue_{\mathrm{ramp1}}(k) + w_{1(2)} sign\left[\rho_2(k) - \rho_{\mathrm{cr}(2)}\right] \\ & + w_{2(2)} queue_{\mathrm{ramp2}}(k) + \ldots + w_{1(n)} sign\left[\rho_n(k) - \rho_{\mathrm{cr}(n)}\right] + w_{2(n)} queue_{\mathrm{ramp}n}(k). \end{aligned} \tag{3.6}$$

Incrementing Eq. (3.6), the following is obtained:

$$\begin{aligned} & w_{1(1)} sign\left[\rho_1(k+1) - \rho_{\mathrm{cr}(1)}\right] + w_{2(1)} queue_{\mathrm{ramp1}}(k+1) + \\ e(k+1) = {} & w_{1(2)} sign\left[\rho_2(k+1) - \rho_{\mathrm{cr}(2)}\right] + w_{2(2)} queue_{\mathrm{ramp2}}(k+1) + \ldots \\ & + w_{1(n)} sign\left[\rho_n(k+1) - \rho_{\mathrm{cr}(n)}\right] + w_{2(n)} queue_{\mathrm{ramp}n}(k+1). \end{aligned}$$

Using the dynamics here,

$$\begin{aligned} e(k+1) = {} & w_{1(1)} sign\left[\rho_1(k) - \rho_{\mathrm{cr}(1)} + \frac{T}{\Delta x_{f1}}\left(-q_1(k) + u_1(k) + q_0(k)\right)\right] \\ & + w_{2(1)}\left[queue_{\mathrm{ramp1}}(k) + \frac{T}{\Delta x_{r1}}\left(r_1(k) - u_1(k)\right)\right] + \ldots \\ & + w_{2(n)}\left[queue_{\mathrm{ramp}n}(n) + \frac{T}{\Delta x_{rn}}\left(r_n(k) - u_n(k)\right)\right]. \\ & + w_{1(n)} sign\left[\rho_n(k) - \rho_{\mathrm{cr}(n)} + \frac{T}{\Delta x_{fn}}\left(-q_n(k) + u_n(k) + q_{n-1}(k)\right)\right] \\ e(k+1) = {} & w_{1(1)} sign\left[\rho_1(k) - \rho_{\mathrm{cr}(1)} + \frac{T}{\Delta x_{f1}}\left(-q_1(k) + q_0(k)\right)\right] \end{aligned}$$

Arranging terms on the right-hand side of the above equation gives

$$\begin{aligned} & + w_{2(1)}\left[queue_{\mathrm{ramp1}}(k) + \frac{T}{\Delta x_{r1}} r_1(k)\right] \\ & + \left[w_{1(1)} sign \frac{T}{\Delta x_{f1}} - w_{2(1)} \frac{T}{\Delta x_{r1}}\right] u_1(k) + \ldots \\ & + w_{1(n)} sign\left[\rho_n(k) - \rho_{\mathrm{cr}(n)} + \frac{T}{\Delta x_{fn}}\left(-q_n(k) + q_{n-1}(k)\right)\right] \end{aligned}$$

$$+w_{2(n)}\left[queue_{\mathrm{ramp}n}(k)+\frac{T}{\Delta x_{rn}}r_n(k)\right]$$

$$+\left[w_{1(n)}sign\frac{T}{\Delta x_{fn}}-w_{2(n)}\frac{T}{\Delta x_{rn}}\right]u_n(k).$$

This difference equation can be written in an organized form as

$$e(k+1)=F(k)+u(k), \tag{3.7}$$

where

$$F(k)=w_{1(1)}sign\left[\rho_1(k)-\rho_{\mathrm{cr}(1)}+\frac{T}{\Delta x_{f1}}(-q_1(k)+q_0(k))\right], \tag{3.8}$$

$$\begin{aligned}
u(k)=&\left[w_{1(1)}sign\frac{T}{\Delta x_{f1}}-w_{2(1)}\frac{T}{\Delta x_{r1}}\right]u_1(k)+\ldots\\
&+\left[w_{1(n)}sign\frac{T}{\Delta x_{fn}}-w_{2(n)}\frac{T}{\Delta x_{rn}}\right]u_n(k)\\
&+w_{2(1)}\left[queue_{\mathrm{ramp}1}(k)+\frac{T}{\Delta x_{r1}}r_1(k)\right]+\ldots\\
&+w_{1(n)}sign\left[\rho_n(k)-\rho_{\mathrm{cr}(n)}+\frac{T}{\Delta x_{fn}}(-q_n(k)+q_{n-1}(k))\right]\\
&+w_{2(n)}\left[queue_{\mathrm{ramp}n}(k)+\frac{T}{\Delta x_{rn}}r_n(k)\right].
\end{aligned} \tag{3.9}$$

In order to obtain Eq. (3.2), the coordinated MIXCROS control law is designed as

$$u(k)=-F(k)-Ke(k). \tag{3.10}$$

Substituting Eq. (3.10) in Eq. (3.7) satisfies the desired dynamics (3.2) in all 2^n regions. In truth, Eq. (3.10) does not accord the control laws, but it provides the condition that the control variables should satisfy. The control law can be designed in a decoupled way or coupled way.

3.7.1 Decoupled Control: D-MIXCROS

The control law can be designed in a decoupled fashion by ignoring the ramp connections totally and treating each ramp as an isolated problem. Therefore, the F (Eq. (3.8)) term in the control law condition equation (Eq. (3.10)) is divided into n components, where n is the number of freeway section with one on-ramp (Fig. 3.1):

$$F(k)=F_1(k)+F_2(k)+\ldots+F_n(k),$$

where

$$F_1(k) = w_{1(1)} sign \left[\rho_1(k) - \rho_{\text{cr}(1)} + \frac{T}{\Delta x_{f1}}(-q_1(k) + q_0(k))\right]$$
$$+ w_{2(1)} \left[queue_{\text{ramp1}}(k) + \frac{T}{\Delta x_{r1}} r_1(k)\right],$$
$$F_n(k) = w_{1(n)} sign \left[\rho_n(k) - \rho_{\text{cr}(n)} + \frac{T}{\Delta x_{fn}}(-q_n(k) + q_{n-1}(k))\right]$$
$$+ w_{2(n)} \left[queue_{\text{rampn}}(k) + \frac{T}{\Delta x_{rn}} r_n(k)\right].$$

Then, using Eq. (3.9), decoupled coordinated control law, namely, D-MIXCROS, for each on-ramp can be derived as

$$u_1(k) = \left(sign(\rho_1(k) - \rho_{\text{cr}(1)}) w_{1(1)} \frac{T}{\Delta x_{f1}} - w_{2(1)} \frac{T}{\Delta x_{r1}}\right)^{-1} (-F_1(k) - Ke_1(k)),$$
$$u_n(k) = \left(sign(\rho_n(k) - \rho_{\text{cr}(n)}) w_{1(n)} \frac{T}{\Delta x_{fn}} - w_{2(n)} \frac{T}{\Delta x_{rn}}\right)^{-1} (-F_n(k) - Ke_n(k)), \quad (3.11)$$

where

$$e(k) = e_1(k) + e_2(k) + \ldots + e_n(k).$$

The error terms are defined as

$$e_1(k) = w_{1(1)} |\rho_1(k) - \tilde{n}_{\text{cr}(1)}| + w_{2(1)} queue_{\text{ramp1}}(k),$$
$$e_n(k) = w_{1(n)} |\rho_n(k) - \rho_{\text{cr}(n)}| + w_{2(n)} queue_{\text{rampn}}(k).$$

The following n decoupled closed-loop dynamics is obtained by the application of Eq. (3.11):

$$e_1(k+1) + Ke_1(k) = 0,$$
$$e_n(k+1) + Ke_n(k) = 0.$$

3.7.2 Coupled Control: C-MIXCROS

By substituting Eq. (3.10) in Eq. (3.9), the following equation is obtained:

$$-F(k) - Ke(k) = \left[w_{1(1)} sign \frac{T}{\Delta x_{f1}} - w_{2(1)} \frac{T}{\Delta x_{r1}}\right] u_1(k) + \ldots$$
$$+ \left[w_{1(n)} sign \frac{T}{\Delta x_{fn}} - w_{2(n)} \frac{T}{\Delta x_{rn}}\right] u_n(k). \quad (3.12)$$

Table 3.1 Calibration parameters for each ramp control implementation

Ramp metering strategy	Calibration parameters for each on-ramp	Total number of calibration parameters for 6 on-ramps
ALINEA	K_R	6
New Control	K_R	6
MIXCROS	K_R, w_1, and w_2	12
METALINE	K_{R1} and K_{R2}	12
D-MIXCROS	K_R, w_1, and w_2	7
C-MIXCROS	K_R, α, w_1, and w_2	13

Using Eq. (3.12), the control effort among the n on-ramps can be distributed as:

$$\alpha_1(-F(k)-Ke(k)) = \left(sign(\rho_1(k)-\rho_{\text{cr}(1)})w_{1(1)}\frac{T}{\Delta x_{f1}} - w_{2(1)}\frac{T}{\Delta x_{r1}}\right)u_1(k),$$

.

.

.

$$\alpha_n(-F(k)-Ke(k)) = \left(sign(\rho_n(k)-\rho_{\text{cr}(n)})w_{1(n)}\frac{T}{\Delta x_{fn}} - w_{2(n)}\frac{T}{\Delta x_{rn}}\right)u_n(k),$$

where

$$\alpha_1+\alpha_2+\ldots+\alpha_n = 1.$$

Therefore, the coupled control laws for the coordinated version of MIXCROS, namely, C-MIXCROS, can be written as

$$u_1(k) = \left(sign(\rho_1(k)-\rho_{\text{cr}(1)})w_{1(1)}\frac{T}{\Delta x_{f1}} - w_{2(1)}\frac{T}{\Delta x_{r1}}\right)^{-1}\alpha_1(-F(k)-Ke(k)),$$

$$u_n(k) = \left(sign(\rho_n(k)-\rho_{\text{cr}(n)})w_{1(n)}\frac{T}{\Delta x_{fn}} - w_{2(n)}\frac{T}{\Delta x_{rn}}\right)^{-1}\alpha_n(-F(k)-Ke(k)).$$

In Table 3.1, the calibration parameters for each ramp metering strategy are presented. As it is evident from this table, local controls, namely, ALINEA and New Control, require the least amount of parameters. On the other hand, coordinated control, namely, D-MIXCROS, has only seven calibration parameters, rendering it advantageous compared with the other coordinated control evaluated, METALINE. The juxtaposition of the performances of each ramp metering strategy is provided in the following section based on the macroscopic simulation environment modeling.

D-MIXCROS is similar to the local version of MIXCROS because it takes each on-ramp system into account separately. However, it differs from the local version because D-MIXCROS uses the same control gain K_R for all the ramps, which ensures a unity in the actions of metered ramps. From a practical view, it is more manageable to implement D-MIXCROS than the local version of MIXCROS on a number of ramps. D-MIXCROS implementation on six on-ramps requires seven parameters to be calibrated, whereas MIXCROS uses 12 calibration parameters (Table 3.1). Hence, it is less complicated to install D-MIXCROS compared with the local version of both MIXCROS and C-MIXCROS on a number of ramps.

In C-MIXCROS, the control effort is apportioned among all on-ramps using distribution factor α_i, which provides the communication between on-ramp systems. With this allocation ratio α_i, the nature of the congestion in each ramp system can be handled meticulously. For example, if the congestion on the second ramp is propagating toward the other on-ramp locations, the allocation ratio, α_2, for the second on-ramp can be reduced, making it less than the other ratios ($\alpha_2 < \alpha_1, \alpha_3$).

3.8 Macroscopic Simulation Model

The proposed coordinated ramp metering strategies, C-MIXCROS and D-MIXCROS, and ALINEA, MIXCROS, and METALINE are tested on six consecutive ramps along a corridor. Each ramp system consists of a 1-lane (1 mi) freeway link and a 1-lane (0.5 mi) ramp link. The simulation duration for each tested case is 300 min.

In both macroscopic and microscopic simulation models, for ALINEA, New Control, and METALINE implementations, a queue threshold of 35 vehicles is used. New Control and METALINE, as well as all versions of MIXCROS, perform satisfactorily without a queue override strategy that shuts off the ramp metering and creates unwanted fluctuations. In ALINEA implementation, for the values of parameter K_R above 240 veh/hr, on-ramp queues are decreased, whereas ramp metering provides no improvement on the downstream traffic conditions in the ramp systems. Thus, the purpose of the control, which is to maintain the downstream freeway section at the set level, is not accommodated. Therefore, the regulator parameter, K_R, is limited between 70 veh/hr and 240 veh/hr for all the ramps.

All the tested ramp metering strategies maintain the freeway outflow close to the capacity while keeping the traffic density below critical density. All the controls except D-MIXCROS and C-MIXCROS experience high fluctuations in the traffic density within the first 50 min of simulation. Queue override tactics employed in these controls mainly induce this problem. That is, these controls use the storage capacity of the on-ramps, which leads to increased traffic flow on the freeway sections. This increased traffic flow results in congestion in downstream locations, causing more restrictive ramp metering rates so as to serve additional throughput from the upstream ramp systems.

Among all tested controls, METALINE has the largest on-ramp queues owing to its restrictive metering. Decreasing freeway demand by 5.26% (Demand Scenario 1) leads to increase in the freeway maximum outflow (throughput) with each ramp metering strategy. With reduced freeway demand, each ramp metering strategy results in approximately the same total travel time. To observe the behavior of

Table 3.2 Overall network results for four demand scenarios

	Base demand		Demand Scenario 1		Demand Scenario 2		Demand Scenario 3	
	A	B	A	B	A	B	A	B
ALINEA	13.73	247.17	13.02	239.58	19.62	239.70	15.21	198.29
New Control	12.70	246.21	12.22	238.36	19.24	239.57	15.29	198.60
MIXCROS	11.92	248.65	12.06	238.44	18.37	238.77	14.91	198.17
METALINE	79.87	314.61	13.44	243.43	459.95	680.33	19.55	201.43
D-MIXCROS	11.92	248.65	12.08	238.43	18.37	238.76	14.91	198.17
C-MIXCROS	11.92	244.56	12.06	238.44	18.37	238.77	14.91	198.18

the controls in the presence of heavy ramp demand (Demand Scenario 2), the ramp demand is increased by 67% compared with the base demand scenario, lowering the freeway demand by 33% because of the limited capacity of the freeway segments. With this demand configuration, all versions of MIXCROS provide superior individual ramp performance results (e.g., increased average freeway downstream flow, speed, and density) compared with all the strategies tested. In Demand Scenario 3, ramp demand is increased only by 33%, whereas freeway demand is lowered by 33%; all ramp metering strategies, both local and coordinated, provide almost the same improvements at the network level. Because of light ramp demand, METALINE is also able to keep the on-ramp queues at reasonable levels with the help of a queue override tactic (Table 3.2). In this table, "A" refers to the total travel time on all 6 ramps (veh.hr/hr) and "B" stands for the total travel time in the network (veh.hr/hr).

3.9 Microscopic Simulation Model

Figure 3.4 shows a screen capture of the PARAMICS model of model of the section of I-295 in South Jersey, created using the available geometric and traffic demand data.

A PARAMICS model of the section of I-295 in South Jersey is created using the available geometric and traffic demand data. The calibrated and validated model of the 11-mile-long 3-lane freeway section includes the junctions of I-295 with Route 38, State HWY 73, State HWY 70, and Berlin Rd. Each on-ramp has 1 lane. Then, an Application Programming Interface (API) is coded to assign green times based on each tested control law to all 4 on-ramps in PARAMICS. In the API file, it is ascertained that the calculated green phase duration is within specified limits (i.e., minimum and maximum values are 2 and 15 s, respectively). Statistics are collected for 3-h simulations from the detectors located downstream and upstream of the ramp and two additional detectors, one at the exit and one at the entrance of the ramp. In the microscopic simulation model, the proposed ramp metering controls (namely, D-MIXCROS and C-MIXCROS), ALINEA, and MIXCROS are

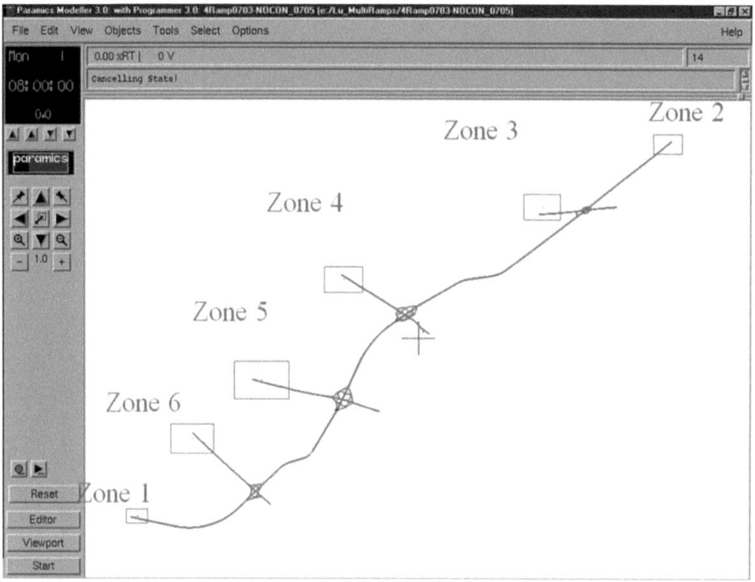

Fig. 3.4 PARAMICS model of the study network

Table 3.3 Congestion levels on each ramp

	1st Ramp (%)	2nd Ramp (%)	3rd Ramp (%)	4th Ramp (%)
1st Demand Level	27	60	68	20
2nd Demand Level	53	39	58	9
3rd Demand Level	0	17	35	24

evaluated and compared with the No Control scenario using three demand scenarios, whose congestion levels are listed in Table 3.3. The congestion level is the percent of time that the downstream link occupancy is greater than the critical occupancy. All simulations are run for 3 h with different seed values for the statistical analysis of the results (which ensures a 95% confidence level).

Ramp metering controls seem to be more effective under certain demand patterns than others. As traffic demand increases, ramp metering tends to more consistently reduce the system travel time. The reason for reduced ramp metering performance is that the third demand scenario represents the low level of congestion on each ramp system. It was also claimed in other studies that the effectiveness of ramp control varies depending on the severity of congestion (Papageorgiou 1988). Table 3.4 summarizes the main findings of this implementation. All controls tested except ALINEA reduced the average travel time regardless of the demand scenario compared with No Control. For the demand scenarios tested, both C-MIXCROS and D-MIXCROS led to maximum improvement for all the performance criteria.

Table 3.4 Overall network results for three demand scenarios

	Demand Scenario 1		Demand Scenario 2		Demand Scenario 3	
	Tot. Travel	Mean Speed (mph)	Tot. Travel Time (veh.hr)	Mean Speed (mph)	Tot. Travel Time (veh.hr)	Mean Speed (mph)
No Control	3723.57	55	4119.18	49.6	3408.15	57.8
ALINEA	3901.73	52.13	4137.14	49.43	3368.08	58.23
Change (%)	4.78	−5.23	0.44	−0.35	−1.18	0.74
MIXCROS	3674.04	55.63	3976.06	51.4	3354.74	58.25
Change (%)	−1.33	1.14	−3.47	3.63	−1.57	0.78
D-MIXCROS	3552.71	57.28	3801.35	54.1	3394.75	57.83
Change (%)	−4.59	4.14	−7.72	9.07	−0.39	0.43
C-MIXCROS	3546.94	57.25	3938.86	52.15	3354.98	58.05
Change (%)	−4.74	4.09	−4.38	5.14	−1.56	0.43

3.10 Conclusions

Evaluation of the new coordinated ramp metering strategy is performed to demonstrate its characteristics and eventually its impact on the ramp system and whole network in two phases. The first phase includes the macroscopic testing of the proposed coordinated ramp metering controls using RMSE (Rutgers Macroscopic Simulation Environment) to compare it with three local (ALINEA, New Control, and MIXCROS) ramp metering controls, as well as one coordinated (METALINE), under the various demand scenarios. The second phase involves evaluating the proposed methodology using a microscopic simulation environment (PARAMICS) under three different demand scenarios.

From these implementations, it is found that the system performs more competently after the implementation of the coordinated version of MIXCROS, namely, C-MIXCROS and D-MIXCROS, compared with other ramp metering controls. As expected, the mainline freeway experiences more favorable traffic conditions when any tested ramp metering control is implemented. However, when the queue thresholds are used in ALINEA, New Control, and METALINE to prevent the ramps from being overloaded, the system benefits of these strategies are reduced. C-MIXCROS and D-MIXCROS significantly improve system performance compared with other controls under various demand conditions, and they are proven to be quite effective in general. Well-tuned parameters are critical to achieve successful ramp metering performance. As opposed to some coordinated ramp metering controls that employ optimization techniques, parameter calibration for C-MIXCROS and D-MIXCROS is comparatively unburdensome.

References

Bogenberger K, May AD. Advanced coordinated traffic responsive ramp metering strategies. In: California PATH Working Paper, UCB-ITS-PWP-99–19, 1999.

Bhouri N, Haj Salem H, Kauppila J. Isolated versus coordinated ramp metering: Field evaluation results of travel time reliability and traffic impact. Transport Res C 2011. Elsevier (Available On-Line, 2011). doi:10.1016/j.trc.2011.11.001.

Chu L, Liu HX, Recker W, Zhang HM. Development of a simulation laboratory for evaluating ramp metering algorithms. In: UCI-ITS-WP-01–23, 2001.

Hasan M, Jha M, Ben-Akiva M. Evaluation Of ramp control algorithms using microscopic traffic simulation. Transport Res C 2002;10(3):229–256.

Hou Z.S., Xu J.X., Yan J.W, An iterative learning approach for density control of freeway traffic flow via ramp metering, Transportation Research Part C, (16) 71–97, 2008.

Kachroo P, Ozbay K. Feedback ramp metering in intelligent transportation systems. New York: Kluwer Academic/Plenum Publishers; 2003.

Kotsialos A, Kosmatopoulos E, Papageorgiou M. Current status of ramp metering. In: European Ramp Metering Project, IST-2002–23110, 2004.

Jacobsen L, Henry K, Mahyar O. Real-time metering algorithm for centralized control. Transport Res Rec TRB 1989;1232. pp. 17–26.

Lipp L., Corcoran L., and Hickman G, *Benefits of central computer control for the Denver ramp metering system*, Transportation Research Board Record 1320, January 1991.

Liu J.C., Kim J., Chen Y., Hao Y., Lee S., Kim T. and Thomadakis M., *An advanced real time metering system (ARMS): the system concept*, Texas Department of Transportation Report Number 1232–24, 1993.

Masher DP, Ross DW, Wong PJ, Tuan PL, Zeidler S, Peracek S. Guidelines for design and operating of ramp control systems. In: Stanford Research Institute Report NCHRP 3–22, SRI Project 3340, Menid Park, CA: SRI; 1975.

May, A. D., Traffic Flow Fundamentals, Prentice-Hall, 1990.

Meldrum D. and Taylor, C. *Freeway traffic data prediction using artificial neural networks and development of a Fuzzy Logic Ramp Metering Algorithm*, Final Technical Report, Washington State Department of Transportation Report No. WA-RD 365.1, Washington, 1995.

Ozbay K, Yasar I, Kachroo P. Comprehensive evaluation of feedback based freeway ramp metering strategy by using microscopic simulation: taking ramp queues into account. In: Transportation Research Record, No. 1867, TRB. Washington, D.C.: National Research Council; 2004. pp. 89–96.

Papageorgiou M. Modelling and real-time control of traffic flow on the southern part of Boulevard Peripherique in Paris. Internal report. INRETS-DART, Arcueil, France, 1988.

Papageorgiou M, Blossville J-M, Hadj-Salem H. Modelling and real-time control of traffic flow on the southern part of Boulevard Peripherique in Paris, Part II: Coordinated on-ramp metering. Transport Res 1990;24A:361–370.

Papageorgiou M, Habib HS, Blosseville J-M. ALINEA: A local feedback control law for on-ramp metering. In: Transportation Research Record, No. 1320, TRB. Washington, D.C.: National Research Council; 1991. pp. 58–64.

Papageorgiou M and Kotsialos A. Freeway Ramp metering: An overview. IEEE Transactions on Intelligent Transportation Systems 2002;3(4): 271–281.

Paesani G., Kerr J., Perovich P. and Khosravi E., *System wide adaptive ramp metering in southern California*, ITS America *7th* Annual Meeting, June 1997.

Stephanedes Y, *Implementation of on-line Zone Control Strategies for optimal ramp metering in the Minneapolis Ring Road*, *7th* International Conference on Road Traffic Monitoring and Control, 1994.

Wei C.H. and Wu K.Y., *Applying an artificial neural network model to freeway ramp metering control*, Transportation Planning Journal, Vol. 25 No. 3, 1996.

Yasar, I. Modeling and PARAMICS based evaluation of new local freeway ramp metering strategy that takes into account ramp queues, M.S. Thesis, Rutgers University 2003.

Yin Y, Liu H, Benouar H. A note on equity of ramp metering. In: IEEE intelligent transportation systems conference, Washington, D.C., USA, October 3–6, 2004.

Yoshino T., Sasaki T and Hasegawa T, *The traffic control system on the Hanshin Expressway*, Interfaces Magazine, Jan./Feb. 1995.

Zhang HM, Recker WW. On optimal freeway ramp control policies for congested traffic corridors. Transport Res B Methodological 1999;33(6):417–436.

Zhang M, Kim T, Nie X, Jin W, Chu L, Recker W. Evaluation of on-ramp control algorithms. In: California PATH Research Report, UCB-ITS-PRR-2001–36, December 2001.

Chapter 4
Solving the Integrated Corridor Control Problem Using Simultaneous Perturbation Stochastic Approximation

Jingtao Ma, Yu (Marco) Nie, and H. Michael Zhang

Abstract Integrating various control measures within a transportation corridor promises to improve the overall operational performance of the entire corridor. In this study, we formulate a corridor traffic control problem that considers two control actions: signal timing and ramp metering, and propose a solution method for the formulated problem. In the formulation, traffic dynamics within a general corridor is modeled on a coherent platform based on the kinematic wave traffic flow model, and the traffic control actions of urban street signals and ramp meters are embedded in the platform. The solution algorithm based on the simultaneous perturbation stochastic approximation (SPSA) scheme is developed to solve the integrated control problem. Numerical experiments show that the SPSA algorithm strikes a better balance between computational efficiency and solution quality compared to other heuristics such as the genetic algorithm (GA) and the hill-climbing algorithm.

J. Ma
Mygistics, Inc., 9755 SW Barnes Rd, Suite 550, Portland, OR 97225, USA
e-mail: jma@mygistics.com

Y. Nie
Department of Civil and Environmental Engineering, Northwestern University,
2145 Sheridan Road, Evanston, IL 60208, USA
e-mail: y-nie@northwestern.edu

H.M. Zhang (✉)
University of California, Davis, Dept. of Civil & Environmental Engineering,
One Shields Avenue, Davis, CA 95616, USA

Tongji University, School of Transportation Engineering, 1239 Siping Road,
Shanghai, China 200092
e-mail: hmzhang@ucdavis.edu

S.V. Ukkusuri and K. Ozbay (eds.), *Advances in Dynamic Network Modeling in Complex Transportation Systems*, Complex Networks and Dynamic Systems 2,
DOI 10.1007/978-1-4614-6243-9_4, © Springer Science+Business Media New York 2013

4.1 Introduction

A transportation corridor is operationally (rather than geographically or organizationally) defined as "a combination of discrete parallel surface transportation networks (e.g., freeway, arterial, transit networks) that link the same major origins and destinations" (Federal Highway Administration 2005). A corridor usually includes various types of facilities (e.g., freeway sections, ramps, and urban streets), which are typically managed by different agencies and jurisdictions. In the current practice, most corridors are operated separately with little consideration to the coordination of individual facilities (Wood 1994; Zhang 2001), although it has long been recognized that integrating the control measures can improve the operational performance of the entire corridor (e.g., (van Zuylen and Taale 2003)).

Two components are fundamental to modeling an integrated corridor control system (e.g., Chang and Stephanedes 1993). The first is the traffic flow model that realistically represents traffic evolution, and the other is the optimization method that generates optimal control plans. Three major categories of traffic flow models have been developed and applied in traffic control studies: the point-queue (P-Q) or vertical queue model, the spatial queue (S-Q) or horizontal queue model and the Lighthill–Whitham–Richards (LWR) model. Most studies, including the classical ones such as Webster's (1958) and later HCM methods, used the P-Q model. In this model the vehicles are assumed to travel at the design speed uniformly along the road section and arrive at the stop line at a constant rate. The vehicles behind the stop line take no physical space and will be discharged at the saturation flow rate during the effective green time. The platoon dispersion model in TRANSYT (Robertson 1969) uses an empirical formula to depict the cyclic flow profiles (CFP) on road sections and thus relaxes the constant arrival assumption, but vehicles are still queued at the stop line. TRANSYT version 8 (Vincent et al. 1980) allows vehicles to join at the end of the stopped queue. Link storage capacity constraint is enforced and no traffic can enter a link if it is occupied by stopped vehicles. This model, known as the spatial queue (S-Q) model, has seen more applications recently in other traffic control studies (Diakaki et al. 2000; Papageorgiou 1995; Shelby 2001).

Both the P-Q and S-Q models can provide good estimates of the *queue size*, i.e., the number of stopped vehicles, particularly under low to medium traffic loads. When the traffic load is high enough to keep the intersection near or over saturated, the traffic densities behind the stop line will be in frequent transitions due to varying arrival rates and intermittent signal services (Stephanopoulos et al. 1979). Shockwaves and acceleration waves, interfaces between two differing traffic states, will be generated in such a complicated way that neither the P-Q nor the S-Q model could capture the spatial extent of queue formations and dissipations. Consequently, *queue lengths* cannot be estimated accurately. In this study, queue length is stated as "the length of the roadway section behind the stop line where traffic conditions are in the congested region of the flow-density curve, i.e., they range from capacity to jammed." (Stephanopoulos et al. 1979). Michalopoulos and

Stephanopoulos (1977) argued that under these circumstances the control action would be dictated by minimizing queue length instead of delay. Stephanopoulous et al. (1979) incorporated the more elaborate LWR model to analyze the complicated queuing phenomena at the signal intersections. In their analysis the linear speed-density relation (Greenshields 1934) and the resulting parabola fundamental diagram (flow-density curve) were used to compute the maximum queue length analytically. Using the triangular fundamental diagram, Helbing (2003) recently derived the formula for queue dynamics and travel time variations with respect to the arrival and departure flow rates, but the derivation simplifies the LWR model into *de facto* spatial queue (S-Q) model. A self-organized control method was later developed for urban signals based on these results (Helbing et al. 2005).

Researchers have made use of finite difference solution schemes to the LWR model, such as the cell transmission model (CTM) (Daganzo 1994; 1995), in traffic control studies. A linear transformation of the CTM model has been carefully designed to study the global optimal ramp metering strategies (Gomez and Horowitz 2004a; 2004b). In an earlier work, Lo (1999; 2001) also modified the original CTM to formulate the signal control problem into a mixed integer program. The program only considered the intersections without turns. He later applied a genetic algorithm (GA)-based solution algorithm to optimize control plans for more general intersection layouts (Lo et al. 2001b).

The optimization methods used to compute the optimal controls plans are highly tied to the underlying traffic flow models. For instance, in (Diakaki et al. 2000; Papageorgiou 1995), the researchers used the store-and-forward approach to depict the flow dynamics of urban streets, ramps, and freeway mainline. This approach is essentially similar to the S-Q model, and the formulated integrated corridor control problem is a linear one with a sparse constraint set, for which highly efficient algorithms exist. But typically the store-and-forward approach requires the control updating time period to be no less than the common cycle length; this feature rules out the possibility of synchronizing the control actions and thus make the model only suitable as a strategic queue-management tool (Papageorgiou 1995). Papageorgiou et al. adopted the high-order flow model in METANET (Messmer and Papageorgiou 1990) and studied integrated ramp metering and variable message sign (VMS) controls, where conjugate gradient algorithms were deployed to solve the integrated control problem (Kotsialos et al. 2002). The same algorithm was applied in (Chang and Stephanedes 1993) as well, where a forward time centered space method was used to model traffic evolution. Similar to the work of (Kotsialos et al. 2002), the resulting system state equations are also twice-differentiable. However, both studies can only guarantee local optima, which can be sensitive to the initial guess of the solution (Chang and Stephanedes 1993).

To summarize, mathematical programming methods (e.g., Chang and Stephanedes 1993, Diakaki et al. 2000, Papageorgiou 1995, and Kotsialos et al. 2002) usually require the traffic flow models to be simplified so that the gradient information can be computed. Unfortunately such a simplification often compromises the fidelity of the underlying traffic flow models. On the other hand, heuristic optimization methods such as the genetic algorithm can search for a near-global optimal control plan while allowing more realistic representation of traffic flow (e.g., Lo et al. 2001b). However, heuristic methods usually need a large number of evaluations of system performance and usually lead to high computational costs.

In this paper, we explore a stochastic approximation technique that can be viewed as a compromise of the above two types of approaches. The proposed simultaneous perturbation stochastic approximation (SPSA) has been used in other fields (Spall 1998) and demonstrated encouraging performances. In this study, an SPSA-based algorithm is developed to compute the time-of-day optimal corridor control plan, while the corridor operational performances under various control plans are evaluated on a CTM-based platform. The platform embeds signal control and ramp metering and can be easily applied to any general traffic corridor network. Numerical examples are used to investigate the effectiveness of the method as compared to other heuristics methods. Practical guidelines of applying the SPSA method are also discussed before we conclude the study.

4.2 Modeling Dynamic Network Flow

This section introduces a network flow model based on cell transmission model (CTM), in which the control actions from traffic signals and ramp meters are embedded.

4.2.1 Flow Dynamics on a General Corridor Roadway Section

The LWR model states the following:

$$\frac{\partial q}{\partial x}+\frac{\partial \rho}{\partial t}=0 \quad q=f(x,\rho,t) \tag{4.1}$$

where q is the flow rate on a road section, ρ is the density, x and t are the space and time variables, respectively. In (Daganzo 1994) Daganzo developed a stable numerical scheme that solves the LWR model. He shows that if the relationship between traffic flow q and density ρ is in the form

$$q=\min\{v\rho,\ q_{\max},\ w(\rho_j-\rho)\} \tag{4.2}$$

where v is the free flow speed, $q_{\max}$ is the maximum flow rate, w is the backward shockwave speed and ρ_j is the jam density, then LWR model can be approximated by a set of difference equations. The model discretizes the entire time horizon T (assignment period) into small intervals t, called the *loading interval* in this paper. Conforming to the loading interval, the model divides every road section of the network into homogeneous segments called cells, in a way that the cell length equals the distance traversed by one typical vehicle at free flow speed in one loading interval. The flows are updated by the following difference equations:

$$y_i(t) = \min\{n_{i-1}(t),\ q_{i,\max},\ \delta(N_i - n_i(t))\} \tag{4.3}$$

and

$$n_i(t+1) = n_i(t) + y_i(t) - y_{i+1}(t) \tag{4.4}$$

where $y_i(t)$, $y_{i+1}(t)$are the number of vehicles entering cell i and $i+1$ at time t, respectively, $n_{i-1}(t), n_i(t), n_{i+1}(t)$ are the numbers of vehicles in the cell $(i-1)$, i and $i+1$ at time t, respectively, $q_{i,\max}$ is the capacity flow into i at t, $N_i - n_i$ is the space available in i, $\delta = w/v$.

Essentially Eq. (4.4) states that the number of vehicles staying in cell i at loading interval $t+1$ is the number of vehicles from interval t plus the incoming vehicles and minus the outgoing vehicles. Daganzo (1995) extended the model to a general network by classifying roadway junctions into basic merges and diverges. Since control actions take places at junctions, we mainly focus on the flow updating rules at general junctions including signalized intersections and metered ramps.

4.2.1.1 Flow Updating at Signalized Urban Intersections

In (Lo 1999), Lo employed CTM to model the flow updates at urban intersections with a few changes. If the flow capacity $q_{\max}$ in Eq. (4.2) is replaced by one that depends on the signal timing variable $g_i(t)$,

$$q_{\max}(t) = \begin{cases} q_{\max} & t \in \text{green} \\ 0 & \text{otherwise} \end{cases}, \tag{4.5}$$

where it switches between $q_{\max}$ (green) and zero (red), the end cell of an intersection approach will serve as a functioning signal, and the flow dynamics still approximates the LWR model. At a typical intersection, traffic is grouped into movements *or* streams. At a generalized intersection (Fig. 4.1), the traffic movements can be decomposed into simple merges and diverges, where different flow updating rules must apply.

4.2.1.2 Signalized Diverges

The diverging flows occur where the traffic stream on a single link splits into left turn, through and right turn movements. Left or right turn bays are common to store the incoming vehicles, and these short sections must also be accommodated in the

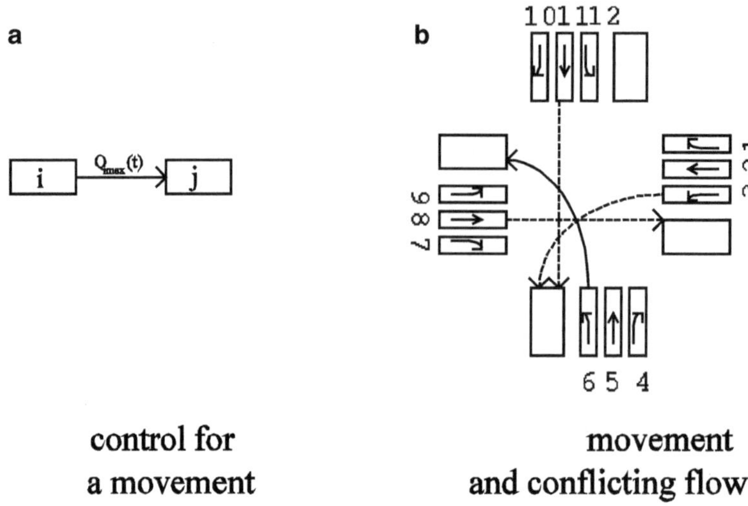

Fig. 4.1 A general representation of cell-based intersection movements

generalized model. Denote the end cell C_s^j of a link j approaching a signalized intersection, the flow conservation equation then reads:

$$n_s(t+1) = \sum_{m=L,R,T} n_{s-1}^m(t) + y_{s-1,s}^m(t) - \sum_{m=L,R,T} y_s^m(t) \tag{4.6}$$

The superscripts of L, R, T denote the left turn, right turn and through movement, respectively. The cell C_{s-1}^j is the preceding cell of C_s^j. The numbers of vehicles into and out of cell s are stated as:

$$y_{s-1,s}^m(t+1) = \min\{n_{s-1}(t),\ q_{s,\max},\ \delta_s(N_s^m - n_s^m)\} \tag{4.7}$$

$$y_{s,s+1}^m(t+1) = \min\{n_s^m(t),\ q_{s,\max}(t),\ \delta_{s+1}(N_{s+1}^m(t) - n_{s+1}^m(t))\},\ m = L,\ R,\ T, \tag{4.8}$$

where the notation naming convention follows (4.3) and (4.4). Note that N_{s+1}^m, $m = L, R, T$ in Eq. (4.7) are the different storage capacities for various movements, ensuring that different sizes of turning bays can be modeled accurately.

4.2.1.3 Signalized Merges

In this study, the right turns are explicitly considered in the signal timing optimization. In this way, the flow updating at intersections is simplified to be the same as a set of coupled consecutive links, which then reads:

$$n_{i+1}(t+1) = n_{i+1}(t) + y_{i+1}(t) - y_{i+2}(t), \tag{4.9}$$

where $(i+1)$ is the start cell index for the downstream link, i.e., the first cell of the downstream link that receives the stream with cell index of i serviced by the signal. The incoming flux $y_{i+1}(t)$ is then determined by the signal timing plan but shares the same updating rules as in (4.3) with $q_{i,\max}$ replaced by $q_{i,\max}(t)$ in (4.5). One may notice that this simplified treatment has also been used in Lo's study (Lo 1999; 2001; 2001b).

The above defined flow dynamics model can conveniently accommodate all four types of signal control actions, namely cycle length $\boldsymbol{C}$, phase sequencing, phase duration $\boldsymbol{g}$ and offset Δ between two adjacent signalized intersections. In this study, the offset is in reference with respect to the start of the analysis horizon; the numerical values of each variable are also calculated in the multiples of the loading interval $\boldsymbol{t}$.

4.2.1.4 Metered Freeway Onramp

Modeling ramp meters has only one control variable to deal with, the metering rate at on-ramp link j at time t. For notational simplification, the ramp subscript j is omitted in the following development. Modifying the demand-supply method for merges (Daganzo 1995), we apply one generic flow updating rule to represent the flow dynamics at a freeway merge section (Jin and Zhang 2003):

$$y_i(t) = \min\{n_i(t), q_{i,\max}(t), \delta(N_i - n_i)\} \tag{4.10}$$

$$D_R^t = \min(D_R^t, R^t) \tag{4.11}$$

$$D^t = D_M^t + D_R^t \tag{4.12}$$

$$S^t = \min(S_M^t, D^t) \tag{4.13}$$

$$f_M^t = \frac{D_M^t}{D^t} S^t \tag{4.14}$$

$$f_R^t = \frac{D_R^t}{D^t} S^t \tag{4.15}$$

where the ramp metering R^t is embedded, and other notations are:

- D_R^t: Ramp demand at time t.
- D^t: Demand upon the beginning cell of the link downstream of the ramp.
- D_M^t: Competing demand on mainline.
- S_M^t: Supply of the beginning cell of the downstream link.
- S^t: Total service flow rate.
- f_R^t: Outflow from ramp.
- f_M^t: Outflow from upstream mainline.

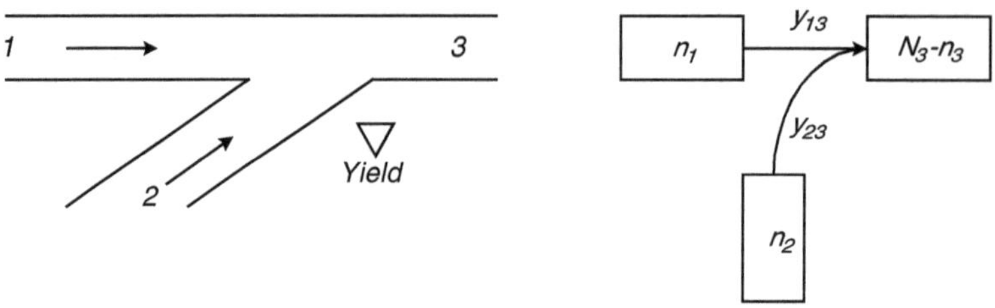

Fig. 4.2 Flow updating by priority control (e.g., Yield Sign)

The modification mainly lies in two aspects: (a) the ramp demand to the merge point is bounded not only by actual demand and the flow capacity but also by the metering rate executed at that time step (4.11); (b) in the overflow or congestion situation, the freeway mainline and ramp flows will be distributed proportionally to their relative demand (4.13)–(4.15) (Jin and Zhang 2003). The ramp metering takes effect in the form of R^t.

4.2.1.5 Priority Rule Controlled Merges and Diverges: All-way STOP and Yield Merge

In addition to the junctions controlled by traffic lights, a large number of junctions are controlled by the rules that drivers must follow in order to go through the junction. Adapted into this study framework, these priority rule-based flow controls can be classified into two categories: all-way stops and yielding merges. Different flow updating mechanisms follow at these two types.

Stop sign control operates on a "first-come-first-serve" basis, where the flow is thus discharged in an ordered manner. At each loading interval, the right-of-way (ROW) is allocated according to the order the flow at each approach arrives. Once the approach gets the ROW, the flow will be discharged according to (4.3) and (4.4) again. However, some exceptions such as the zero-demand approach at certain interval must be handled separately. One flow updating algorithm for this all-way stop sign has been developed in (Ma 2008) and will not be detailed here.

The updating rule for competing flows at yield sign control is essentially merges under priority rules (Fig. 4.2).

Under a yield sign, the yielding flow will only be able to take the remainder of the available space at each loading interval. At any loading interval t, the flow updates at a yield sign will be specified by:

$$y_{13}^t = \min\{n_1^t,\ N_3 - n_3^t,\ q_{1,\max}\} \tag{4.16}$$

$$y_{23}^t = \min\{n_2^t,\ \max\{N_3 - n_3^t - y_{13}^t,\ 0\},\ q_{2,\max}\} \tag{4.17}$$

where the flow on approach 1 has the priority and the flow on 2 has to yield to flow 1.

One may note that the above flow dynamics model cover most vehicle traffic network layout sufficiently. To gain maximum computing performance, however, some network links could be simplified with either the P-Q or the S-Q models as in the introduction. A polymorphic link flow dynamics model has been developed and implemented within the same framework (Nie et al. 2008). This framework allows for heterogeneous link flow models: for example, if some links serve merely as flow exchange between more important facilities, these links can be modeled using P-Q or S-Q models; but if the queuing dynamics is critical to the subarea e.g., closely spaced controlled junctions within the same interchange area, it is better to apply the above flow dynamics model.

4.2.2 *Traffic Demand Input and Vehicle Routing*

In the model, the traffic demand is given externally at any source node j:

$$Q_j(t) = \sum_r \sum_s D^{r,s}(t), \forall (r,\ s) \in \{(R,\ S)\} \tag{4.18}$$

where $Q_j(t)$ is the sum of the time-dependent demands entering the source node j.

Because our main focus is the development of optimal control plans, the discussion of users' route choice behavior is reduced to a minimum. In the later numerical examples, only one pre-determined time-dependent shortest path is utilized for any path flow $D^{r,s}$. The network flow pattern and the resulting performance measure will only be determined by the specified control plan. Nevertheless, one needs to note that the proposed framework can be easily adapted to study the interaction between users' route choice and control strategies (e.g., Chen 1998, Yang and Yagar 1995).

4.2.3 *Minimizing Total Delay for Integrated Corridor Control*

The performance of a control plan is often evaluated through delays and the number of stops. In this study, we select minimizing delay as the control objective. The fundamental diagram depicts two regions that traffic flow status can fall into, the free flow region and the forced flow region. Once the flow falls in the forced flow region, the vehicles will not operate at the free flow speed any more, and delays are incurred to the vehicles. In the model, the total delay is the accumulation of the delay at the cell level, while the latter is conveniently expressed as the following:

$$d_i(t) = d(n_i(t) - y_i(t)) = (n_i(t) - y_i(t)) \bullet t \tag{4.19}$$

where $d_i(t)$ is the delay occurring at cell i during time interval t, $n_i(t)$ and $y_i(t)$ are the numbers of vehicles in i at t and the number of vehicles that can go out of

i at t, and $\bullet$ represents dot product. When the loading interval t is a unit time one, the delay can simply be numerically represented by $n_i(t) - y_i(t)$. That is, the delay is the time that $n_i(t) - y_i(t)$ vehicles are forced to stay in cell i during time step t (because CTM dictates the movement of vehicles from only one cell into the next one at each time step). The objective function is then the summation over all cells and the overall time horizon:

$$D \min(C, \Delta, g, R) = \min \sum_t \sum_i d_i(t) \tag{4.20}$$

where $D(\cdot)$ is the total delay of the system, and $(\boldsymbol{C}, \Delta, \boldsymbol{g}, \boldsymbol{R})$ is the vector of the cycle, offset, phase duration of all street signals, and $\boldsymbol{R}$ is the vector of metering rates of all ramp meters in concern.

4.2.4 Practical Control Constraints

In this study, we aim at computing the control plan for the corridor network with time-of-day control devices. In practice, the traffic controls usually enforce some physical constraints, including the maximum and minimum duration of the cycle length and green duration for any phase, and the max/min metering rates as follows:

$$C_{i,\ \min} \le C_i \le C_{i,\ \max} \tag{4.21}$$

$$g_{i,\ \min} \le g_i \le g_{i,\ \max} \tag{4.22}$$

$$R_{i,\ \min} \le R_i \le R_{i,\ \max} \tag{4.23}$$

The cycle length constraint for any intersection then reads:

$$\sum_{h=1}^{N} g_h^j = C^j - NL \tag{4.24}$$

It states that the sum of the effective green duration of the phases $h = 1 \ldots N$ at intersection j has to be equal to the available green time $C^J - NL$, i.e., the cycle length deducted by the loss time of all phases.

Note that the choice of the loading interval length can impact both the computing performance and the solution quality. The network representation by cells requires a minimum length unit to adequately characterize the junction layout e.g., turn bays or taper lanes (usually smaller a few hundred feet). At the same time, the control parameters even in their simplified forms (green time and red time durations including loss time) are typically set at the 0.1 s level. The authors have developed a numerical solution schema for further dividing the traffic demand into particles and thus allowing for finer resolution of both network representation and control parameters; however, it was concluded that the loading interval of 1 s or 2 s provides a satisfying balance between the computing overhead and the solution quality.

4.3 The SPSA Method and SPSA-based Integrated Control

Optimization techniques based on stochastic perturbation apply in virtually all engineering areas where a closed-form solution to the problem is not available, or the input information into the model could be contaminated with noise. One of these techniques is the simultaneous perturbation stochastic approximation (SPSA) method that uses only the objective function information to compute approximated gradient information. This method has been used in many areas such as industrial quality control, neural network training, sensor placement, and configuration (Spall 1998). In the formulated corridor control problem (4.1)–(4.24), the complexity of the traffic dynamics model precludes direct computation of the gradient information and heuristic method is thus considered more suitable. Because of its high computational efficiency proved in other studies, SPSA method is employed in this study to solve the integrated corridor control problem. The introduction of SPSA method here draws largely on the theoretical development of the SPSA method in (Chin 1994; Sadegh 1997; Sadegh and Spall 1998).

4.3.1 Introduction of the SPSA Method

For a general SPSA procedure, the general objective function $L(\theta)$ as $\boldsymbol{D}(\boldsymbol{C}, \Delta, \boldsymbol{g}, \boldsymbol{R})$ in (4.20) is a scalar-valued performance measure of the system, and θ is a continuous-valued p-dimensional vector of parameters, i.e., $(\boldsymbol{C}, \Delta, \boldsymbol{g}, \boldsymbol{R})$ in the corridor control context. It could happen that noises ϵ occur when measuring the system performance measure $z(\theta)$:

$$z(\theta) = L(\theta) + \epsilon \tag{4.25}$$

As a matter of fact, the SPSA method is mostly superior in the context of optimization with noisy measurements of the system of interest.

The SPSA method starts from an initial guess of θ (one feasible solution) and by a sequential "simultaneous perturbation" over the solution path, the approximation of the gradient $\varphi(\theta) \equiv \frac{\partial L(\theta)}{\partial \theta}$ will converge to zero, under several regularity conditions over the sequence.

Assume that $L(\theta)$ is differentiable over θ and the minimum is obtained at a zero point of the gradient, i.e.,

$$\varphi(\theta) = \left| \frac{\partial L(\theta)}{\partial \theta} \right|_{\theta=\theta^*} = 0 \tag{4.26}$$

The recursive updating of θ takes the standard form:

$$\hat{\theta}_{k+1} = \hat{\theta}_k - a_k \hat{\varphi}(\hat{\theta}_k) \tag{4.27}$$

where the gain sequence$\{a_k\}$ must satisfy certain regularity conditions.

The perturbation is performed upon evaluating $\varphi(\hat{\theta}_k)$. First define a p-dimensional mutually independent mean-zero random variable vector $\Delta_k \in R^p\{\Delta_{k1}, \cdots, \Delta_{kp}\}$ satisfying certain conditions, the most important of which is that $E(|\Delta_{ki}^{-1}|)$ is bounded above by some constant α_1, $E(|\Delta_{ki}^{-1}|) \leq \alpha_1$. An optimal distribution of Δ_k is symmetric Bernouli (Sadegh 1997), i.e., $P(\Delta_{ki} = \pm 1) = \frac{1}{2}$. Furthermore, $\{\Delta_k\}$ is a mutually independent sequence that is also independent of $\hat{\theta}_0, \hat{\theta}_1, \cdots, \hat{\theta}_k$. Let

$$z_k^{(+)}(\theta_k) = L(\hat{\theta}_k + c_k \Delta_k) + \epsilon_k^{(+)} \tag{4.28}$$

$$z_k^{(-)}(\theta_k) = L(\hat{\theta}_k - c_k \Delta_k) + \epsilon_k^{(-)} \tag{4.29}$$

where c_k is a positive scalar satisfying the regularity conditions, and $z_k^{(+)}(\theta_k)$, $z_k^{(-)}(\theta_k)$ are the measurements of the system under the perturbation $\hat{\theta}_k + c_k \Delta_k$ and $\hat{\theta}_k - c_k \Delta_k$, respectively.

The approximation of the gradient will then become:

$$\hat{\varphi}_k(\hat{\theta}_k) = \frac{z^{(+)k} - z_k^{(-)}}{2c_k} \begin{bmatrix} \Delta_{k1}^{-1} \\ \vdots \\ \Delta_{kp}^{-1} \end{bmatrix} \tag{4.30}$$

Spall (1992) showed that by recursively updating θ_k, the gradient will converge to a zero point. The basic recursive form (4.27) and gradient approximation (4.30) ensure that the approximation will settle down at a local minimum at least.

4.3.2 Regularity Conditions Assuring Convergence

Five assumptions are made upon the gain sequence a_k to ensure θ_k to converge almost surely to at least a local optimum θ^*. We refer to (Spall 1992) for the full derivation. A brief description of the assumptions (called "regularity conditions") is given below:

A1: $a_k,\ c_k > 0 \forall k$; $a_k \to 0,\ c_k \to 0$ as $k \to \infty$; $\Sigma_{k=0}^{\infty} a_k = \infty$; $\Sigma_{k=0}^{\infty} \left(\frac{a_k}{c_k}\right)^2 = 0$;

A2: For some $\alpha_0,\ \alpha_1,\ \alpha_2 > 0$ and $\forall k$, $E\epsilon^{(\pm)^2} \leq \alpha_0$, $EL(\hat{\theta} \pm \bar{\Delta}_k)^2 \leq \alpha_1$, $E\Delta_{kl}^{-2} \leq \alpha_2$, $l = 1, \cdots, p$;

A3: $||\hat{\theta}_k|| < \infty, \forall k$;

A4: θ^* is an asymptotically stable solution of the differential equation $\mathrm{d}x(t)/\mathrm{d}t = -\varphi(x)$.

A5: Let $D(\theta^*) = \{x_0 : \lim_{t\to\infty} x(t|t_0) = \theta^*\}$ where $x(t|x_0)$ denotes the solution to the differential equation of A4 based on initial conditions x_0, there exists a compact set $S \subseteq D(\theta^*)$ such that $(\hat{\theta}_k) \in S$ infinitely often for almost all sample points.

The gain sequences of $\{a_k\}$ and $\{c_k\}$ generally take the power functions:

$$a_k = \frac{a}{(1+A+k)^{\alpha}}, \qquad c_k = \frac{1}{(1+k)^{\gamma}} \tag{4.31}$$

where k is the iterator, and A is a constant introduced to stabilize the optimization process.

It is argued (Sadegh and Spall 1998) that the asymptotically optimal values of α and γ are 1 and $\frac{1}{6}$, respectively. But Spall (1998) pointed out that $\alpha < 1.0$ usually produces a better finite-sample performance. Hence another set of values of 0.602 and 0.101 that are the lowest allowable to satisfy the regularity conditions (**A1–A5**) were recommended.

It is observed that for most engineering problems these conditions are almost automatically satisfied with only **A3** being hard to verify for general cases (Spall 1992). In the corridor control problem, the violation of **A3** implies that the transportation system leads to a complete gridlock. As this could be partly avoided by placing the practical constraints over the control (4.19)–(4.21), it does not impose difficulties in the solution as indicated in the numerical example.

4.3.3 Constrained Optimization via Stochastic Approximation

The SPSA procedure introduced above is suitable for solving unconstrained optimization problems, which is not directly applicable to our optimal corridor control problem. Sadegh (1997) proposed a projection method to restrict $\theta_k \in R^p$ at each iteration k to fall in the feasibility range of the control variables by simply replacing any violating $\hat{\theta}_k$ with the nearest point $\theta_k \in G(\theta)$ where $G(\theta)$ is the feasibility set of the control vector:

$$\hat{\theta}_{k+1} = P(\theta_k - a_k \hat{g}_k(\hat{\theta}_k)) \tag{4.32}$$

The perturbed vectors $\hat{\theta}_k - c_k\Delta_k$ and $\hat{\theta}_k + c_k\Delta_k$ in evaluating the loss function (4.28)–(4.29) will also be projected such that these two perturbed vectors lie in the feasibility range. By forcing another restriction (Assumption 1) over the constraints, SPSA can still converge to a Karash–Kuhn–Tucker point almost surely (*a.s.*) [Proposition 1 in Sadegh (1997)].

The set $G = \{\theta : f_i(\theta) \leq 0, i = 1, \ldots, s\}$ is nonempty and bounded, and the functions $q_i(\theta), i = 1, \ldots, s$, are continuously differentiable. At each $\theta \in \mathrm{col}(G)$ where *col* denotes the boundary; the gradients of the active constraints are linearly independent. Furthermore, there exists an $\xi < 0$ such that the set $G^- = \{\theta : f_i(\theta) \leq r, i = 1, \ldots, s\}$ is nonempty for $\xi \leq r < 0$ (i.e., the set G has a nonempty interior).

Because the parameter vector θ may have various numerical magnitudes (e.g., the ramp metering rate $\mathbf{R}$ is measured in hundreds of vehicles per hour and the green duration g is measured in seconds), they have to be normalized during the decaying process. The following normalization process is then introduced:

$$g_i^n = \frac{g_i - g_{i,\min}}{g_{i,\max} - g_{i,\min}} \tag{4.33}$$

where g_i^n can be any control variable with the physical boundary in (4.19)–(4.21). The following proposition examines whether the normalization process would affect the performance of SPSA.

Proposition 1. A normalized version of the projection method in constrained SPSA can assure a convergence to at least a local optimum a.s.

Proof: It is trivial to verify the nonemptiness of the control feasibility set $G(\theta)$ since any points that fall in the box constraints (4.21)–(4.23) will fulfill the conditions. Since all constraints including the box constraints and summation constraint (4.24) are all linear, the following equation holds:

$$\frac{\partial f_i(\theta)}{\partial \theta_j} = 1 \mathrm{or} - 1,\ i = 1, \ldots,\ s,\ j = 1, \ldots,\ q \tag{4.34}$$

As $g_{i,\max}$, $g_{i,\min}$ are constants, the linear transformation (4.33) does not change the above argument; then Assumption 1 for the control feasibility set after the linear transformation still holds.

With the assumptions A1–A5 and the above verification of Assumption 1, we conclude that after the linear transformation (4.33) as $k \to \infty$ with $\hat{g}_k(\hat{\theta}_k) = \hat{g}_k^{SP}(P_k(\hat{\theta}_k))$, $\hat{\theta}_k \to \hat{\theta}^*$ ■

4.4 Solution Algorithm

The iterative SPSA solution algorithm for the time-of-day corridor control has the following steps.

SPSA Algorithm for Integrated Network Traffic Control

Step 1: Initialization and Coefficient Selection.

1.0 Set iterator k =0
1.1 Generate the control vector and normalize it via (4.33) as θ^N
1.2 Pick an initial feasible solution of θ_0^N
1.3 Select nonnegative coefficients a, c, A, α and γ

Step 2: Simultaneous Perturbation.

Generate a p-dimensional random perturbation vector Δ_k, *where each component is mutually independent Bernoulli* ± 1 *distributed with probability of 1/2 for each* ± 1 *outcome.*

Step 3: Loss Function Evaluation by Dynamic Network Loading (DNL).

3.1 Perturb the normalized control vector with $\hat{\theta}_k \pm c_k \Delta_k$;
3.2 Project the perturbed control vectors onto $G(\theta)$ *from (4.32);*
3.3 Transform the projected control vector back to the real valued control variables;
3.4 Evaluate the system performances by loading the demand onto the network under both set of control variables and obtain (4.28)–(4.29);

Step 4: Gradient Approximation.

Calculate the approximated gradient from (4.30).

Step 5: Control Update.
Update $\hat{\theta}_k$ *with (4.27).*

Step 6: Convergence Check.
If the convergence criteria are met, stop. Otherwise, set $k = k+1$ *and go to step 2.*

As with any other heuristic method, the selection of appropriate parameters including the gain sequence a_k and c_k is crucial. A few selection guidelines are discussed after the numerical experiments.

4.5 Numerical Examples

4.5.1 A Simple Network to Investigate the Convergence of SPSA

A typical diamond interchange is first constructed, where the freeway traffic travels from 1 to 2 and the surface street traffic 3–4 can go both ways (Fig. 4.3). One ramp meter and two intersection signal groups control the traffic flow. Both ramp meter and signal controllers are assumed pre-timed, and the phasing diagram is shown in Fig. 4.3. For illustrative purpose, only fifteen minutes of demand is set up and the hourly trip rates are also shown.

One may notice that only two independent control variables are present in the sample network, corresponding to the normalization procedure in (4.33), namely the green ratio g_1 for phase 1 (the green ratio for phase 2 will be $1 - g_1$ if we omit the loss time for the time being) and the metering rate R. The maximum/minimum green time is set to be 50 and 10 s respectively, and the range for the metering rates is set to be 300–1,500 vph. To locate the possible global optimal control plan, an exhaustive search through the feasibility solution space is performed. In this

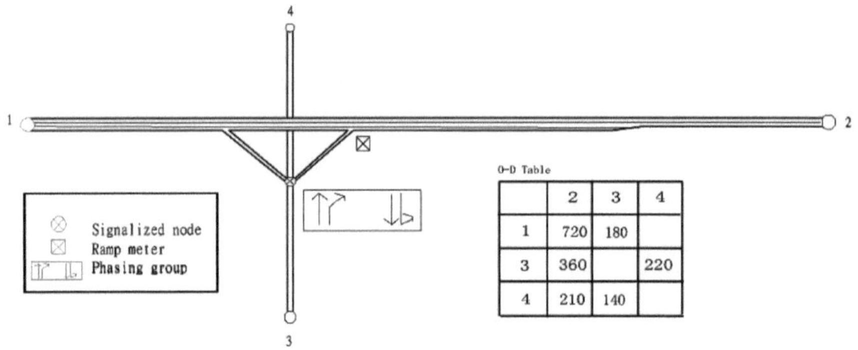

Fig. 4.3 Geometric layout and demand of sample network I

example, the network loading interval t is 1 s; thus the increment of the phase duration is set equal to t, while an increment of 20 vph is selected to scan the range of metering rates. Therefore, the exhaustive search goes through a total number of 2,400 $(40 \times 60) g_i - R$ combinations, and the contour of the objective function (total travel time) under various combinations is plotted in Fig. 4.3a). The contour implies only one global optimal solution in the search space at $(g_1, R) = (0.61, 1500)$, with a total travel time of 312 veh-hr. In the example, the optimal metering rate is the upper bound of the feasible range, that is, allowing as many flows as possible into the freeway mainline during this 15-min period.

Two SPSA processes with different initial feasible "guesses" (θ_0) are experimented and plotted in Fig. 4.4. The first starts with $(g_1, R) = (0.30, 300)$ and stabilizes itself at $(g_1, R) = (0.58, 1473)$; the second starts with $(g_1, R) = (0.80, 600)$ and stabilizes itself at $(g_1, R) = (0.57, 1498)$. Generally both processes can reach the near-global optimal solution along a different path. Figure 4.3b shows that the convergence process of SPSA has a feature of quick drop at the first few iterations; after less than forty iterations in the example, both processes reach near-global optima. This example confirms the ability of the constrained SPSA method can be applied to study the corridor control problem.

4.5.2 A Real Network to Investigate the Effectiveness of the SPSA Algorithm

4.5.2.1 Network Background and Preliminary Work

The other network is a real one of SR-81 corridor at Fort Worth, Texas. A DynaSmart-P network has been developed elsewhere (Mahamassani et al. 2004) and is converted into the CTM representation. The geometric layout is illustrated in Fig. 4.5.

Due to the differences in the network representation [e.g., the travel demand releasing mechanisms in DynaSmart-P are different from the network dynamics

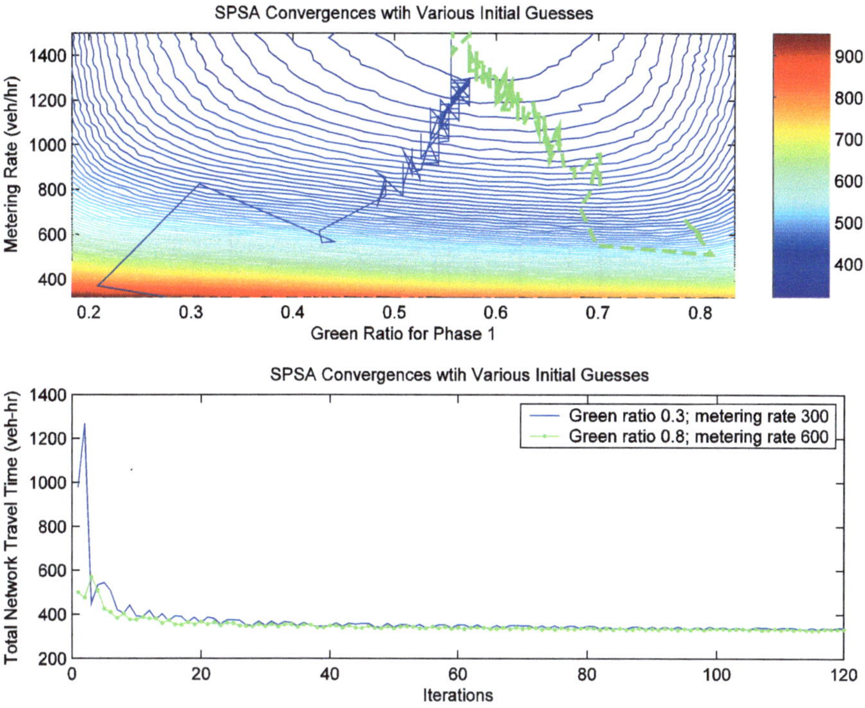

Fig. 4.4 SPSA convergence process under various initial feasible solutions (network I)

model characterized by Eq. (4.1)–(4.18)] and lack of further data support, the network was slightly modified in the conversion. The most important modification is the controller type changes. In the original network, the signals are most vehicle-actuated controllers. Since herein only time-of-day corridor control is considered, all controllers are assumed pre-timed. The phasing sequence and phase diagrams for each of the signal controllers are inherited from the original settings.

Selecting an appropriate initial control setting θ_0 is the first step to compute the optimal control plan. The preliminary experimentations indicate that the constructed network is heavily loaded if the controls are not properly set. An arbitrary control plan then cannot act as θ_0 because the performance index cannot be evaluated if a gridlock happens under the control plan. A "good" control plan that at least allows the traffic flows smoothly through the network must be found before the SPSA optimization process could start.

The signal timing design procedure in HCM is followed to compute a feasible control plan. First the demand is loaded onto the network and the network flow pattern is obtained. Then the cycle length and green splits at each intersection

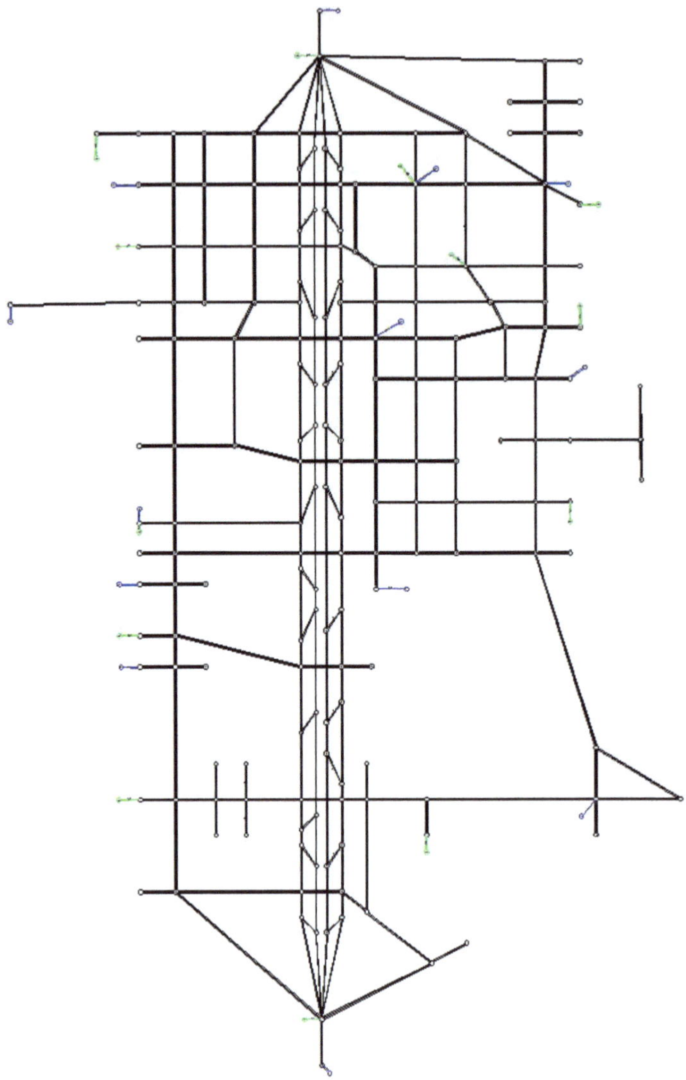

Fig. 4.5 Geometric layout of Dallas Fort Worth network

is computed under the "equi-saturation" logic (Chin 1994). The resulting timing plans and an initial offset of zero for each intersection and a no-meter [$R = q_{max}$ in Eq. (4.11)] solution does not cause a gridlock. This HCM control plan is taken as θ_0 for the successive SPSA optimization process.

4.5.2.2 SPSA Application in the Dallas Network

A two-level procedure is taken to compute the optimal corridor control plan for the Fort Worth network. First the green splits and metering rates at each intersection or metered ramp are optimized; based on this and an adjusted common cycle length, the offsets are then computed. The first level of optimization has a total of 209 control variables encoded as the vector $\boldsymbol{\theta}$. For the second level, the offset for each intersection is all referenced to the start of the study period, and altogether 46 decision variables are encoded.

Two more well-documented heuristics methods are also implemented in this study to compare their relative computational performances. For the first level, the genetic algorithm (GA) in (Lo et al. 2001b) is extended to accommodate ramp metering controls. The "hill-climbing" method in TRANSYT (Robertson 1969) is used to compute the offset of the second level of optimization. On the one hand, choosing these two methods for benchmarking SPSA was mainly because of their legacy value and acceptance in the community of traffic control researchers and practitioners; on the other hand, these two methods each can assist in better understanding SPSA searching when compared to intuitive rules (hill-climbing) or random generation of candidate solutions (GA).

The selection of parameters is important to the GA algorithm as well. The most important parameters in this algorithm include population size at each generation, and mutation rate. In this study, we apply a real-value gene-coding scheme instead of the commonly used binary-coding scheme (e.g., Sadegh and Spall 1998 and Lo 2001b). The real-value coding scheme is considered more efficient and accurate (Wright 1991). However, no empirical formula is available to estimate an appropriate population size as in the case of binary coding; a trial-and-error process then has to be used to come up with the following GA parameters: population size is 50; mutation rate 0.1; a predefined maximum generation number of 80. The convergence process with these two algorithms is shown in Fig. 4.6.

The computation time is mostly consumed by the evaluation of the system performance, that is, the dynamic network loading (DNL) as in Step 3 of the SPSA algorithm. For example, a single DNL process for the half-hour demand of Dallas Forth Worth network generally takes 7–10 s on an up-to-date PC (Pentium-4 3G CPU, 1G RAM). Then the total number of $z(\boldsymbol{\theta})$ evaluations determines the amount of computational resources when computing the optimal control plan. In Fig. 4.5, the total network travel time (TNTT) for GA is averaged over each generation; while the TNTT in SPSA process are sampled every 20 DNL evaluations. It can be seen that SPSA only needs about 350 performance evaluations to reach a stable solution, while GA needs 3,200 evaluations. It also illustrates that the objective function value can get a very sharp drop in early SPSA iterations, and this advantage can be utilized in other optimization applications to perform a quick search for a good starting solution. However, SPSA was slightly outperformed by GA in terms of the stable solutions they reached. SPSA does not jump out of an inferior "stable" solution (323.1 veh-hr) in the later process, while GA-based optimization obtained a better stable solution (317.2 veh-hr) in terms of the TNTT.

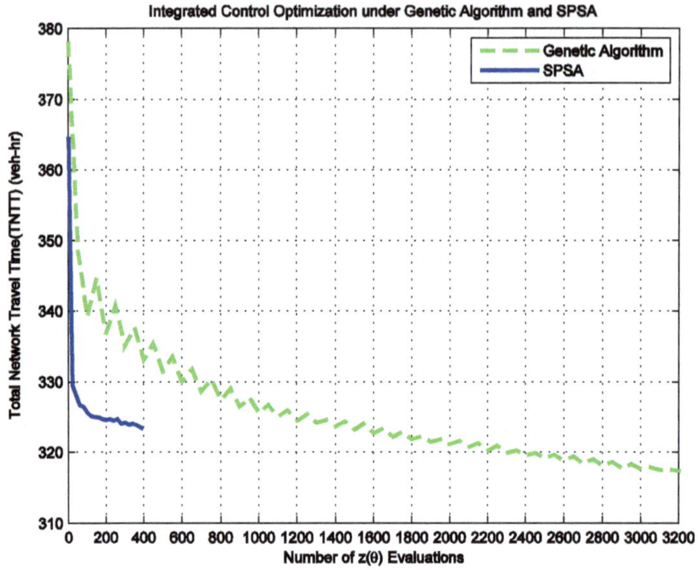

Fig. 4.6 Computational performances of SPSA- *vs.* GA-based optimization of green splits/ metering rates for network II

The second level, offset optimization, is also conducted using both the SPSA algorithm and genetic algorithm based on the green splits obtained from their corresponding first level of optimization. The longest cycle length calculated from the critical intersection is used as the common cycle length. The cycle length and phase durations of the rest intersections are scaled accordingly. For the purpose of comparison, the classic "hill-climbing" algorithm is also implemented to compute the offset for each intersection.

The hill-climbing method proceeds as a sequence of adjusting the offset at each intersection. First a step size is selected; the adjustments are then performed by a line search to find an improved global objective function that is also computed from network loading. The adjustments are incremental by the selected step size as long as the search improves the objective function. If the search degrades the objective function, the direction of adjustments will be reversed and continued at the same step size. In this way, a better offset is achieved for the intersection. Then the search proceeds to the remaining intersections. An optimization decision is made for each of several step sizes.

The optimization results of GA, SPSA, and hill-climbing methods are shown in Fig. 4.7. While all three methods can reduce the TNTT further by adjusting the offset for each intersection, hill-climbing method can only reach an inferior solution compared to the other two. Furthermore, the SPSA method outperforms both GA and hill-climbing methods using less DNL evaluations. The solutions and performances at both levels are summarized in Table 4.1. It is interesting to note that the optimal solutions after the successive optimization processes from GA and

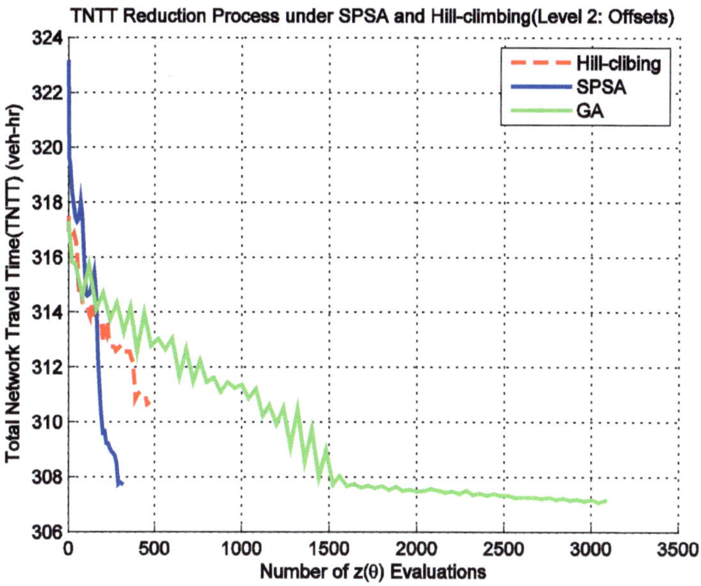

Fig. 4.7 Computational performances of SPSA- *vs*. Hill-climbing based optimization of offset for network II

SPSA are now very close (TNTT-GA 306.9 vs. TNTT-SPSA 307.5); it implies that various (local) optima could exist when searching for the optimal corridor control plan for a real network. Even though no method can guarantee a global optimal solution to the formulated corridor control problem, SPSA and GA can reach stable solutions that are comparable to each other.

4.5.2.3 Integrated Corridor Control under Different Network Traffic Loads

Corridor networks serve as the backbone for surface transportation systems and are usually heavily used. Integration of control measures as proposed in this study can improve the operational performance as illustrated in the above numerical examples. Yet, it is unknown whether integrated control can perform equally well under various traffic loads. For this purpose, the above experimentation process is repeated under another set of traffic demand. The new demand is uniformly decreased to be 80% of the original, which is considered a congested network. The following table indicates the extent to which the network performances are improved (Table 4.2).

It is clearly seen that the improvement is higher under the decreased demand, which may represent the low to medium network load. It may imply that integrated control may see diminishing benefits when the network becomes more congested.

Table 4.1 Performance comparison of various optimization methods

	Level 1				Level 2			
Method	# of $z(\theta)$ Evaluations	$z(\theta_0)$	$z(\theta^*)$	Improvement (%)	# of $z(\theta)$ Evaluations	$z(\theta_0)$	$z(\theta^*)$	Improvement (%)
SPSA	388	364.4	323.1	11	327	323.1	307.5	4.8
Genetic algorithm	3,200	377.9	317.2	16	2,800	317.2	306.9	3.2
Hill-climbing	N/A	N/A	N/A		470	317.2	311.1	1.0

Table 4.2 Network operational performance improvements under different traffic loads

Level of	Greensplits and Metering Rates			Offsets			Overall
Demand (%)	$z(\theta_0)$	$z(\theta^*)$	$\pm\%$	$z(\theta_0)$	$z(\theta^*)$	$\pm\%$	Improvement
100	364.4	323.1	11%	323.1	308.2	4%	15%
80	313.2	256.8	18%	256.8	250.6	2%	20%

4.6 Guidelines for Selecting SPSA Parameters

Selection of appropriate parameters for the gain sequence a_k and c_k is important to the performance of SPSA process. Spall (1998) provided a few guidelines for the choice of the related parameters, i.e., α, γ, a, A and c.

With the Bernoulli ± 1 distribution for Δ_k, c can be set at a level approximately equal to the standard deviation of the measurement noise in $z(\theta)$ so that the magnitude of the approximated gradient $\hat{g}_k(\hat{\theta}_k)$ does not go excessively large. In our study, the system performance evaluation is deterministic from (4.1) to (4.24); in this case, c can be some small positive number. In our experimentations with various networks in the normalization scheme, it is found that 0.05 provides acceptable results, namely the change in each element of θ in the initial iterations is in the magnitude of around 5%.

It is also suggested that a "stability constant" A should be used for the sequence of a_k when large noises or variations of system performance measures are observed. A useful guideline for choosing A is to set to 10% (or less) of the maximum number of expected or allowed iterations. Meanwhile, Spall (1998) also suggested running a few preliminary replications of $\hat{g}_0(\hat{\theta}_0)$, and choosing a such that $\frac{a}{(A+1)^{\alpha}}$ times the magnitudes of $\hat{g}_0(\hat{\theta}_0)$ should be approximately equal to the smallest change in θ. It is found that a larger a could lead to faster convergence to the optimal solution, but it may also run into the risk of reaching infeasible solutions (gridlock in our corridor control context). Following the above guidelines, we have found that the initial changes of 3–4 s in phase duration or offset values can generally provide smoother SPSA convergence. One may also notice from the above numerical experiments particularly with the real world network that some extra efforts could be necessary to get the initial feasible solution for different corridor networks.

4.7 Conclusions

An integrated corridor control problem is formulated and solved in this study. Based on the cell transmission model, the platform can capture the queuing phenomena within a general corridor network under all traffic conditions. Urban signal control and ramp metering are embedded in the platform generically. A new heuristics solution algorithm is developed using the simultaneous perturbation stochastic approximation method. The algorithm can compute a near-global optimal control plan more efficiently compared to other heuristics methods, even if it may not guarantee global optima. Our results indicate that SPSA can be used to solve integrated corridor control problems for large-scale networks. Numerical experiments also indicated that integrated corridor control can perform well under various traffic loads but appears to see higher improvements under medium congestion.

Acknowledgment We very much appreciate Professor James C. Spall at Applied Physics Laboratory at Johns-Hopkins University for his valuable comments in the early development of this work. Mr. Xuesong Zhou from University of Texas at Austin generously shared the Dallas Fort Worth network data used in this study. We would also like to thank all anonymous reviewers for their comments that lead to the final version. This work is partially funded by Caltrans under TO 5300 and TO 4136. The views are the authors' alone.

References

Chang K, Stephanedes YJ. Optimal control of freeway corridors. ASCE J Transport Eng. 1993;119(4):504–514.

Chen OJ. Integration of dynamic traffic control and assignment. Ph.D Thesis, Massachusetts Institute of Technology, Department of Civil and environmental Engineering. Cambridge, Massachusetts, USA. 1998.

Chin DC. Comparative study of stochastic algorithms for system optimization based on gradient approximations. IEEE Trans Syst Man Cybern B 1994;27;244:249.

Daganzo CF. The cell transmission model, part II: network traffic. Transpn Res B 1995;29:79:93.

Daganzo CF. The cell transmission model: a dynamic representation of highway traffic consistent with the hydrodynamic theory. Transpn Res B 1994;28:269:287.

Diakaki C, Papageorgiou M, McLean T. Integrated traffic-responsive urban corridor control strategy in Glasgow, Scotland: application and evaluation. Transport Res Rec. 2000;1727:101:111.

Federal Highway Administration (FHWA) 2005. Integrated Corridor Management System (ICMS) Work Plan. http://www.itsdocs.fhwa.dot.gov/icms/icmsworkplan.htm. accessed 2005.

Gomez G, Horowitz R. Globally optimal solutions to the onramp metering problem - part I. IEEE on ITS'04, Washington D.C. USA. 2004a.

Gomez G, Horowitz R. Globally optimal solutions to the onramp metering problem - part II. IEEE on ITS'04, Washington D.C. USA. 2004b.

Greenshields BD. A study of traffic capacity. Proc Highway Res Board 1934;14:448:477.

Helbing D. A section-based queuing-theoretical traffic model for congestion and travel time analysis in networks. J Phys Math Gen 2003;36:L593–L598.

Helbing D, Lämmer S, Lebacque J-P. Self-organized control of irregular or perturbed network traffic in optimal control and dynamic games. Dortrecht: Springer; 2005. p. 239–274.

Jin W, Zhang HM. On the distribution schemes for determining flows through a merge. Transpn Res B. 2003;37(6):521–540.

Kotsialos A, Papageorgiou M, Mangeals M, Haj-Salem H. Coordinated and integrated control of motorway networks via nonlinear optimal control. Transpn Res C 2002;10:65–84.

Lo HK. A cell-based traffic signal formulation: strategies and benefits of dynamic timing plans. Transport Sci. 2001a;35(2):148–164.

Lo HK. A novel traffic signal control formulation. Transport Res A 1999;33A:433–448.

Lo HK, Chang E, Chan YC. Dynamic network traffic control. Transpn Res B 2001b;35B:721–744.

Mahamassani H, Sbayti H, Zhou X. DYNASMART-P: Intelligent Transportation Network Planning Tool, Version 1.0 User Guide. Maryland Transportation Initiative, University of Maryland, College Park. MD 20742. 2004.

Ma J. An Efficiency-Equity Solution to the Integrated Transportation Corridor Control Problem, PhD Thesis, Department of Civil and Environmental Engineering, University of California at Davis, Davis, CA, USA. 2008.

Messmer A, Papageorgiou M. METANET: a macroscopic simulation program for motorway networks. Traffic Eng Contr. 1990;31:466–473.

Michalopoulos PG, Stephanopoulos G. Oversaturated signal systems with queue length constraints – I single intersection. Transpn Res. 1977;11(6):413–421.

Nie Y., Ma J, Zhang HM. A polymorphic dynamic network loading model. Comput Aided Civ Infrastruct Eng. 2008;23(2):86–103.

Papageorgiou M. An integrated control approach for traffic corridors. Transpn Res C. 1995;3(1):19–30.

Robertson DI. TRANSYT: a traffic network study tool. Technical report 253, Transport Road Research Laboratory, UK. 1969.

Sadegh P. Constrained optimization via stochastic approximation with a simultaneous perturbation gradient approximation. Automatica 1997;33:889–892.

Sadegh P, Spall JC. Optimal random perturbations for multivariate stochastic approximation using a simultaneous perturbation gradient approximation. IEEE Trans Automat Contr. 1998;43(3):1480:1484.

Shelby S. Design and evaluation of real-time adaptive traffic signal control algorithms. Ph.D. thesis, University of Arizona, System and Industrial Engineering Department, December 2001.

Spall JC. Implementation of the simultaneous perturbation algorithm for stochastic optimization. IEEE Trans Aero Electron Syst. 1998;34(3):817:823.

Spall JC. Multivariate stochastic approximation using a simultaneous perturbation gradient approximation. IEEE Trans Automat Contr. 1992;37(3):332:341.

Stephanopoulos G, c PG, Stephanopoulos G. Modeling and analysis of traffic queue dynamics at signalized intersections. Transpn Res A 1979;13(3):295–307.

van Zuylen H, Taale H. Urban networks with ring roads: a tri-level optimization. In: Proceedings of 83th TRB annual meeting, Washington D.C. 2003.

Vincent RA, Mitchell AI, Robertson DI. User guide to TRANSYT version 8. Technical Report TRRL Laboratory Report 888, TRRL Department of the Environment, Crowthorne, Berkshire, UK. 1980.

Webster FV. Traffic signal settings. Technical Report 39, Transport Road Research Laboratory, Crowthorne, Berkshire, UK. 1958.

Wood K. Urban traffic control, systems review. Project report 41, Transport Road Research Laboratory, Crowthorne, Berkshire, UK 1994.

Wright AH. Genetic algorithms for real parameter optimization. Foundations of genetic algorithms. San Mateo, California: Morgan Kaufmann; 1991.

Yang H, Yagar S. Traffic assignment and traffic control in saturated road networks. Transpn Res B 1995;29(2):125–139.

Zhang HM, et al. Evaluation of on-ramp control algorithms. Research Report UCB-ITS-PRR-2001–36, University of California, Davis. 2001.

Chapter 5
Analyses of Arterial Travel Times Based on Probe Data

Isaac Kumar Isukapati, George F. List, Stacy Eisenman, Jeffrey Wojtowicz, and William Wallace

Abstract This paper presents an analysis of arterial travel times based on AVI (automatic vehicle identification) data from vehicles that were equipped with toll tags. The source is a six-month experiment conducted on a small arterial network in upstate New York. Data were collected using wireless, solar-powered toll tag readers. The paper explores and examines trends by time of day, day of the week, and as affected by weather and other conditions. The results point toward the value of using such data for travel time prediction, travel time reliability monitoring, incident detection, and overall performance monitoring.

5.1 Introduction

Travel times have always been an important aspect of transportation systems (Berry 1952; Greenshields 1934; Greenshields et al. 1947). People and shippers want to know how long it will take to get from point A to B and whether the trip will be completed on time. Technological advances in the past two decades

I.K. Isukapati • G.F. List (✉) • S. Eisenman
Department of Civil, Construction, and Environmental Engineering,
North Carolina State University, 208 Mann Hall, 2501 Stinson Drive,
Campus Box 7908, Raleigh, NC 27695–7908 USA
e-mail: ikisukap@ncsu.edu; gflist@ncsu.edu

J. Wojtowicz • W. Wallace
Rensselaer Polytechnic Institute, a: Department of Civil and Environmental Engineering,
b: Department of Industrial and Systems Engineering, 110 Eighth Street, Troy, NY 12180 USA
e-mail: wojtoj@rpi.edu; wallaw@rpi.edu

S.V. Ukkusuri and K. Ozbay (eds.), *Advances in Dynamic Network Modeling in Complex Transportation Systems*, Complex Networks and Dynamic Systems 2,
DOI 10.1007/978-1-4614-6243-9_5,

have made it possible to more effectively monitor trip times and their variability, especially AVI and AVL technology. As the paper demonstrates, useful insights can be developed today by studying the trip times that these technologies provide. In fact, observability has risen to the point where the Federal Highway Administration has issued a notice of proposed rulemaking in which it expects States to "monitor, in real-time, the traffic and travel conditions of the major highways of the United States and to share these data with State and local governments and with the traveling public. This rule establishes minimum parameters and requirements for States to make available and share traffic and travel conditions information via real-time information programs." Hence, not only has it become possible to monitor the effects of recurring and nonrecurring congestion, but also to report the findings to the public. This leads to challenging questions that transportation researchers have been trying to answer for a long time: "Can the reliability or unreliability of travel times be predicted or observed?"

Traditionally, travel times have been studied "indirectly" from field-based observations of counts and occupancy transformed into spot speeds or observed in snippets via license plate surveys and other techniques. But with the advent of AVI (automatic vehicle identification) and AVL (automatic vehicle location) technologies, it has become much easier to observe travel times directly.

The purpose of this study was to see what could be learned about arterial travel times based on data from AVI-equipped vehicles (automatic vehicle identification, often used for toll tags). It presents the findings from six months of data collected on a small arterial network in upstate New York.

5.2 Prior Work

The oldest analyses of travel times made use of license plate surveys (Oppenlander 1976; Schaefer 1988; Shuldiner 1996; Williams 1986). Oppenlander (1976) developed a methodology for sample size determination applicable to travel time studies for both license-plate and test-car techniques, whereas Williams (1986) created a license-plate survey technique to develop travel time measures and other metrics for traffic monitoring. The latter provided information to the system operator on topics such as the number of vanpools, origin and destinations, travel and trip time lengths, and traffic assignment. Schaefer (1988) provided guidance on the mathematical and statistical considerations that should be employed when conducting license-plate matching surveys.

Shuldiner (1996) used video technology for reading and processing vehicle license plate images. Vehicle license-plate images were acquired by video camcorders. They concluded that video and machine vision analysis is an effective means of conducting a wide variety of traffic engineering and traffic management studies such as understanding travel time and micro-scale origin–destination pattern analyses. List (1993) used handheld tape recorders to study the travel time improvements from time-based signal coordination. Observers with accurate clocks recorded the times when vehicle license plates were observed at various locations and then point-to-point (P2P) travel times were derived from these data by matching the license plate numbers. The sample sizes were small and matching the license plate numbers was a challenge. Observers would mistake one letter for another ("E's" for "F's") and they would only capture a few of the characters in the license plate (e.g., only the first or last few characters).

The ADVANCE project in Chicago was one of the first studies where travel times were characterized using probes (Boyce et al. 1991; 2002). Vehicles equipped with dead reckoning equipment recorded travel times across northeast Chicago, principally between Evanston and O'Hare airport. These travel times were then used to help motorists determine the best paths to choose in traversing this largely arterial-based urban network. The project was moderately successful, although it pushed the limits of the then-available technology.

After ADVANCE, List et al. were the next researchers to use probes to share travel time data. They placed 200 probes in service in a peer-server-peer ATIS system in upstate New York (Demers et al. 2006a). The main focus was on journey-to-work trips for people traveling to a cluster of business in a single area. The experiment demonstrated that such a system could and would be used to help drivers determine the best routes to use through congested and incident-impacted networks.

At this point, many examples exist of using AVL equipment to monitor travel times although there have been many other efforts to explore use GPS equipment in transportation (Bertini et al. 2005; Byon et al. 2006; Pan et al. 2007; Quiroga and Bullock 1998). Byon et al. (2006) developed GISTT (GPS-GIS Integrated System) for Travel Time Surveys. This system enables one to match GPS data with spatial map features using GIS for monitoring traffic conditions on specific links. Pan et al. (2007) proposed GPS-based methodology for collecting historical travel time data that includes link travel time and information on intersection signal delay, for an arterial. Furthermore, they developed a post-trip map-matching algorithm to project GPS data onto an arterial network.

Taxis have been used to collect traffic data in such wide-ranging locations as Japan, Germany, and Malaysia. In Berlin, the taxis served as floating cars and their travel time observations were processed into traffic information that other services then offered to clients (Pan et al. 2007). In addition, trucking companies use AVL technologies to manage their truck fleets (Morris et al. 1998). Information about the truck movements are retrieved in real time; and the dispatcher switches load assignments and re-routes the trucks to maximize quality of service while minimizing the impacts of congestion. Transit operators use AVL technology to track buses and improve system performance. They improve the system's ability to keep the buses on time, alert riders to bus arrival times and locations, and make it easier for the system to meet rider demands.

Furthermore, the buses are sometimes used as probes to evaluate traffic conditions on arterials (Berkow et al. 2008; Bertini et al. 2005; Chakroborty and Kikuchi 2004; Hellinga and Fu 2002). Berkow et al. (2008) developed techniques for constructing the shape of the congested regime in time and space along urban arterial, combining signal system detectors and buses as probe vehicles. Chakroborty and Kikuchi (2004) developed a model for predicting travel times for automobile using transit vehicles as probes. Hall and Nilesh(2000) compared automobile and transit vehicle trajectories to explore alternative methods for detecting congestion on arterials.

Yamamoto et al. (2006) studied the variability of travel time estimates using probe vehicle data. Hellinga and Fu (2002) developed insights into reducing the bias on link travel time estimates from probe data. Cetin et al. (2005) suggested the factors affecting minimum number of probes required for reliable travel time and Quiroga and Bullock (1998) created guidelines for determining the sample size for travel time studies.

AVI technology is as old as AVL, but it has only more recently been used to monitor travel times. The use of AVI technology became more popular when toll tags were put into service by agencies such as the New York State Thruway and companies such as Mark IV (Vavra 1999). With AVI probes, as with AVL, it becomes possible to monitor point-to-point (P2P) travel times on both freeways and arterials. Li et al. (2006) used automatic vehicle identification (AVI) data to gain insight into travel time variability and its causes.

Freeways have been studied heavily in terms of understanding the travel time distributions. However, most of the investigations have had an objective of predicting future (or current) travel times. Arterials have been studied far less in terms of understanding the travel time distributions, where prediction has been an objective (Lin et al. 2003). Wasson et al. (2008) suggested media access control (MAC) matching for estimating travel time in real time. There were also several efforts to create probe-based real-time route guidance systems (Demers et al. 2006a, 2006b; Fontaine and Smith 2005; List et al. 2005a, 2005b; List and Demers 2006; Ma and Koutsopoulos 2012; Dion and Rakha 2006). This paper focuses on observations about arterial travel times from AVI probe data.

5.3 The Data

The data were obtained during a field test of a portable, wireless, solar-powered tag reader. Studying travel times was not the main intent. However, in this analysis, the data have been repurposed for that objective.

The test of the portable readers was conducted for six months during the fall of 2007 and the winter of 2008. The location was North Greenbush Road (US-4) in North Greenbush, NY, just east of Albany as shown in Fig. 5.1. North Greenbush Road passes by Hudson Valley Community College (located just above reader #5); a minor league baseball field (next to HVCC); the Rensselaer Technology Park (just west of reader #1); and other residential and commercial areas.

Five tag readers were stationed at the locations shown. (A sixth tag reader, #4, not shown, was moved from place to place and collected very little data.) Data were captured using wireless, solar-powered tag readers, mounted on side of the road. An encrypted tag ID, the tag reader ID, and a time stamp were collected with every tag read. The percentage of vehicles with toll tags ranged from 22.5% to 30%.

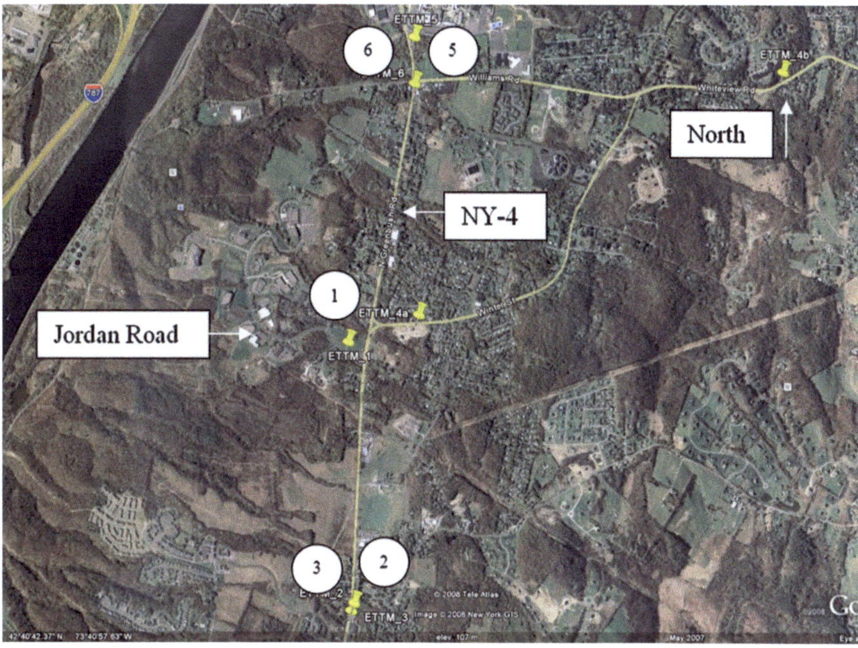

Fig. 5.1 Reader deployment locations

Table 5.1 Breakdown of the data records

	Reader						
Item	#1	#2	#3	#5	#6	Total	Percent
Duplicates	2,000	582	2,041	9	799	5,431	0.87
Multiple records	151	142	840	98	0	1,231	0.20
Test records	61	54	19	21	36	191	0.03
Records after cleaning	76,991	162,802	184,607	55,814	137,910	618,124	98.90
Total records	79,203	163,580	187,507	55,942	138,745	624,977	100.00

More than 620,000 tag reads were collected from more than 54,000 different tags. Each data point represents a specific tag being observed at one of the five locations on a given day at a specific time. More than 39,000 (about 72%) of the tags were seen more than once. Over 1,200 of the tags were seen more than 100 times. Table 5.1 indicates that more than 98% of these observations were useful. The others were duplicates or testing observations.

Except for reader #1, just one direction of travel was recorded by each reader. For example, reader #6 was used to capture data for vehicles traveling southbound on Route 4; reader #5 was used for vehicles traveling north. Reader #1 was used for vehicles going in both directions on Jordan Road (into and out of the Rensselaer Technology Park, which the road serves). The reader (trailer) was placed in the median just west of the intersection.

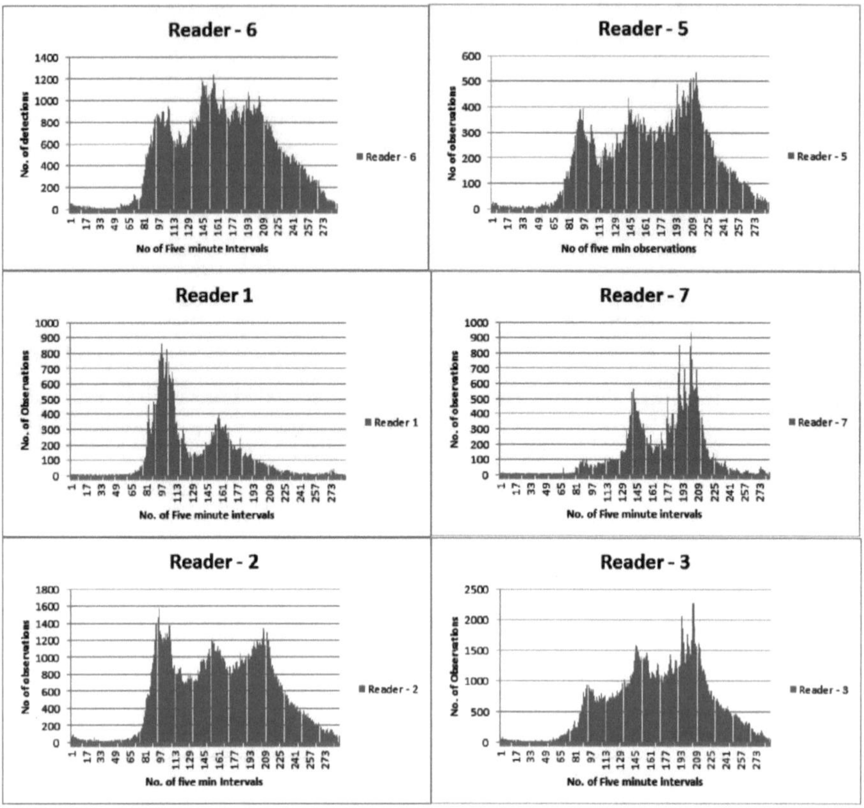

Fig. 5.2 Number of observed tag-reads at individual readers (5-min time bins)

To simplify analysis of the data collected by reader #1, a virtual tag reader #7 was created in the database for the outbound (eastbound) reads on Jordan Road. Thus, in the discussions that follow, the reader #1 tag reads are for vehicles entering the tech park (going west on Jordan Road), while the reader #7 tag reads are for vehicles leaving the tech park (going east on Jordan Road).

Figure 5.2 plots the number of tag-reads observed in each 5-min bin at four of the readers for the entire data set. (The patterns for readers 2 and 3 are the same as 5 and 6.) Readers 5 and 6 show pronounced AM, midday, and PM peaks on top of a general level of activity that begins in early morning and diminishes late at night. Reader 1 shows a peak at 8 AM and another around 12:30–1:00 PM, whereas Reader 7 has peaks around 12:00 Noon and 5:00 PM. It can be inferred that people start their work day at Tech Park around 8 AM; some of them go out for lunch around Noon (first peak at Reader-7) and come back to Tech Park between 12:30 and 1:00 PM (second peak at Reader-1); and then they leave Tech Park at the end of workday around 5:00 PM (second peak at Reader-7).

Table 5.2 Breakdown of the from/to times observed

From\To	1	2	3	5	6	7	Total
1	2,914	2,236	10,368	196	896	20,944	37,554
2	18,251	27,205	49,782	20,457	32,374	4,699	152,768
3	3,695	87,203	33,782	6,083	28,406	2,468	161,637
5	561	8,098	5,248	13,363	22,050	590	49,910
6	6,922	13,913	65,926	6,301	30,762	7,034	130,858
7	4,604	3,438	13,005	5,294	8,831	3,787	38,959
Total	36,947	142,093	178,111	51,694	123,319	39,522	571,686

To create trip times between reader pairs, the roughly 620,000 tag reads were sorted by tag ID, date, and time. Trip times $tk(i,j) = tk(j) - tk(i)$ were then computed for each successive pair of reads (i,j) observed for vehicle *k*—where *tk(i)* and *tk(j)* are in chronological order. More than 570,000 such *tk(i, j)* times were computed. As shown in Table 5.2, they cover all 36 of the possible (i,j) combinations including those where *i* and *j* are the same.

The reader might wonder why the (i,j) pairs of tag reads are not just for the pairs that make sense, such as (6,1), (6,3), and (7,3) for vehicles going southbound. The answer is: (1) the readers were not in operation all of the time, (2) low-power readers were used—intentionally—so some tags were not seen when they passed the readers, and (3) the network is an open network, so there are paths the vehicles can follow that do not involve passing the readers that might seem logical. What is true is that the (i,j) pairs do represent the sequence of tag reads observed for the vehicles.

Hence, the challenge is one of interpretation. For example, the times for which *i* and *j* are the same, arise because (1) double reads a couple seconds apart (these were intentionally not scrubbed), (2) missed intermediate reads (which means the time between the reads tends to be long), and (3) reads (unintentional) for vehicles going in the opposite direction. Moreover, the observations for readers that are adjacent—(1,7), (7,1), (2,3), (3,2), (5,6), and (6,5)—arose either because a tag was read at nearly the same time by both readers—which happened infrequently—or the vehicle was observed first leaving the network—e.g., passing reader #5 going northbound out of the network—and then subsequently re-entering the network—e.g., traveling southbound at reader #6. These latter *tk(i, j)* times tend to be quite long.

Six of the (i,j) pairs represent origin–destination (OD) pairs for trips through the network that fit the classical manner in which that word is used:

- 6–3: the southbound through move on Route 4
- 6–1: the southbound right turn into the tech park
- 7–3: the southbound right out of the tech park
- 2–5: the northbound through move on Route 4
- 2–1: the northbound left turn into the tech park
- 7–5: the northbound left turn out of the tech park

These are the (i,j) studied in detail in the rest of the paper.

5.4 Analysis #1: All of the Trip Times

This analysis looks at all of the *tk(i, j)* values for the six OD pairs. The main question is: how short are the trip times; how long; and how are they distributed? What are the 15th, 50th, 85th, and 95th percentiles, the mean, and other values, and how do those percentiles vary from one day to another? Are they consistent from one day to another or do they change? Are the values affected by weather and other factors? Also, how do the trip times vary from one day to another. (It is important to keep in mind that the AVI system does not have a direct way to separate travel times from trip times because it is not possible to directly distinguish between vehicles that have made nonstop trips and those that have stopped due to intermediate stops or incidents.)

Figure 5.3 plots all of the (i, j) times for (6,3)—the southbound OD pair on Route 4. The x axis shows the time of day when the first tag read occurred—i.e., at reader #6. Almost 66,000 data points are plotted. Notice that the y-axis is in days. Some of these "trip" times are very long. The largest is 192 days—effectively the tag was seen at the beginning of the experiment and then not again until the end. In sharp contrast, the 95th percentile is 1.9 days, the 80th percentile is 8.7 min, and the 50th percentile is only 3.3 min (the distance is 2.1 miles). Clearly, most of the "trip" times really are trip times in the classical sense of the word. However, some of them are very long—because there is no restriction on how much time can transpire between the tag reads. The purpose in presenting these data is to make sure the readers of the paper understand how wide ranging the (i, j) times can be.

Since the focus of this analysis is on travel times—not trip-making patterns—while Fig. 5.2 is helpful in understanding the nature of the data, a more detailed

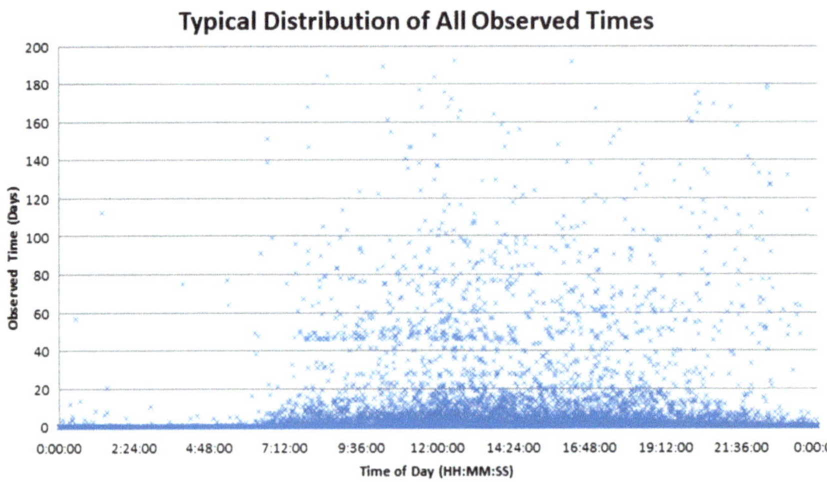

Fig. 5.3 Typical distribution of all *(i, j)* times observed

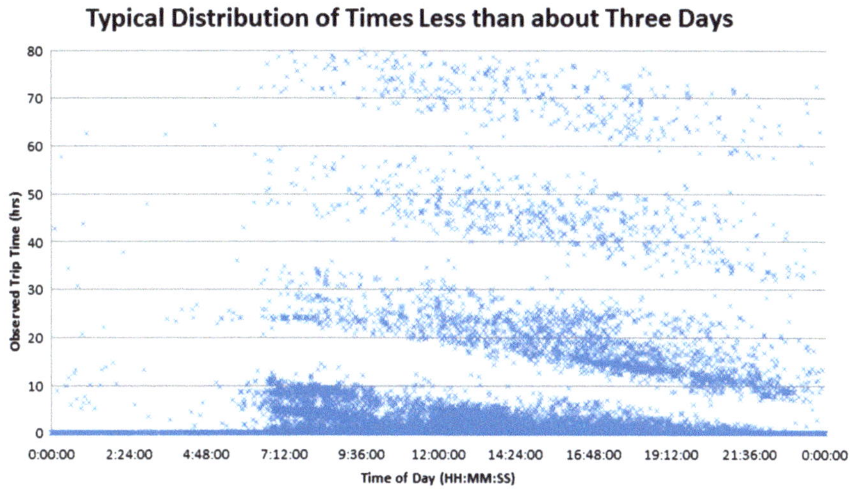

Fig. 5.4 Typical multi-day pattern in the *(i, j)* times observed

analysis is needed to understand travel times. Dropping down first to about three days, first, Fig. 5.4 shows the roughly 63,000 times that were less than 80-h long for reader pair (6,3). The x-axis again uses the time of the observation at reader #3.

Three patterns are immediately evident. The first is that in the early hours of the morning very few people are making trips. Very few observations exist and the trip times tend to be very short (near 0).

The second pattern is a triangular cluster of trip times that descends from a maximum of about 12 h at 7 am down to zero at midnight. Moreover, there are dense clusters within this cluster: one between 7 and 10 am that has times descending from 5 h to 3 reflecting people who work a half day and then go home or to lunch; a second again from 7 to 10 am that descends from 10 h down to 8 reflecting people who work the whole day without leaving. There is a third subcluster from 1 until 3 pm that descends from 5 h down to about 3 h reflecting people who work the afternoon and/or return to work from lunch.

The third large pattern is the descending bands of times greater than a day that start about 7 am and extend until midnight. The first band starts at between 24 and 36 h at 7 am and descends to 7–16 h just before midnight. The second starts at 48–60 h and descends to 30–48 h. The reason why these bands exist is mostly missed reads. The second read (in this case at reader #3) that should have occurred did not; it did not happen until the next day. Another reason, less likely, is that the tag was somewhere in the vehicle where it could not be detected. The third is that the vehicle made a trip that used other local roads and did not take it past another reader (in this case, reader #3) until the next day. The fourth is that the vehicle was not moving between the first and second reads. The bottom line is that the bands of observations that span multiple days are mostly noise and can be ignored in the context of analyzing travel times.

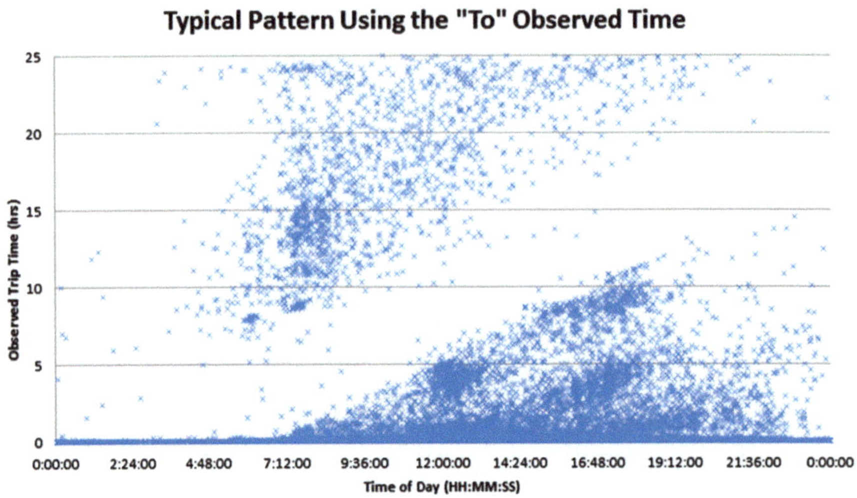

Fig. 5.5 Typical diurnal pattern based on the "To" time

Another important point to make is that pattern in the plot depends on whether the "from" time or the "to" time is used on the x-axis to plot the data. Figure 5.5 uses the "to" time instead of the "from" time that was used in Fig. 5.3—that is, the time when the vehicle was observed passing the "from" reader while instead. Only the times less than 24 h are plotted. While the lower-triangular daily pattern is again evident, it is flipped horizontally.

Focusing now on only the much shorter trip times, a fairly severe truncation is helpful. IT also helps to focus to trip rates—trip times per unit distance—rather than trip times themselves so that data from different (i, j) pairs can be compared. For reader pair (6,3) this means dividing by 2.1 miles, for example. If the cutoff rate is set at 4 min/mile or 15 miles per hour, then about 85% of observations with a trip time less than a day are kept.

Figure 5.6 shows the diurnal distribution of the trip rates less than 4 min/mile (speeds down to 15 mph). This includes weekends and holidays. For this (i, j) pair the minimum rate is just above 1 min/mile (the speed limit is 45 mph or 1.33 min/mile; 55 mph is 1.1 min/mile and 50 mph is 1.2 min/mile).

It is clear that the spread in trip rates is much larger during the day than it is at night. This makes sense: congestion is present as well as signal delay. The spread during the morning peak is about 1.2–2.5 miles/min; it drops to about 1.2–1.7 min at about 10 am; climbs back to 1.2–2.0 min by noontime; stays there until about 5 pm; and then drops to 1.2–1.5 min around midnight.

Figure 5.7 shows the difference between the daily patterns on the weekdays and the weekends. The weekdays look very much like the pattern shown in Fig. 5.4, while the weekends have no morning peak and the rates in general are lower.

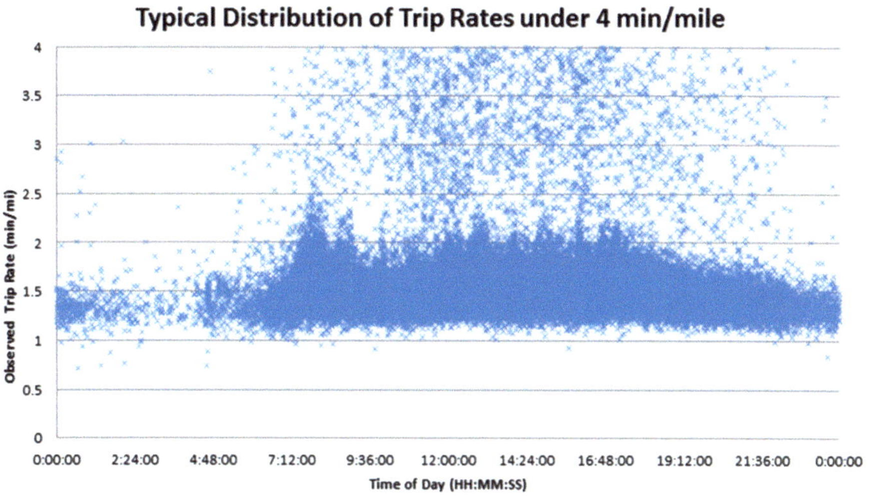

Fig. 5.6 Typical trip rates by time of day

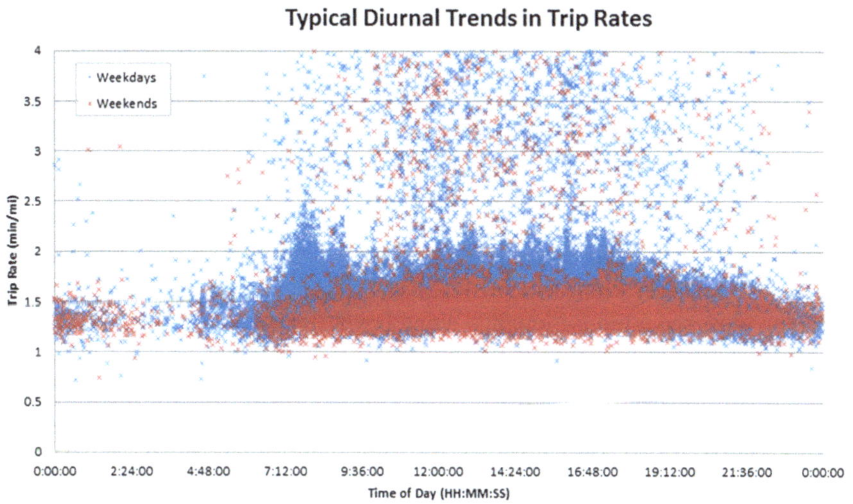

Fig. 5.7 Typical trip rates on weekdays and weekends

Now that a general understanding of the trip times has been obtained by studying Figs. 5.2–5.6, attention can be shifted to an analysis of the travel times. More analysis is needed to see what the travel time trends are. One way to do this is to use cumulative histograms. A cumulative histogram is like a cumulative probability density function except that the actual counts are used—the total is not normalized to 100%.

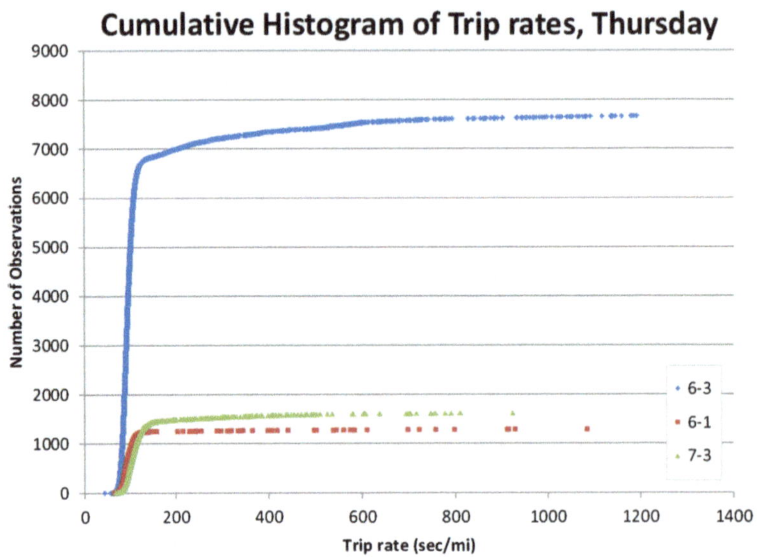

Fig. 5.8 Typical cumulative histograms of the trip rates for a day

Figure 5.8 shows typical cumulative histograms based on the southbound movements 6–3 (through), 6–1 (southbound right), and 7–3 (eastbound right on Jordan Drive). For each day and each OD pair (τ-CHs), the observed trip rates have been sorted by day, OD pair and travel rate; and the number of vehicles n having trip rates less than or equal to τ has been computed. The maximum trip rate considered was 20 min/mile (3 mph). The value of the trip rate τ is plotted on the x-axis and the number of vehicles n with a trip rate less than or equal to τ is plotted on the y-axis.

These plots show the information about travel times that is of interest, but the scaling focuses too much on the long trip times. The shape of the τ-CHs near the origin is hard to see; yet it is that portion which is of greatest interest in differentiating between trip times and travel times. Using log–log plots is better (logarithmic scales on both axes).

Figure 5.9 presents log–log plots of the τ-CHs. Notice that the τ-CHs for the northbound ODs on Thursdays show similar values for both τ and n. The northbound left into Jordan Road (2–1) has the smallest value of τ, followed by the northbound left (7–5) and then the through (2–5). This order is the same on Sundays, but the locus of the τ-CH in terms of the n values are different; the northbound left has fewer vehicles (less than 100) while the northbound through is upwards of 300.

Figure 5.10 shows the τ-CHs for the southbound flows. Notice that on Thursdays, the values of τ for the southbound right (6–1) and eastbound right (7–3) are the shortest and nearly identical while the τ values for the southbound through are larger. The n values for the two right turns (6–1 and 7–3) are nearly identical while the n value for the southbound through is almost an order of magnitude larger.

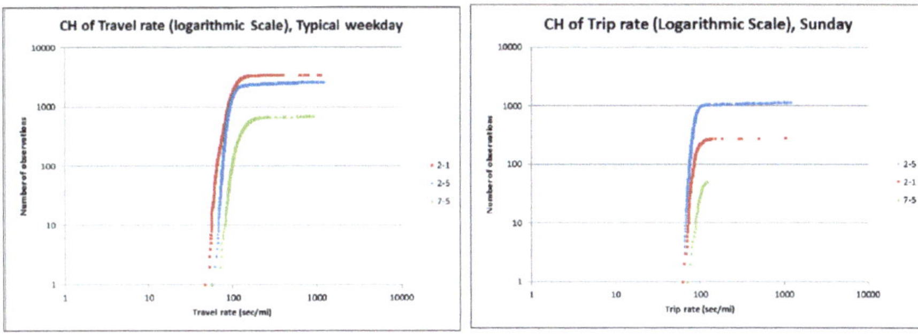

Fig. 5.9 Typical τ-CHs for northbound trips for a weekday and a weekend day

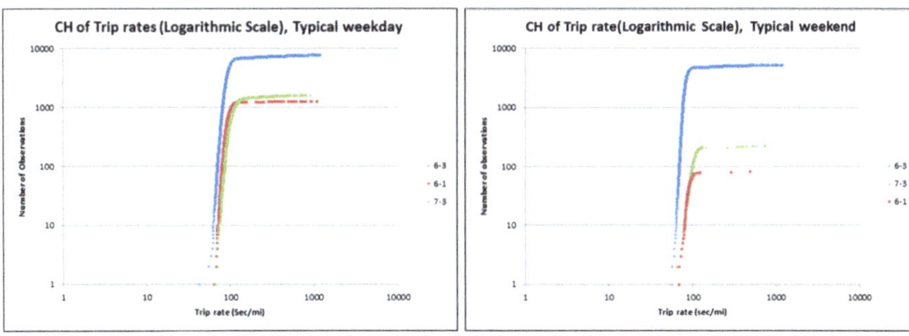

Fig. 5.10 Typical τ-CHs for southbound trips for a weekday and a weekend

On Sundays the same trends for the τ values still pertain, but the right turn volumes are an order of magnitude smaller (about 100) than on Tuesdays, while the through is half the Thursday value (about 5,000 instead of 10,000).

It is also useful to study the percentiles of the trip rates and see how they vary from day to day. These percentiles, and their ratios to other metrics like the mean, are commonly used to characterize travel time reliability. Figure 5.11 plots trends in the 10th, 50th (median), and 95th percentiles for the Tuesday and Sunday data. (The trends in the mean are also plotted). The sequential number of the Tuesday (or Sunday) is plotted in the x direction and the value of the percentiles is plotted in the y direction. It is apparent that:

- The 10th and 50th percentiles are quite consistent except on days when something is amiss (e.g., the 2nd Tuesday for movement 2–1).
- The 95th percentile varies widely. It follows a pattern very different from the 10th and 50th percentiles. It appears to be noisy and random. It could be that the people making stops have the most influence on this value, not the behavior of the arterial. The signals may also be having an effect; and that effect may also vary widely from one day to another.

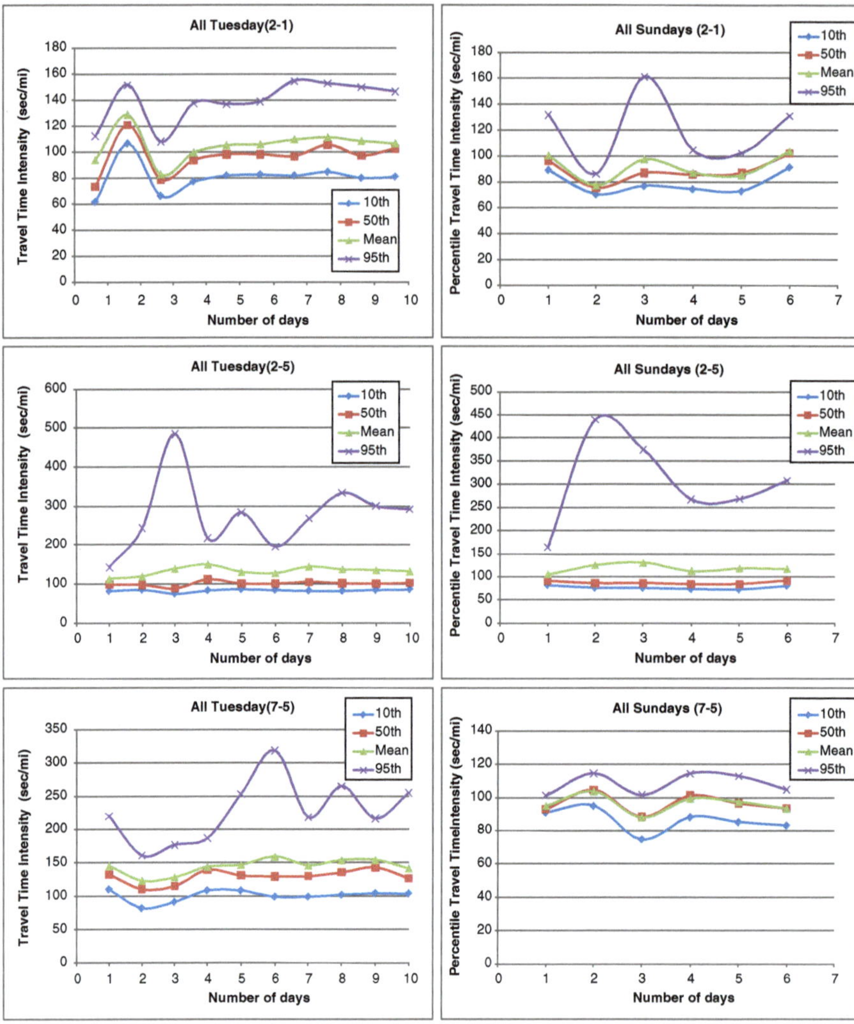

Fig. 5.11 Percentile trip rates (NB)

It is very important to recognize that these τ-CHs are based on observations of individual vehicles, not average travel times for groups of vehicles—as is the case for freeway detectors—across short time intervals—such as 5 min. The popular "travel time index" as it is often called is based on the 95th percentile of these observed average travel times divided by the average of those average travel times. The difference is akin to the contrast between the statistics relating to the random variable x versus statistics related to its mean $\bar{x}$.

Tables 5.3 and 5.4 present numerical data for the northbound OD pairs on Tuesdays and Sundays shown in Fig. 5.10 plus the southbound data.

Table 5.3 Summary of statistics for a typical weekday

Typical weekday									
Day	O–D Pair	Dist (mile)	Shortest TT	10th%	50th%	Mean	85th%	95th%	Max TT
Tues (SB)	6–3	2.1	55	80	93	124	112	264	1,485
	6–1	1.1	58	80	92	104	105	120	1,098
	7–3	1	63	91	110	140	133	288	1,917
Tues (NB)	2–5	2.2	63	83	101	142	140	316	1,420
	2–1	1.1	56	77	95	96	113	120	125
	7–5	1.3	67	98	129	145	172	249	1,108

Table 5.4 Numerical data for the Tuesdays and Sundays

Typical weekend									
Day	O–D Pair	Dist (mile)	Shortest TT	10th%	50th%	Mean	85th%	95th%	Max TT
Sun (SB)	6–3	2.1	56	74	83	113	94	233	1,485
	6–1	1.1	68	80	89	97	100	112	479
	7–3	1	56	87	101	131	117	351	1,767
Sun (NB)	2–5	2.2	62	76	87	130	102	361	1,407
	2–1	1.1	62	75	87	97	107	120	1,005
	7–5	1.3	72	85	98	98	109	115	122

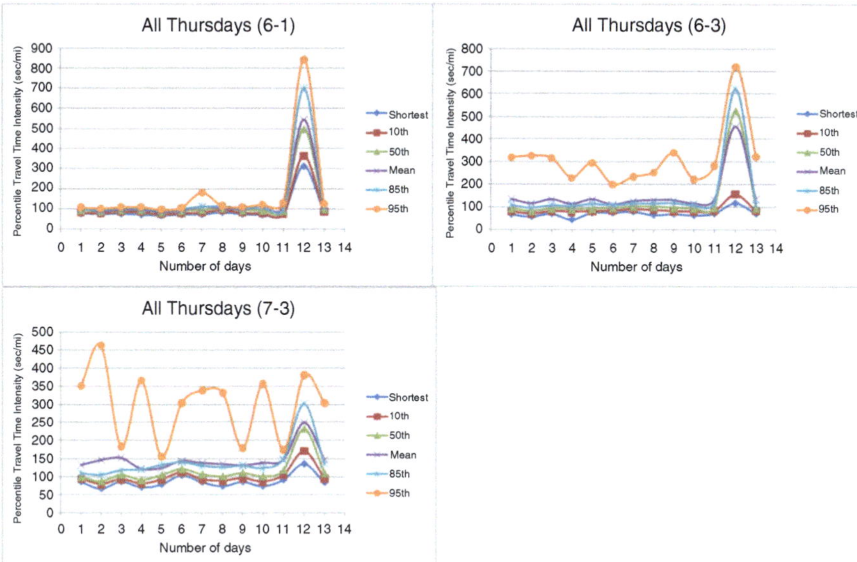

Fig. 5.12 Percentile travel time graphs for SB

Figure 5.12 shows trends in the 10th, 50th, and 95th percentiles for the southbound OD pairs (6–3, 6–1, and 7–3) for the Thursdays (instead of Tuesdays) during the experiment. (Remember that 6–1 is the inbound right onto Jordan Road and 7–3 is the outbound right.) The dramatically higher percentile values on the 12th Thursday immediately stand out. It turns out there was a snowstorm on this day and traffic was snarled; so perhaps low percentiles can be used to identify incidents and changes in network performance due to other factors.

Table 5.5 shows the numerical values for the southbound OD pairs for the Thursday data presented in Fig. 5.12.

5.5 Analysis #2: Trends within Days

This analysis focuses on trends within individual days. It asks: how do the short-run τ-CHs vary as a function of congestion, weather impacts, etc. Of course, as Fig. 5.5 shows, there is a tendency for the long-valued trip times to become increasingly more prevalent across the course of a given day; so this trend needs to be taken into account; but what other trends do the data show?

To do this analysis the trip time observations are sorted by OD pair and then placed in chronological order based on the "D" time (for the entire five months). This allows us to look at trends in the trip rates over time.

Table 5.5 Summary of trip rate statistics for Thursdays for the southbound OD pairs

O–D Pair	Date	Shortest TT	10th%	50th%	Mean	85th%	95th%
	26-Jul	75	82	92	100	100	107
	2-Aug	72	78	86	87	95	103
	9-Aug	74	85	93	95	105	109
	4-Oct	69	84	92	94	104	109
	11-Oct	65	73	82	83	92	97
	18-Oct	69	78	88	90	99	105
6-1 (1.1 mile)	25-Oct	71	83	94	110	111	181
	1-Nov	80	95	102	103	112	115
	8-Nov	72	81	91	97	103	110
	15-Nov	68	77	90	108	108	121
	29-Nov	74	76	88	102	97	127
	13-Dec	311	363	497	544	700	841
	20-Dec	80	85	96	109	109	125
6-3 (2.1 mile)	26-Jul	67	79	92	134	109	319
	2-Aug	56	70	81	115	97	325
	9-Aug	69	80	92	132	107	316
	4-Oct	43	78	90	112	103	228
	11-Oct	73	80	91	133	114	293
	18-Oct	74	84	95	110	108	199
	25-Oct	79	90	100	125	115	232
	1-Nov	63	86	100	129	115	253
	8-Nov	68	81	96	129	118	339
	15-Nov	63	80	93	115	109	222
	29-Nov	70	82	97	131	118	282
	13-Dec	116	157	524	456	622	719
	20-Dec	70	83	96	134	113	322
7-3 (1 mile)	26-Jul	86	94	100	132	109	351
	2-Aug	67	79	88	146	105	463
	9-Aug	87	93	106	152	118	182
	4-Oct	71	80	92	121	120	366
	11-Oct	79	92	106	124	132	156
	18-Oct	104	111	122	143	140	304
	25-Oct	85	92	107	138	130	338
	1 Nov	74	88	102	134	126	332
	8-Nov	86	95	111	131	130	178
	15-Nov	74	86	102	138	124	356
	29-Nov	92	104	119	147	149	174
	13-Dec	136	170	232	249	302	381
	20-Dec	84	95	110	147	135	303

The first subanalysis looks at variations in the τ-CHs for every half hour. The fifty most recent trip rate observations are used to do this. Since 50 observations are being used, each one is a two-percentile point in the τ-CH. For completeness, all the trip times observed are examined first, not just those with trip rates under 1,200 s/mile (refer back to Fig. 5.3).

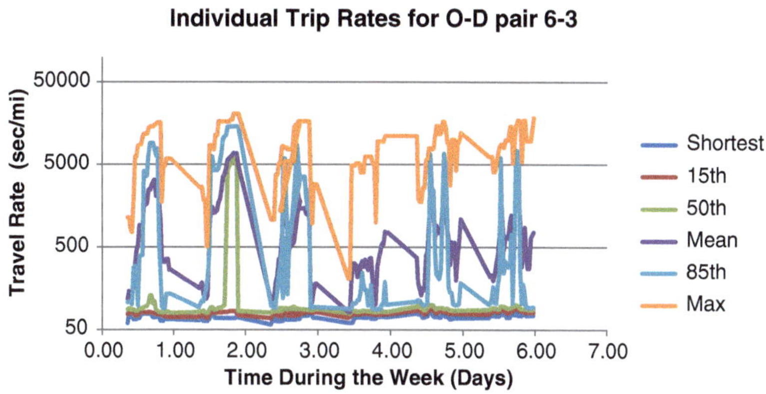

Fig. 5.13 Trip rate percentiles across a week

Across a single week, there is a clear pattern in these trip rates. As Fig. 5.13 shows, the values for the shortest trip rate, the 15th and 50th percentile hold steady while the values for the mean, 85th, 95th, and maximum vary widely. Our conclusion is that the values up to the 50th percentile reflect travel rates while the other percentiles reflect trip rates.

The next subanalysis involves seeing what happens within individual days. Rather than using Fig. 5.13 to do this, where the results are present but hard to see, a second figure is created. Figure 5.14 shows trends in the τ-CHs for August 1, 2007. Each τ-CH is labeled to show the half hour to which it pertains (e.g., 8:30, 9:00, etc.) and the "D" time of the oldest observation in the set of 50.

The top-left plot shows τ-CHs for every half hour from 8:30 to 12:00 noon; the top right one shows τ-CHs from 12:30 to 16:00; etc.; the last τ-CH pertains to 23:00, encompassing observations from 20:40 to 23:00.

Notice that the first τ-CH is labeled 5:59–8:30 am. This means the τ-CH pertains to the most recent 50 observations seen at 8:30 am on this day and the earliest of these was observed at 5:59 am. It is unfortunate that the 50 observations span more than the half hour from 8:00 to 8:30 am, but the density of AVI-equipped probes is not high enough to do that. The extent to which the observations are reused decreases during the peak periods. For example, the second τ-CH is based on observations from 7:37 to 9:00 am, so the observations reused are those from 7:37 to 8:30 am.

The τ-CHs show that the minimum trip rate remains fairly consistent at about 150 s. This means some vehicles always traverse the network without significant delay; 150 s/mile is equivalent to about 24 mph. As the morning peak progresses, the percentage of vehicles that experience this minimum trip rate becomes smaller, dropping to only about half (20–25 out of 50) midday. Rather than this being a change in the trip rate for the OD pair, this more likely reflects an increase in the percentage of vehicles that are stopping between the two readers; so these observations are not evidence of a slowdown in the travel, it shows changes in

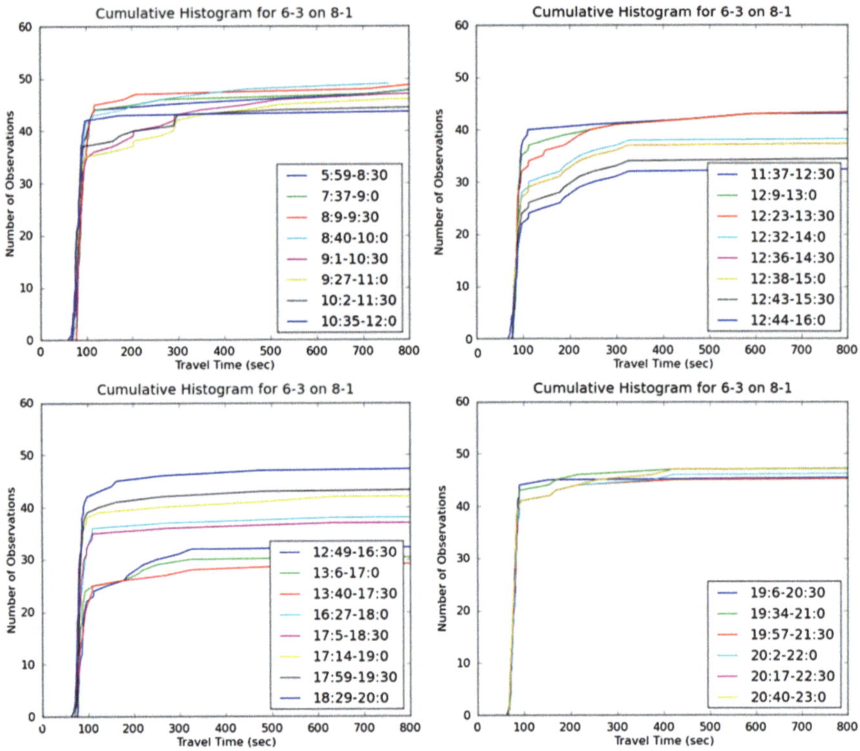

Fig. 5.14 Trends in the $T - Chs$ across a day (All Data)

activity patterns. There are still vehicles that can achieve the minimum trip rate (consistent with the findings for the Thursday analysis presented before). The number of vehicles that have these large trip rates seems to peak at about 3:00 pm (12:38–15:00 and 12:43–15:30) and then by 6:00 pm (e.g., 16:27–18:00) there are far fewer. Later in the evening, nearly all of the trip rates are at or near the minimum value.

Another subanalysis asks what happens if the trip rates greater than 1,200 s/mile are omitted. Figure 5.15 presents the results.

The same trends are evident but the breakpoint where the distribution tails off to larger values is much higher; the long trip times are a much smaller percentage of the total. It is also apparent that the span of time involved in the 50 observations changes. In the early hours and late at night, the spans change considerably, but in the middle of the day, they do not change much. Notice that the span of observations at 14:00 changes only from 12:32 in the case of using all the observations to 12:29 when only those under 1,200 s/mile are employed.

The third subanalysis asks: can incidents and other events be spotted using these τ-CH plots? The answer seems to be "yes." Figure 5.16 shows the τ-CHs for OD pair 6–3 on December 13, 2007, a day when there was a snowstorm. In this instance, all

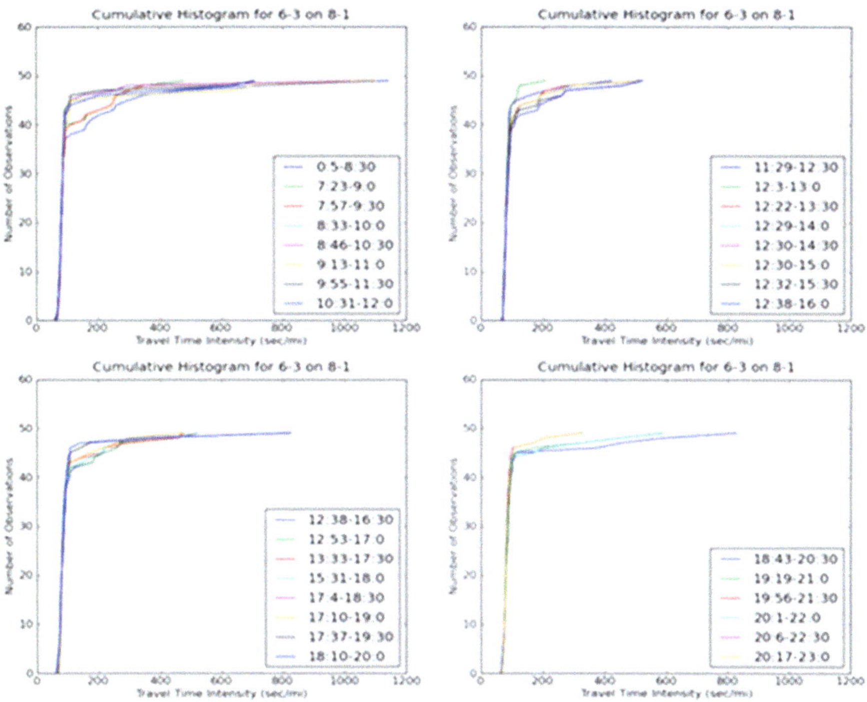

Fig. 5.15 Trends in the $T - Chs$ across a day ($\mathrm{T} \leq 1,200\,\mathrm{s/mile}$)

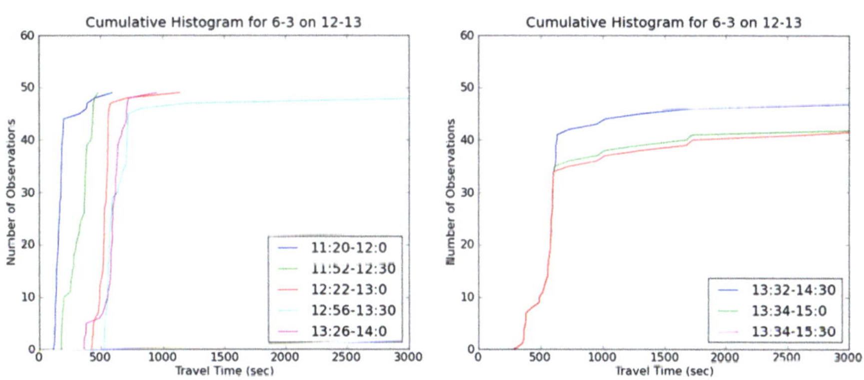

Fig. 5.16 Trends in the $T - Chs$ on a snow day (All Data)

of the data are employed so that truncation does not filter out our ability to see what happened. It is possible to see that the minimum trip rate increases substantially, the long trip rates are all but absent—probably because far fewer people stop at intermediate locations; and the τ-CHs are stretched out across a greater range of τ values—maybe showing that the drivers are being more risk averse because of the

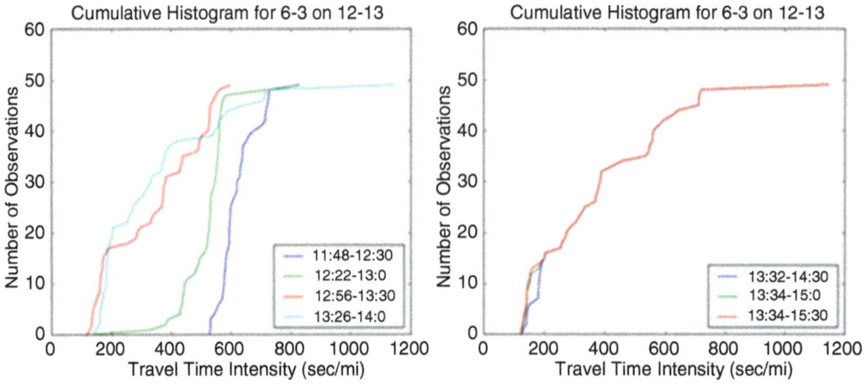

Fig. 5.17 Trends in the *T-Chs* on a snow day ($T \leq 1,200\,\text{s/mile}$)

slippery, snowy conditions. This suggests that incident detection algorithms based on tracking the low-percentile values of the τ-CHs may have promise.

As with the previous investigation, it is interesting to see how these τ-CHs change if the 1,200 s/mile trip rate cutoff is applied. The result is shown in Fig. 5.17. The same trends are present, but the tailing off is less apparent and are less because the long trip intensity observations are omitted. (Please note that the scales are very different on the horizontal axis when comparing the graphs.)

The final subanalysis in this "trends within days" study examines the relationship between the half-hourly τ-CHs and the histogram for the average trip times. To be clear about the difference, Fig. 5.12 shows the patterns in the shortest, 15th, 50th, etc. percentiles for the individual vehicle trip rates across the week of August 1; while the focus here is on the histogram for the averages of each of those half-hourly distributions across the week. This is akin to comparing the cumulative histograms for individual speeds observed across all lanes of a freeway count station within each 5-min period of a week with the histogram of the averages for those 5-min observations.

Figure 5.18 shows the histogram for the average values of τ observed across the week of August 1, 2007 without the 1,200 s/mile filter being applied. Notice that the values are quite large. The average of these averages is 2,137 s/mile, which shows the significant impact of the large trip rates—equivalent to 1.7 mph—and the 95th percentile is 3,848 s/mile—equivalent to 0.93 mph! The insight is that these averages, which include all of the observations, are not particularly useful in terms of indicating what the travel times might be; the long trip times have too significant an impact. In contrast, the 50th percentile values for these CHs during the same week (not shown in the figure) have an average τ of 289 s/mile—equivalent to 12.4 mph—and a 95th percentile of 126 s/mile—equivalent to 28.4 mph—and less than the average—which is evidence of the impact of the very large values—out of the 182 observations of these 50th percentile values across the week, almost all of them are in the range of 80–100 s/mile, but seven of them are in the range of 4,500–6,000 s/mile.

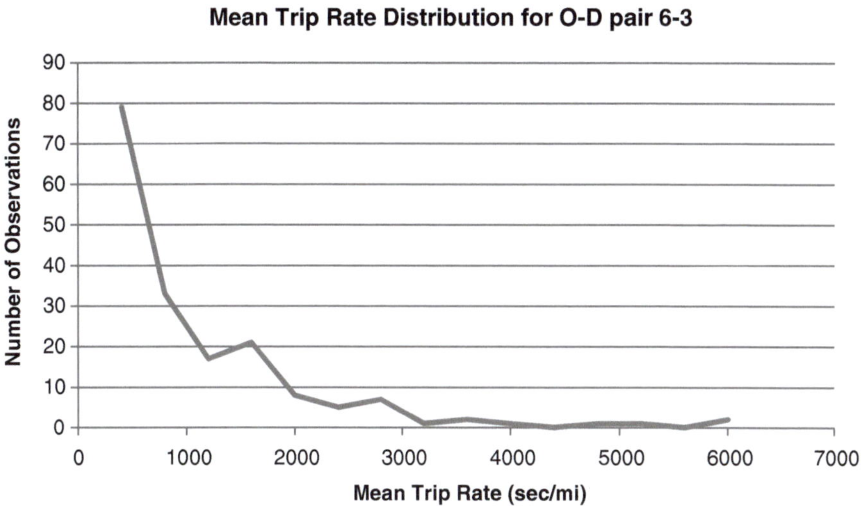

Fig. 5.18 Histogram for the mean trip rates observed during the week of August 1, 2007, O–D pair 6–3

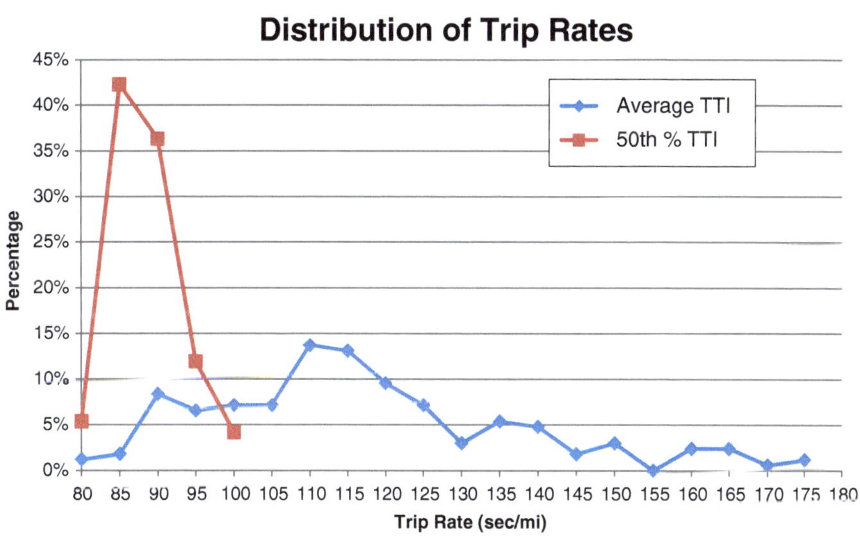

Fig. 5.19 Histogram for the mean trip rates observed during the week of August 1, 2007, O–D pair 6–3 ($T \leq 1{,}200\,\text{s/mile}$)

In contrast, Fig. 5.19 shows the distribution of the average trip times when the 1,200 s/mile filter is applied. As would be expected, all the values are much smaller. In fact, the average of these averages is only 114 s/mile, more than an order of

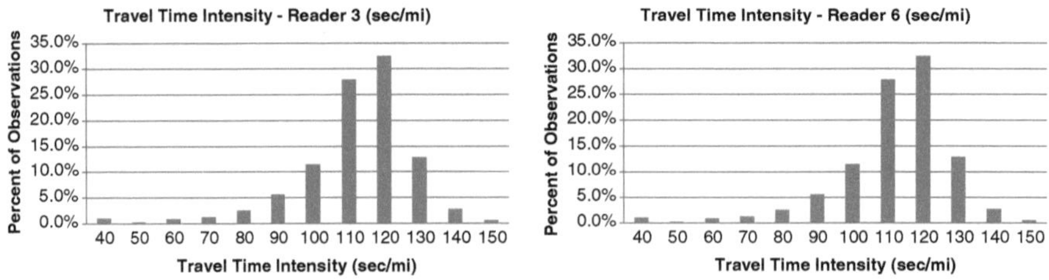

Spot" trip rates from spot speed data collected at readers 3 and 6

Fig. 5.20 "Spot" trip rates from spot speed data collected at readers 3 and 6

magnitude smaller than the result when all the observations are employed. This is equivalent to 31.6 mph, a value which is far more likely to be reflective of the space mean speeds that occur between readers 6 and 3—and the 95th percentile is 157 s/mile—equivalent to 22.9 mph. That being said, the 50th percentile still offers a compelling alternative to the mean; it has an average of 86 s/mile—equivalent to 41.9 mph and a 95th percentile of 95 s/mile—equivalent to 37.9 mph—both of which are probably even more typical of the space mean speeds between readers 6 and 3.

As a closing subanalysis, spot speeds that were collected adjacent to Readers 3 and 6 were inverted to create "spot" trip rates. They ranged from 90 to 130 s/mile at those locations as shown in Fig. 5.20, values that are slightly larger than, but consistent with the 50th percentile values shown in Fig. 5.18. Whether these spot rates can be used to gain insights about travel times or travel time reliability is an open question; but they are comparable to the distance-based rates observed from the toll tag data.

5.6 Key Findings and Conclusions

The most important finding is that the AVI data are very valuable. The data reveal important trends in the trip times and travel times. They can be used to estimate travel times, watch for incidents, monitor travel time reliability, etc., albeit with some care. The data are clearly not as valuable as AVL data, where the entire vehicle trajectories can be observed, but compared with point detectors where the status of the network can only be observed at the instrumented locations, AVI data are much better; they provide an opportunity for a major improvement in system observability.

The next observation is that the arterial trip times are richly varied; with a clear daily pattern in the longest trip times (at least for this network). Whether this is a tendency for all arterial trip times is, of course, unknown, but it seems likely that many arterials may have similar patterns. These trip times were expected, but not their significance.

The presence of these long trip times makes it important to understand how to both filter them and use them. It is not just a filtering issue. The data do need to be filtered so that travel times can be discerned (instead of trip times). However, the long values may be useful in detecting incidents; so not filtering the data is a useful idea. It may be that two data processing algorithms are needed—one to estimate travel times and the other to look for evidence of incidents.

It is also important to understand how the time of revelation affects these thoughts. The trip times only become observable when the second reader is passed—when the trip ends, not when it starts. Effectively, a detection lag exists; one that cannot be overcome without more readers. Incidents might happen before they are observable because the second tag read has not yet occurred. On the other hand, expected downstream reads (from upstream observations) could be compared with actual downstream reads to get an early sense that something may have happened.

These observations suggest that AVL systems will add significant value over AVI-based systems. They will provide better information about travel times than AVI systems. With AVL systems, it will be easier to filter out the travel times for the extended trips.

Incident detection and characterization will also be more difficult with AVI systems. Since with AVI systems the whole trajectory is not known, it will be difficult to distinguish between vehicles that are delayed due to incidents and those that have simply made roadside stops. It will also be difficult to tell exactly where the incident has occurred (if one has) because the only information available will be travel times between AVI readers. In contrast, with AVL systems, it will be easy to spot vehicles that have made a roadside stop and distinguish them from ones involved in or affected by incidents. It will also be a lot easier to see where the incident has occurred because it will be possible to see the AVL vehicles that have been stopped by the incident.

The trends in the percentiles for the trip time density functions suggest the following:

- The 10th and 50th percentile values are likely to be very consistent.
- But the 95th percentile values vary dramatically.
- Moreover, the mean travel times vary more than the 50th percentile times because they are affected by, and are sensitive to, the long trip times.
- This means travel time metrics predicted on ratios to the mean will be confounded by the variations in the mean.
- It would likely be better to use metrics predicted on the 50th percentile or the free flow travel time.

It may be that the 95th percentile will not be the best metric to use as an indicator of travel time reliability. A lower percentile, like the 10th or 50th, might be better. All of the percentiles are affected when there are incidents; and the lower percentiles seem to be affected only when there are incidents, whereas the 95th percentile has a lot of volatility, seemingly caused by other factors.

That having been said, the 95th percentile values can perhaps still be used to provide guidance to travelers about what travel times to expect; but that 95th percentile needs to be based on the travel times, not the trip times. Based on the plots reviewed so far, one can make statements like: half of the travelers will need X minutes, so allowing that much time will be fine if you are someone who drives faster than about half of the population; but if you want to be sure you are not late, you should allow Y minutes. You are likely to be quite early, but you will not be late.

A final observation is that the movements are slightly different in trends they portray. Some OD pairs show very stable values, as with the ones involving right turns where the impacts of signal timing or congestion is likely to be low. The ones with left turns show more variation. This finding has not been stressed heavily in the results presented, but it can be seen in the differences between the southbound OD pairs—that involve two right turns—and the northbound OD pairs—that involve two left turns.

Acknowledgements The authors are deeply indebted to the Paul Manuel and the other individuals at Mark IV Industries who helped us to create the wireless, solar-powered tag readers, and for making it possible for us to experiment with them in the field. We are also deeply indebted to Brian Menyuk, at NYSDOT, who helped us make arrangements to deploy the detectors in the field. Also critical was Tony Annese, of Annese and Associates, who helped created the software for managing the data flows, and to Michael Schauer, and others, of USDOT, for helping to transform the vision of an experiment into a reality. The authors are solely responsible for the analyses conducted and the conclusions drawn.

References

Berkow M, Wolfe M, Monsere C Bertini R. Using signal system data and buses as probe vehicles to define the congested regime on arterials. In: Proceedings of the 87th Annual Meeting of the Transportation Research Board, 2008.

Berry DS. Evaluation of techniques for determining overall travel time. Highway Res Rec Proc. 1952; 31: 429–439.

Bertini RL, Cameron GJ, Peters J. Evaluating traffic signal improvements using archived transit AVL data. ITE J 2005; 75(2); 69–75.

Boyce D. A memoir of the ADVANCE project. ITS J 2002; 7(2): 105–130.

Boyce DE, Kirson A, Schofer JL. Design and implementation of ADVANCE. The Illinois dynamic navigation and route guidance demonstration program. In: Vehicle Navigation Information Systems Conference Proceedings Part 1 (of 2), 1991.

Byon Y-J, Shalaby A, Abdulhai B. Travel time collection and traffic monitoring via GPS technologies. In: Proceedings of the IEEE Intelligent Transportation Systems Conference, 2006.

Cetin M, List GF, Zhou Y. Factors affecting minimum number of probes required for reliable estimation of travel time. Transport Res Rec. 2005; 1917; 37–44.

Chakroborty P, Kikuchi S. Using bus travel time data to estimate travel times on urban corridors. Transport Res Rec 2004; 1870; 18–25.

Demers A, List GF, Wallace WA, Lee E, Wojtowicz J. Probes as path seekers: a new paradigm. Transport Res Rec 2006a; 1944: 107–114.

Demers A, List GF, Wojtowicz J, Kornhauser A, Wallace WA, Lee EE, Salasznyk P. Experimenting with real-time ATIS: stepping forward from ADVANCE. In: Proceedings of the 9th International Conference on Applications of Advanced Technology in Transportation, 2006b.

Dion F, Rakha H. Estimating dynamic roadway travel times using automatic vehicle identification data for low sampling rates. Transport Res B 2006; 40(9): 745–766.

Fontaine MD, Smith BL. Probe-based traffic monitoring systems with wireless location technology. Transport Res Rec. 2005; 1925: 3–11.
Greenshields BD. A study of traffic capacity. Proc Highway Res Board 1934; 14: 468.
Greenshields BD, Schapiro D, Erickson EL. Traffic performance at urban intersections, Technical Report No. 1, Bureau of Highway Traffic, 1947.
Hall R, Nilesh V. Buses as a traffic probe demonstration project. Transport Res Rec. 2000; 1731: 96–103.
Hellinga BR, Fu L. Reducing bias in probe-based arterial link travel time estimates. Transport Res C 2002; 10(4): 257–273.
Kwon J, Coifman B, Bickel P. Day-to-day travel-time trends and travel-time prediction from loop-detector data. Transport Res Rec. 2000; 1717: 120–129.
Lin W, Kulkarni A, Mirchandani P. Arterial travel time estimation for advanced traveler information systems. In: Proceedings of the 82nd Annual Meeting of the Transportation Research Board, 2003.
Li R, Rose G, Sarvi M. Using automatic vehicle identification data to gain insight into travel time variability and its causes. Transport Res Rec. 2006; 1945: 24–32.
List GF. Signal hardware demonstration project, final report, prepared for the New York State Energy Research and Development Authority, 1993.
List GF, Demers A. Estimating highway facility performance from AVL data. In: Proceedings of the fifth international symposium on highway capacity and quality of service, pp. 319–328, 2006.
List GF, Wallace WA, Demers A, Salasznyk P, Lee E, Wojtowicz J. Field experiment with a wireless GPS-based ATIS system. In: Proceedings of the 12th World Congress on ITS, 2005a.
List GF, Demers A, Wallace WA, Lee E, Wojtowicz J. ATIS via wireless probes: smart vehicles for smart travelers. INFORMS Annual Meeting, 2005b.
Ma X, Koutsopoulos HN. Estimation of the automatic vehicle identification based spatial travel time information collected in Stockholm. IET Intell Transport Syst. 2012; 4(4): 298–306.
Morris AG, Kornhauser AL, Kay MJ. Urban freight mobility: collection of data on time, costs, and barriers related to moving product into the central business district. Transport Res Rec. 1998; 1613: 27–32.
Oppenlander JC. Sample size determination for travel time and delay studies. Traffic Eng. 1976; 25–28.
Pan C, Lu J, Wang D, Ran B. Data collection based on global positioning system for travel time and delay for arterial roadway network. Transport Res Rec 2007; 2024: 35–43.
Quiroga CA, Bullock D. Determination of sample sizes for travel time studies. ITE J 1998; 68(8): 92–98.
Quiroga CA, Bullock D. Travel time studies with global positioning and geographic information systems: an integrated methodology. Transport Res C 1998; 6(1/2): 101–127.
Rice J, Van Zwet E. A simple and effective method for predicting travel times on freeways. IEEE Trans Intell Transport Syst. 2004; 5(3): 200–207.
Schaefer MC. License plate matching surveys: practical issues and statistical considerations. ITE J, 1988; 58(7): 37–42.
Shuldiner PW. Acquisition and analysis of license plate data using video and machine vision technology. In: Proceedings of the National Traffic Data Acquisition Conference, pp. 435–454, 1996.
Vavra TG. Evaluating EZ-Pass: Using conjoint analysis to assess consumer response to a new tollway technology. Market Res. 1999; 11(2): 5–16.
Wasson JS, Sturdevant JR, Bullock DM. Real-time travel time estimates using media access control address matching. ITE J 2008; 78(6): 20–23.
Williams J. Survey and analysis of vanpooling in metropolitan Washington, D.C. Transport Res Rec. 1986; 1082: 15–22.
Wojtowicz J, Murrugarra RI, Bertoli B, Wallace WA, Manuel P, He W, Body C. RFID technology for AVI: field demonstration of a wireless solar powered E-ZPass tag reader. In: Proceedings of the 15th World Congress on Intelligent Transport Systems, 2008.

Yamamoto T, Liu K, Morikawa T. Variability of travel time estimates using probe vehicle data. In: Proceedings of the Fourth International Conference on Traffic and Transportation Studies, pp. 278–287, 2006.

Chapter 6
A Multibuffer Model for LWR Road Networks

Mauro Garavello and Benedetto Piccoli

Abstract This paper introduces a new model for describing intersections in road networks, whose load dynamics is governed by the Lighthill–Whitham–Richards model. More precisely we define a solution for intersections using a multibuffer, i.e. a set of buffers, one for each outgoing road. We compare the obtained dynamics with those of some models previously introduced in the literature. In particular, we are able to respect the preferences of drivers and to not block the intersection when only one outgoing road is full. This improves some weaknesses of previous models.

6.1 Introduction

The modeling of traffic at a macroscopic level is nowadays a well-established approach in the transportation engineering community, which has its roots in the fundamental model proposed independently by Lighthill and Whitham (1955), and Richards (1956). Their work introduced to the traffic community the kinematic wave theory, which enables one to reconstruct macroscopic features of traffic flow, in particular tracing backward queues propagation. The model, consisting in a single conservation law for car density, is referred to as LWR model and is based on expressing the average velocity as function only of the car density. Greenshields

M. Garavello
Dipartimento di Scienze e Tecnologie Avanzate, Università del Piemonte Orientale "A. Avogadro", viale T. Michel 11, 15121 Alessandria, Italy

Dipartimento di Matematica e Applicazioni, Università di Milano Bicocca, via R. Cozzi 53, 20125 Milano, Italy
e-mail: mauro.garavello@unimib.it

B. Piccoli (✉)
Department of Mathematical Sciences and Center for Computational and Integrative Biology Rutgers University - Camden, 311 N 5th Street, Camden, NJ 08102, USA
e-mail: piccoli@camden.rutgers.edu

S.V. Ukkusuri and K. Ozbay (eds.), *Advances in Dynamic Network Modeling in Complex Transportation Systems*, Complex Networks and Dynamic Systems 2,
DOI 10.1007/978-1-4614-6243-9_6, © Springer Science+Business Media New York 2013

(1935) empirically measured a relation between the density and the flow of vehicles, now known as the fundamental diagram. There is a wide literature of studies on the fundamental diagram; for a review, see Garavello and Piccoli (2006). The numerics for such model was addressed in various papers; see Lebacque (1996). The supply demand approach there proposed is equivalent to the classical Godunov numerical scheme for general conservation laws (Godunov 1959). Notice also that such approach is intimately related to the cell transmission model of Daganzo (1994), which can be seen as a discretization of the LWR model.

More recently, a growing attention has been devoted to extensions of the same model to networks; see, for instance, (Chitour and Piccoli 2005; Coclite et al. 2005; Garavello and Piccoli 2005; Helbing et al. 2007; Holden and Risebro 1995). The interest was also motivated by other applications: data networks (D'apice et al. 2006), supply chain (D'Apice and Manzo 2006; Göttlich et al. 2005), air traffic management (Sun et al. 2007). Here we focus on the LWR model on a network, but the results are of use to other research domains.

To define a dynamics on the whole network, one first considers Riemann problems at nodes, which are Cauchy problems with constant initial data on each arc. Notice that the only conservation of cars is not sufficient to determine a unique dynamics. Thus one has to prescribe solutions for every initial data and we call the relative map a Riemann solver at nodes. Then it is possible to construct approximate solutions, via wave front tracking (see Garavello and Piccoli 2009), using classical self-similar entropic solutions for Riemann problems inside arcs and an assigned Riemann solver at nodes.

Various ways to define solutions at intersections were proposed in the literature; see, for instance, (Chitour and Piccoli 2005; Coclite et al. 2005; Garavello and Piccoli 2006; Herty et al. 2009; Marigo and Piccoli 2008).

This paper introduces a new model for describing dynamics at junctions in road networks. Due to finite speed of waves, we can reduce to the case of a simple road network, composed by a single junction with an arbitrary number of incoming and outgoing roads. On each road, the evolution of the car traffic is governed by the LWR model.

In the same spirit as (Göttlich et al. 2006; Herty et al. 2007, 2009), we suppose that, inside the crossroad, there are some buffers, with finite size. More precisely, we assume that there is a buffer in front of each outgoing road, so that the number of buffers equals that of outgoing roads. The basic idea behind this construction is that a car exiting an incoming road enters the buffer associated with its desired destination and then it passes to the corresponding outgoing road by a FIFO policy. Since the number of buffers is equal to the number of outgoing roads, our model is able to capture the preferences of drivers. On the contrary, the model proposed in Herty et al. (2009) contains only one buffer inside the junction and so all the cars enter the same buffer losing the information about their origins and destinations.

In Sect. 6.2, in order to justify the study, we illustrate, with a simple example, the main differences of our approach with respect to those introduced in Coclite et al. (2005) and Herty et al. (2009). In Sect. 6.3, after introducing the basic assumptions and definitions about conservation laws, we recall the construction of the Riemann solver introduced in Coclite et al. (2005). The latter is defined for a model without buffers and makes use of a traffic distribution matrix together with maximization of the through flux. In Sect. 6.5, we describe in detail the Riemann solver with multibuffer.

Moreover we provide an analytic comparison of our model with an ODE-PDE ones defined in Herty et al. (2007) for supply chains and networks and with that of Herty et al. (2009). The paper ends with Sect. 6.7, which contains the conclusions.

6.2 Model Justification

In this section, we present some examples to show the behaviors of solutions of different models in literature. More precisely, we consider a junction J with a single incoming road I_1 and two outgoing roads I_2 and I_3. The initial loads of the roads are given by

$$\rho_{l,0} = \begin{cases} \sigma, & \text{if } l = 1, \\ 1, & \text{if } l = 2 \end{cases} \qquad \text{and} \qquad \sigma < \rho_{3,0} < 1, \tag{6.1}$$

where $\rho_{\max} = 1$ denotes the maximal possible density in the roads, $f(\rho)$ is the flux when density is ρ, while $\sigma \in (0,1)$ is the critical density between the free and congested traffic flow. In the following paragraphs, we consider three different types of solutions at junctions: the first one, denoted by $\mathcal{RS}_{\text{CGP}}$, is that introduced in Coclite et al. (2005), the second one was proposed by Herty et al. (2009) and, finally, the last one is that introduced in this paper. The last two models use buffers to describe the dynamics; in this case we also assume that buffers are initially empty.

The solution $\mathcal{RS}_{\text{CGP}}$ introduced in Coclite et al. (2005). It is simple to see that the solution at the junction J, with respect to $\mathcal{RS}_{\text{CGP}}$, is given by the triple (ρ_1, ρ_2, ρ_3) defined by

$$\begin{aligned} \rho_1(t,x) &= \begin{cases} \sigma, & \text{if} \quad x < \bar{\lambda}_1 t, \\ 1, & \text{if} \quad \bar{\lambda}_1 t < x < 0, \end{cases} \\ \rho_2(t,x) &= 1 \\ \rho_3(t,x) &= \begin{cases} 0, & \text{if} \quad 0 < x < \bar{\lambda}_2 t \\ \rho_{3,0}, & \text{if} \quad x > \bar{\lambda}_2 t \end{cases} \end{aligned}$$

where $\bar{\lambda}_1 = -\frac{f(\sigma)}{1-\sigma}$ and $\bar{\lambda}_2 = \frac{f(\rho_{3,0})}{\rho_{3,0}}$; see Fig. 6.1. A complete description of $\mathcal{RS}_{\text{CGP}}$ is done in Sect. 6.4. This example shows that, if an outgoing road is full, then no car crosses the junction. This implies that the road I_3 empties and in the incoming road I_1 it appears a shock with negative speed, connecting σ with the maximum possible density. It is questionable that the model does not allow any car going to the empty road I_3.

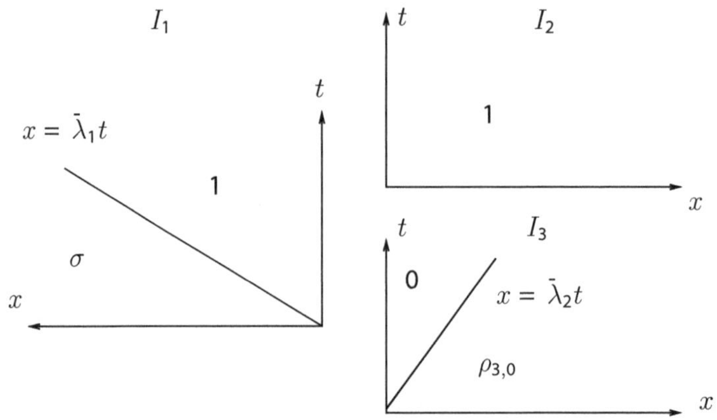

Fig. 6.1 The solutions to the Riemann solver $\mathcal{RS}_{CGP}$ in I_1, I_2 and I_3

The solution with buffer of Herty et al. (2009). In this part we describe the solution at J using the model of a junction with a buffer, as introduced in Herty et al. (2009). For a detailed description of this model, we refer to Herty et al. (2009). Assume that the capacity of the buffer μ is greater than or equal to $f(\sigma)$. For $0 < t < \bar{t}$ $\left(\bar{t} = \frac{r_{\max}}{f(\sigma) - f(\rho_{3,0})}\right)$ the solution is given by

$$\begin{aligned} \rho_1(t,x) &= \sigma \\ \rho_2(t,x) &= 1 \\ \rho_3(t,x) &= \rho_{3,0} \\ r(t) &= [f(\sigma) - f(\rho_{3,0})]\, t \end{aligned}$$

while, if $t > \bar{t}$,

$$\begin{aligned} \rho_1(t,x) &= \begin{cases} \sigma, & \text{if} \quad x < \bar{\lambda}_1 (t - \bar{t}), \\ \rho_{3,0}, & \text{if} \quad \bar{\lambda}_1 (t - \bar{t}) < x < 0, \end{cases} \\ \rho_2(t,x) &= 1 \\ \rho_3(t,x) &= \rho_{3,0} \\ r(t) &= r_{\max}; \end{aligned}$$

see Fig. 6.2. Here $r(t)$ denotes the load of the buffer at time t.

Note that in this case the queue in the buffer increases and when it reaches the maximum value, then a shock with negative speed appears in I_1 and connects the states σ and $\rho_{3,0}$.

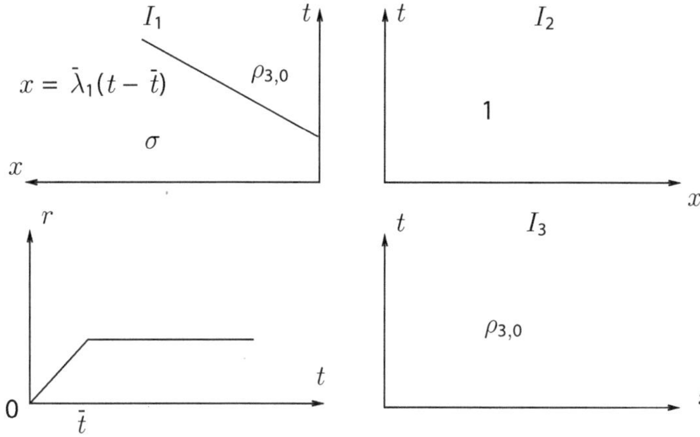

Fig. 6.2 The solutions to the Riemann solver of Herty et al. (2009) in I_1, I_2, and I_3 and the load of the buffer

Notice that all cars will finally reach road $\rho_{3,0}$ for every time. Therefore, the presence of a unique buffer erases the original will of drivers. This phenomenon is a drawback from modeling point of view.

The solution with multibuffer. Assume that the capacities of the buffers are given by $\mu_2 = \mu_3 \geq f(\sigma)$. Call $(\alpha_{21}, \alpha_{31})$ the traffic distribution matrix. For simplicity, we further assume that $\alpha_{31} f(\sigma) < f(\rho_{3,0})$. Define $\bar{t} = \frac{r_2^{\max}}{\alpha_{21} f(\sigma)}$. Then, for $0 < t < \bar{t}$, the solution is given by

$$
\begin{aligned}
\rho_1(t,x) &= \sigma \\
\rho_2(t,x) &= 1 \\
\rho_3(t,x) &= \begin{cases} \rho_{3,0}, & \text{if} \quad x > \bar{\lambda}_3 t \\ \bar{\rho}_3, & \text{if} \quad 0 < x < \bar{\lambda}_3 t \end{cases} \\
r_2(t) &= \alpha_{1,2} f(\sigma) t \\
r_3(t) &= 0
\end{aligned}
$$

where $\bar{\rho}_3 < \sigma$, $f(\bar{\rho}_3) = \alpha_{31} f(\sigma)$ and $\bar{\lambda}_3 = \frac{f(\rho_{3,0}) - f(\bar{\rho}_3)}{\rho_{3,0} - \bar{\rho}_3}$. For $t > \bar{t}$, the solution is given by

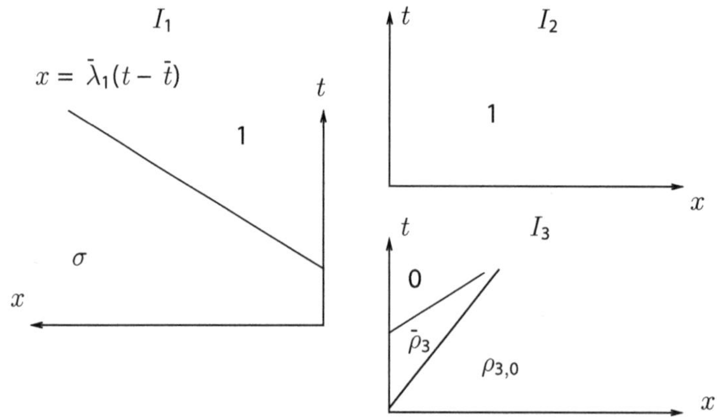

Fig. 6.3 The solutions to the Riemann solver with multibuffer in I_1, I_2, and I_3

$$\begin{aligned}
\rho_1(t,x) &= \begin{cases} \sigma, & \text{if} \quad x \le \bar{\lambda}_1(t-\bar{t}) \\ 1, & \text{if} \quad \bar{\lambda}_1(t-\bar{t}) < x \le 0 \end{cases} \\
\rho_2(t,x) &= 1 \\
\rho_3(t,x) &= \begin{cases} 0, & \text{if} \quad 0 \le x < \tilde{\lambda}_3(t-\bar{t}) \\ \bar{\rho}_3, & \text{if} \quad \tilde{\lambda}_3(t-\bar{t}) \le x < \bar{\lambda}_3 t \\ \rho_{3,0}, & \text{if} \quad \bar{\lambda}_3 t < x \end{cases} \\
r_2(t) &= r_2^{\max} \\
r_3(t) &= 0
\end{aligned}$$

where $\bar{\lambda}_1 = -\frac{f(\sigma)}{1-\sigma}$ and $\tilde{\lambda}_3 = \frac{f(\bar{\rho}_3)}{\bar{\rho}_3}$; see Figs. 6.3 and 6.4. Here $r_j(t)$ denotes the load of the j-th buffer at time t.

Notice that cars flow to road I_3 meanwhile the buffer of road I_2 is not yet full, then they stop. Finally, the situation is intermediate between the two above solvers. More precisely, the flow is neither stopped immediately nor continued for all time. Moreover, cars going through the junction are not redirected, but travel towards the outgoing roads or buffers according to drivers' preferences, expressed by the traffic distribution matrix.

The fact that the inflow stops after some time, even if one exiting road may still allow some outflow, is a limitation due to the use of the Lighthill–Whitham–Richards model for multiple lanes. However, such choice describes in an accurate way the links in highways with respect to the junctions in urban networks; see Newell (1993a,b,c). We summarize the key results of this section in Table 6.1.

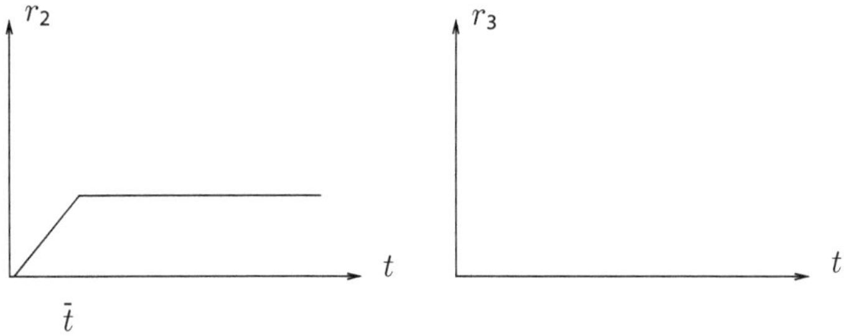

Fig. 6.4 The solutions to the Riemann solver with multibuffer for the buffers r_2 and r_3

Table 6.1 Comparison between different Riemann solvers according to the example of Sect. 6.2

	$\mathcal{RS}_{CGP}$	$\mathcal{RS}$-buffer	$\mathcal{RS}$-multibuffer
Stopping of the flow	Yes: at $t = 0$	No	Yes: when $r_2 = r_{max}$
Redirection of cars	No	Yes	No
Number of buffers	0	1	2
Asymptotic densities	$\rho_1 = 1$ $\rho_2 = 1$ $\rho_3 = 0$	$\rho_1 = \rho_{3,0}$ $\rho_2 = 1$ $\rho_3 = \rho_{3,0}$	$\rho_1 = 1$ $\rho_2 = 1$ $\rho_3 = 0$
Asymptotic buffers' loads		$r = r_{max}$	$r_2 = r_{max}$ $r_3 = 0$

6.3 Basic Definitions and Notations

Consider a junction J with n incoming roads $I_1, \dots, I_n$ and m outgoing roads $I_{n+1}, \dots, I_{n+m}$. We model each incoming road I_i ($i \in \{1, \dots, n\}$) of the junction with the real interval $I_i =]-\infty, 0]$. Similarly we model each outgoing road I_j ($j \in \{n+1, \dots, n+m\}$) of the junction with the real interval $I_j = [0, +\infty[$. On each road I_l ($l \in \{1, \dots, n+m\}$) we consider the partial differential equation

$$(\rho_l)_t + f(\rho_l)_x = 0, \tag{6.2}$$

where $\rho_l = \rho_l(t,x) \in [0, \rho_{max}]$ is the *density* of cars, $v_l = v_l(\rho_l)$ is the *velocity* of cars, and $f(\rho_l) = v_l(\rho_l)\,\rho_l$ is the *flux*. Hence the datum is given by a finite collection of functions ρ_l defined on $[0, +\infty[\times I_l$. For simplicity, we put $\rho_{max} = 1$. On the flux f we make the following assumption

($\mathcal{F}$) $f : [0,1] \to \mathbb{R}$ is piecewise smooth, concave (i.e., almost everywhere $f'' \leq 0$), $f(0) = f(1) = 0$ and there exists a unique a point of maximum $\sigma \in]0,1[$.

Definition 6.3.1. A function $\rho_l \in C([0,+\infty[; L^1_{\text{loc}}(I_l))$ is an entropy-admissible solution to (6.2) in the road I_l if the following holds.

1. For every function $\varphi : [0,+\infty[\times I_l \to \mathbb{R}$ smooth with compact support in$]0,+\infty[\times (I_l \setminus \{0\})$

$$\int_0^{+\infty} \int_{I_l} \left(\rho_l \frac{\partial \varphi}{\partial t} + f(\rho_l) \frac{\partial \varphi}{\partial x} \right) \mathrm{d}x\mathrm{d}t = 0. \tag{6.3}$$

2. For every $k \in \mathbb{R}$ and every $\tilde{\varphi} : [0,+\infty[\times I_l \to \mathbb{R}$ smooth, positive with compact support in $]0,+\infty[\times (I_l \setminus \{0\})$

$$\int_0^{+\infty} \int_{I_l} \left(|\rho_l - k| \frac{\partial \tilde{\varphi}}{\partial t} + sgn(\rho_l - k)(f(\rho_i) - f(k)) \frac{\partial \tilde{\varphi}}{\partial x} \right) \mathrm{d}x\mathrm{d}t \geq 0. \tag{6.4}$$

We now want to describe preferences of drivers. This is done defining a traffic distribution matrix, whose coefficients represent percentages of incoming fluxes which distribute to each outgoing road. Consider the set

$$\mathcal{A} := \left\{ A = \{\alpha_{ji}\}_{\substack{i=1,\dots,n \\ j=n+1,\dots,n+m}} : \begin{array}{c} 0 < \alpha_{ji} < 1 \; \forall i,j, \\ \sum\limits_{j=n+1}^{n+m} \alpha_{ji} = 1 \; \forall i \end{array} \right\}. \tag{6.5}$$

Here the coefficient α_{ji} indicates the portion of cars coming from incoming road I_i which goes to outgoing road I_j.

Let $\{e_1, \dots, e_n\}$ be the canonical basis of $\mathbb{R}^n$. For every $i = 1, \dots, n$, we denote $H_i = \{e_i\}^{\perp}$. If $A \in \mathcal{A}$, then we write, for every $j = n+1, \dots, n+m$, $\alpha_j = (\alpha_{j1}, \dots, \alpha_{jn}) \in \mathbb{R}^n$ and $H_j = \{\alpha_j\}^{\perp}$. Let $\mathcal{K}$ be the set of indices $\mathbf{k} = (k_1, \dots, k_\ell)$, $1 \leq \ell \leq n-1$, such that $0 \leq k_1 < k_2 < \dots < k_\ell \leq n+m$ and for every $\mathbf{k} \in \mathcal{K}$ define

$$H_{\mathbf{k}} = \bigcap_{h=1}^{\ell} H_{k_h}.$$

Writing $\mathbf{1} = (1, \dots, 1) \in \mathbb{R}^n$ and following Coclite et al. (2005) we define the set

$$\mathfrak{N} := \left\{ A \in \mathcal{A} : \mathbf{1} \notin H_{\mathbf{k}}^{\perp} \text{ for every } \mathbf{k} \in \mathcal{K} \right\}. \tag{6.6}$$

Notice that if $n \geq m$, then $\mathfrak{N} = \emptyset$. The matrices of $\mathfrak{N}$ will give a unique solution to the Riemann problem at J.

Remark 1. If $n \geq m$, or more generally $A \notin \mathfrak{N}$, one can resort to right of way parameters to determine a unique solution. The construction is similar to the one used later in Sect. 6.5. For a detailed description, we refer the reader to Chitour and Piccoli (2005). An alternative approach, for incoming roads which share junction area but not necessarily lead to same outgoing roads as it happens in T-junctions, we refer to Marigo and Piccoli (2008). Finally, for traffic data with source–destination patterns, a complete theory is available in Garavello and Piccoli (2005).

Define also the set:

$$\mathcal{P} = \left\{ P = \{p_{ji}\}_{\substack{i=1,\ldots,n \\ j=n+1,\ldots,n+m}} : 0 < p_{ji} < 1 \ \forall i,j,\ \sum_{i=1}^{n} p_{ji} = 1 \ \forall j \right\}.$$

A matrix $P \in \mathcal{P}$ represents priority coefficients among incoming roads to enter each outgoing road.

6.4 The Riemann Problem at J Without Buffers

In this section, we recall the concept of Riemann problem at the junction and the solution, proposed for traffic, in Coclite et al. (2005). Fix $\rho_{1,0},\ldots,\rho_{n+m,0} \in [0,1]$, then the corresponding Riemann problem at J is given by:

$$\begin{cases} \frac{\partial}{\partial t}\rho_l + \frac{\partial}{\partial x} f(\rho_l) = 0, \\ \rho_l(0,\cdot) = \rho_{l,0}, \end{cases} \qquad l \in \{1,\ldots,n+m\}, \tag{6.7}$$

namely the Cauchy problem with initial data constant on each road.

To define the dynamics on the whole network, we need to determine solutions at junctions. In particular, we want to describe the solution to Riemann problems at J. This is achieved by describing solutions via a map which to initial conditions associates boundary data for all roads of the junction. More precisely, we define:

Definition 6.4.1. A Riemann solver $\mathcal{RS}$ is a function

$$\begin{array}{rccc} \mathcal{RS}: & [0,1]^{n+m} & \longrightarrow & [0,1]^{n+m} \\ & (\rho_{1,0},\ldots,\rho_{n+m,0}) & \longmapsto & (\bar\rho_1,\ldots,\bar\rho_{n+m}) \end{array}$$

satisfying:

1. For every $i \in \{1,\ldots,n\}$, the classical Riemann problem

$$\begin{cases} \rho_t + f(\rho)_x = 0, & x \in \mathbb{R}, t > 0, \\ \rho(0,x) = \begin{cases} \rho_{i,0}, & \text{if } x < 0, \\ \bar\rho_i, & \text{if } x > 0, \end{cases} \end{cases}$$

is solved with waves with negative speed.

2. For every $j \in \{n+1,\dots,n+m\}$, the classical Riemann problem

$$\begin{cases} \rho_t + f(\rho)_x = 0, & x \in \mathbb{R}, t > 0, \\ \rho(0,x) = \begin{cases} \bar{\rho}_j, & \text{if } x < 0, \\ \rho_{j,0}, & \text{if } x > 0, \end{cases} \end{cases}$$

   is solved with waves with positive speed.
3. $\sum_{i=1}^{n} f(\bar{\rho}_i) = \sum_{j=n+1}^{n+m} f(\bar{\rho}_j)$.

Remark 2. In the above definition the first two conditions ensure that boundary value problems on each road are solved in a strong sense. This means that weak solutions will indeed achieve the prescribed boundary value as a trace. This is not the case for general weak solutions to boundary value problems, see, for instance, Bardos et al. (1979).
Condition 3. then guarantees conservation of cars through the junction, imposing equality between total incoming and outgoing fluxes.

We need another property to ensure that a Riemann solver is well defined. Indeed, it may happen that a value attained by $\mathcal{RS}$ is not a fixed point for $\mathcal{RS}$ itself. Therefore, one may need to reapply $\mathcal{RS}$ thus not giving rise to a well-defined procedure. We then define:

Definition 6.4.2. We say that a Riemann solver $\mathcal{RS}$ satisfies the consistency condition if, for every $(\rho_1,\dots,\rho_{n+m}) \in [0,1]^{n+m}$, then

$$\mathcal{RS}\left(\mathcal{RS}(\rho_1,\dots,\rho_{n+m})\right) = \mathcal{RS}(\rho_1,\dots,\rho_{n+m}).$$

We are now ready to give the definition of solution at the junction:

Definition 6.4.3. Given a Riemann solver $\mathcal{RS}$, a solution to the Riemann problem (6.7) is a collection of functions $(\rho_1,\dots,\rho_{n+m})$ such that:

1. For every $l \in \{1,\dots,n+m\}$, the function ρ_l is an entropy-admissible solution to (6.2) in the road I_l, in the sense of Definition 6.3.1.
2. For every $l \in \{1,\dots,n+m\}$ and for a.e. $t > 0$, the function $x \mapsto \rho_l(t,x)$ has a version with bounded total variation.
3. For every $l \in \{1,\dots,n+m\}$, $\rho_l(0,x) = \rho_{l,0}$ for a.e. $x \in I_l$.
4. For a.e. $t > 0$, it holds

$$\mathcal{RS}\left(\rho_1(t,0-),\dots,\rho_{n+m}(t,0+)\right) = \left(\rho_1(t,0-),\dots,\rho_{n+m}(t,0+)\right).$$

There are some general properties which hold for all Riemann solvers. To describe the latter, introduce the following sets:

1. For every $i \in \{1,\dots,n\}$, define

$$\Omega_i = \begin{cases} [0, f(\rho_{i,0})], & \text{if } 0 \le \rho_{i,0} \le \sigma, \\ [0, f(\sigma)], & \text{if } \sigma \le \rho_{i,0} \le 1. \end{cases} \tag{6.8}$$

2. For every $j \in \{n+1,\dots,n+m\}$, define

$$\Omega_j = \begin{cases} [0, f(\sigma)], & \text{if } 0 \le \rho_{j,0} \le \sigma, \\ [0, f(\rho_{j,0})], & \text{if } \sigma \le \rho_{j,0} \le 1. \end{cases} \tag{6.9}$$

3. For every $l \in \{1,\dots,n+m\}$, define

$$\gamma_l^{\max} = \max \Omega_l. \tag{6.10}$$

Proposition 6.4.1. The following statements hold.

1. For every $i \in \{1,\dots,n\}$, an element $\bar{\gamma}$ belongs to Ω_i if and only if there exists $\bar{\rho}_i \in [0,1]$ such that $f(\bar{\rho}_i) = \bar{\gamma}$ and point 1 of Definition 6.4.1 is satisfied.
2. For every $j \in \{n+1,\dots,n+m\}$, an element $\bar{\gamma}$ belongs to Ω_j if and only if there exists $\bar{\rho}_j \in [0,1]$ such that $f(\bar{\rho}_j) = \bar{\gamma}$ and point 2 of Definition 6.4.1 is satisfied.

The proof is trivial and hence omitted. Here we recall the construction of the Riemann solver, introduced for traffic in Coclite et al. (2005). For simplicity in this paper, we denote it with the symbol $\mathcal{RS}_{\text{CGP}}$.

1. Fix a matrix $A \in \mathfrak{N}$ and consider the closed, convex and not empty set

$$\Omega = \left\{ (\gamma_1,\cdots,\gamma_n) \in \prod_{i=1}^{n} \Omega_i : A \cdot (\gamma_1,\cdots,\gamma_n)^{\mathrm{T}} \in \prod_{j=n+1}^{n+m} \Omega_j \right\}. \tag{6.11}$$

2. Find the point $(\bar{\gamma}_1,\dots,\bar{\gamma}_n) \in \Omega$ which maximizes the function

$$E(\gamma_1,\dots,\gamma_n) = \gamma_1 + \cdots + \gamma_n, \tag{6.12}$$

and define $(\bar{\gamma}_{n+1},\dots,\bar{\gamma}_{n+m})^{\mathrm{T}} := A \cdot (\bar{\gamma}_1,\dots,\bar{\gamma}_n)^{\mathrm{T}}$. Since $A \in \mathfrak{N}$, the point $(\bar{\gamma}_1,\dots,\bar{\gamma}_n)$ is uniquely defined.
3. For every $i \in \{1,\dots,n\}$, set $\bar{\rho}_i$ either by $\rho_{i,0}$ if $f(\rho_{i,0}) = \bar{\gamma}_i$, or by the solution to $f(\rho) = \bar{\gamma}_i$ such that $\bar{\rho}_i \ge \sigma$. For every $j \in \{n+1,\dots,n+m\}$, set $\bar{\rho}_j$ either by $\rho_{j,0}$ if $f(\rho_{j,0}) = \bar{\gamma}_j$, or by the solution to $f(\rho) = \bar{\gamma}_j$ such that $\bar{\rho}_j \le \sigma$. Finally, define $\mathcal{RS}_{\text{CGP}} : [0,1]^{n+m} \to [0,1]^{n+m}$ by

$$\mathcal{RS}_{\text{CGP}}(\rho_{1,0},\dots,\rho_{n+m,0}) = (\bar{\rho}_1,\dots,\bar{\rho}_n,\bar{\rho}_{n+1},\dots,\bar{\rho}_{n+m}). \tag{6.13}$$

Remark 3. The previous rules 1 and 2 of the construction of $\mathcal{RS}_{\text{CGP}}$ correspond to conditions (A) and (B) in Coclite et al. (2005). In particular, 1 describes the preferences of drivers, while 2 implies that drivers behave as to maximize the sum of incoming fluxes. Clearly these are modeling conditions and their choice is somehow

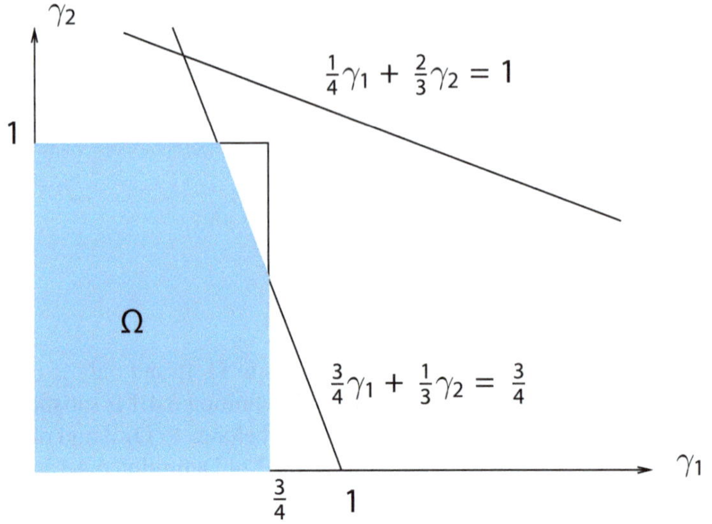

Fig. 6.5 The set Ω of Example 1

arbitrary. In D'Apice and Piccoli (2008), there are some other possible Riemann solvers at J, some of which not based on maximization procedures.

Let us now illustrate with an example the Riemann solver $\mathcal{RS}_{CGP}$ in the case of two incoming and two outgoing roads.

Example 1. Let J be a junction with two incoming roads I_1 and I_2 and two outgoing ones I_3 and I_4. Fix a distribution matrix $A \in \mathfrak{N}$ such that $\alpha_{31} = 1/4$, $\alpha_{32} = 2/3$, $\alpha_{41} = 3/4$, and $\alpha_{42} = 1/3$. Let us assume that the flux $f(\rho)$ is equal to $4\rho(1-\rho)$ and the initial conditions for the Riemann problem (6.7) are given by

$$\rho_{1,0} = \frac{1}{4}, \qquad \rho_{2,0} = 1, \qquad \rho_{3,0} = \frac{1}{2}, \qquad \rho_{4,0} = \frac{3}{4}.$$

We easily deduce that

$$\Omega_1 = \left[0, \frac{3}{4}\right], \qquad \Omega_2 = [0,1], \qquad \Omega_3 = [0,1], \qquad \Omega_4 = \left[0, \frac{3}{4}\right]$$

and so

$$\Omega = \left\{(\gamma_1,\gamma_2) \in \left[0, \frac{3}{4}\right] \times [0,1] : 0 \leq \frac{1}{4}\gamma_1 + \frac{2}{3}\gamma_2 \leq 1, 0 \leq \frac{3}{4}\gamma_1 + \frac{1}{3}\gamma_2 \leq \frac{3}{4}\right\};$$

see Fig. 6.5. Therefore the point of maximum in Ω for the function E, defined in (6.12), is given by $(\bar{\gamma}_1, \bar{\gamma}_2) = \left(\frac{5}{9}, 1\right)$ and consequently $(\bar{\gamma}_3, \bar{\gamma}_4) = \left(\frac{29}{36}, \frac{3}{4}\right)$. Finally we have

$$\bar{\rho}_1 = \frac{1}{2} + \sqrt{\frac{1}{4} - \frac{5}{36}}, \quad \bar{\rho}_2 = \frac{1}{2}, \quad \bar{\rho}_3 = \frac{1}{2} + \sqrt{\frac{1}{4} - \frac{29}{144}}, \quad \bar{\rho}_4 = \frac{3}{4}.$$

6.5 Riemann Solver with Multibuffer

In this section we introduce a new way to solve the Riemann problem at the junction J. We imagine that the junction J is composed of n incoming roads, m outgoing roads and m different buffers in front of each outgoing road; see Fig. 6.6. Let us index buffers by $j \in \{n+1, \ldots, n+m\}$. Then a function $r_j(t)$, which represents the load of the j-th buffer at time $t > 0$, is associated with each buffer. Moreover, for every $j \in \{n+1, \ldots, n+m\}$, the numbers $\mu_j > \sum_i \alpha_{ji} f(\sigma)$ and $r_j^{\max} > 0$ denote, respectively, the maximum number of cars, which can enter or exit the j-th buffer per unit of time, and the maximum number of cars which can be stored in the j-th buffer; hence we deduce the constraints $0 \le r_j(t) \le r_j^{\max}$ for every $j \in \{n+1, \ldots, n+m\}$.

Remark 4. Notice that the assumption $\mu_j > \sum_i \alpha_{ji} f(\sigma)$ means that buffers have large capacities w.r.t. maximal road flux. This is a necessary assumption to obtain a consistent solution to Riemann problems at junctions. Such situation is illustrated in Example 2.

If one of the buffers is full, then we impose that no car passes through the junction. If instead all the buffers are not full, then every car enters into the buffer associated with the desired destination and, finally, it enters into the outgoing road by FIFO policy.

Remark 5. Note that if one of the buffers is full, then it is necessary to not allow cars to cross J, otherwise the preferences of the drivers may not be satisfied. Indeed every buffer should receive a nonzero percentage of the flux entering the junction J. However, if some buffer is full, then it cannot receive any car; therefore, the only way to respect the constraints, imposed by the matrix A, is to block the intersection.

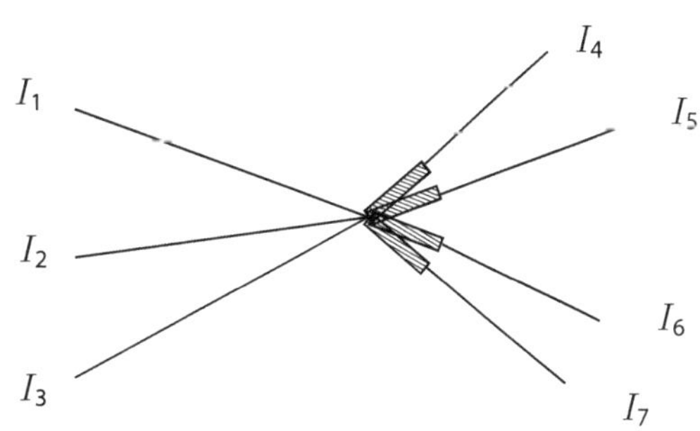

Fig. 6.6 A junction with multibuffer

Let $A \in \mathfrak{N}$, $P \in \mathcal{P}$ be the matrices of the preferences of drivers and of priority coefficients among incoming roads.

We define a Riemann solver with multibuffer

$$\begin{array}{rcl} \mathcal{RS} : [0,1]^{n+m} \times \prod_{j=n+1}^{n+m} \left[0, r_j^{\max}\right] & \longrightarrow & [0,1]^{n+m} \times [0, nf(\sigma)]^m \\ (\rho_{1,0}, \dots, \rho_{n+m,0}, r_{n+1}, \dots, r_{n+m}) & \longmapsto & \left(\bar{\rho}_1, \dots, \bar{\rho}_{n+m}, f_{n+1}^{\text{in}}, \dots, f_{n+m}^{\text{in}}\right) \end{array} \tag{6.14}$$

as follows.

1. For every $i \in \{1, \dots, n\}$ and $j \in \{n+1, \dots, n+m\}$, define

$$\gamma_{i,j}^{\max} = \alpha_{ji} \gamma_i^{\max},$$

   i.e., the maximum flux, which can exit from I_i and enter into I_j.
2. For every $j \in \{n+1, \dots, n+m\}$, define

$$\hat{f}_j^{\text{in}} = \min\left\{ \sum_{i=1}^{n} \gamma_{i,j}^{\max}, \mu_j \right\} \qquad \text{and} \qquad \hat{f}_j^{\text{out}} = \min\left\{ \gamma_j^{\max}, \mu_j \right\},$$

   respectively, the maximum flux which can enter into the j-th buffer and which can enter into the road I_j.
3. If $\min_{j=n+1}^{n+m} \left\{ r_j^{\max} - r_j \right\} > 0$ (i.e., no buffer is full), then define the convex set:

$$\mathbb{R}^{n \times m} \supset K = \left\{ \{k_{ji}\}_{\substack{i=1,\dots,n \\ j=n+1,\dots,n+m}} : 0 \le k_{ji} \le \gamma_{i,j}^{\max}, \sum_{i=1}^{n} k_{ji} = \hat{f}_j^{\text{in}} \, \forall j \right\},$$

   and the point $\tilde{Q} \in \mathbb{R}^{n \times m}$ by setting $\tilde{q}_{ji} = p_{ji} \hat{f}_j^{\text{in}}$. Let $Q = \text{proj}_K(\tilde{Q}) \in K$, where proj_K is the orthogonal projection over the convex set K. Finally, for every $i \in \{1, \dots, n\}$, we set

$$\bar{\gamma}_i = \sum_{j=n+1}^{n+m} q_{ji},$$

   where q_{ji} are the components of Q and represent the number of cars going from I_i to the j-th buffer.
4. If $\min_{j=n+1}^{n+m} \left\{ r_j^{\max} - r_j \right\} = 0$ (i.e., at least one buffer is full), we set $\bar{\gamma}_i = 0$ for every $i \in \{1, \dots, n\}$.
5. For every $j \in \{n+1, \dots, n+m\}$, we set

$$\bar{\gamma}_j = \begin{cases} \hat{f}_j^{\text{out}}, & \text{if } r_j > 0, \\ \min\{\hat{f}_j^{\text{in}}, \hat{f}_j^{\text{out}}\}, & \text{if } r_j = 0. \end{cases}$$

6. For every $j \in \{n+1,\dots,n+m\}$, define

$$f_j^{\text{in}} = \begin{cases} \sum_{i=1}^{n} q_{ji} & \text{if } \min_{j=n+1}^{n+m} \left\{ r_j^{\max} - r_j \right\} > 0, \\ 0 & \text{if } \min_{j=n+1}^{n+m} \left\{ r_j^{\max} - r_j \right\} = 0. \end{cases}$$

7. For every $i \in \{1,\dots,n\}$, define $\bar{\rho}_i$ either by $\rho_{i,0}$ if $f(\rho_{i,0}) = \bar{\gamma}_i$, or by the solution to $f(\rho) = \bar{\gamma}_i$ such that $\bar{\rho}_i \geq \sigma$.
8. For every $j \in \{n+1,\dots,n+m\}$, define $\bar{\rho}_j$ either by $\rho_{j,0}$ if $f(\rho_{j,0}) = \bar{\gamma}_j$, or by the solution to $f(\rho) = \bar{\gamma}_j$ such that $\bar{\rho}_j \leq \sigma$.
9. Define

$$\mathcal{RS}(\rho_{1,0},\dots,\rho_{n+m,0},r_{n+1},\dots,r_{n+m}) = (\bar{\rho}_1,\dots,\bar{\rho}_{n+m},f_{n+1}^{\text{in}},\dots,f_{n+m}^{\text{in}}).$$

Remark 6. Let us comment on the various steps to define the Riemann solver with multibuffer. Steps 1 and 2 define maximal fluxes from incoming roads to buffers and from buffers to outgoing roads. In Step 3, if all buffers are not full, we define fluxes from incoming roads by projecting a vector representing priorities over the set of admissible fluxes. If, on the contrary, at least one buffer is full, then in Step 4 we simply set all fluxes from incoming roads to vanish. Then Step 5 determine fluxes from buffers to outgoing roads, which values depend on the status of the buffer: empty or non empty. Step 6 defines fluxes entering buffers. Finally, Steps 7 and 8 describe how to compute the boundary values both for incoming and outgoing roads.

Remark 7. The Riemann solver with multibuffer can be slightly modified in order to cover also the case of junctions with traffic lights. For example, putting some time-dependent switches in the coefficients of the distribution matrix A realizes this aim. Moreover it is also possible to consider a different flux function for each road of the junction; the only constraint is that each flux function satisfies hypothesis $(\mathcal{F})$. Finally, as in Garavello and Piccoli (2009), it is possible to consider Riemann solvers at J, which depend on the time evolution of certain parameters. This permits to treat the case of varying the capacity of the links. All these extensions are just technical; hence for clarity we prefer to not consider them in the paper.

Example 2. Consider a junction with one incoming road I_1 and two outgoing roads I_2 and I_3 and distribution matrix $(\alpha_{21},\alpha_{31})$. For initial conditions we assume $0 < \rho_{1,0} = \rho_{2,0} = \rho_{3,0} < \sigma$ and empty buffers, while capacities satisfy $\mu_2 < \alpha_{21} f(\rho_{1,0})$ while $\mu_3 > f(\sigma)$. Applying the Riemann solver with multibuffer we get:

$$\hat{f}_2^{\text{in}} = \mu_2, \qquad \hat{f}_3^{\text{in}} = \alpha_{31} f(\rho_{1,0}),$$

thus

$$\mathcal{RS}(\rho_{1,0},\rho_{2,0},\rho_{3,0},0,0) = (\bar{\rho}_1,\bar{\rho}_2,\bar{\rho}_3,\mu_2,\alpha_{31} f(\rho_{1,0})),$$

where $f(\bar{\rho}_1) = \mu_2 + \alpha_{31} f(\rho_{1,0})$, $f(\bar{\rho}_2) = \mu_2$, $f(\bar{\rho}_3) = \alpha_{31} f(\rho_{1,0})$. Now if we apply $\mathcal{RS}$ to the initial conditions $(\bar{\rho}_1, \bar{\rho}_2, \bar{\rho}_3, 0, 0)$ we get

$$\mathcal{RS}(\bar{\rho}_1, \bar{\rho}_2, \bar{\rho}_3, 0, 0) = (\hat{\rho}_1, \bar{\rho}_2, \hat{\rho}_3, \mu_2, \alpha_{31} f(\sigma)),$$

where $f(\hat{\rho}_1) = \mu_2 + \alpha_{31} f(\sigma)$, $f(\hat{\rho}_3) = \alpha_{31} f(\sigma)$. This proves that if μ_j is below $\sum_i \alpha_{ji} f(\sigma)$, then the Riemann solver lacks of consistency.

We give the following definition of solution to Riemann problems with multibuffer.

Definition 6.5.1. Let $r_{n+1,0}, \ldots, r_{n+m,0}$ be the initial loads of the buffers. A solution to the Riemann problem (6.7) with multibuffer is given by

$$\begin{aligned} (\rho_1, \ldots, \rho_{n+m}) &\in C\left([0, +\infty[; \textstyle\prod_{l=1}^{n+m} L^1_{\mathrm{loc}}(I_l)\right) \\ (r_{n+1}, \ldots, r_{n+m}) &\in W^{1,\infty}([0, +\infty[; \mathbb{R}^m) \end{aligned}$$

such that:

1. For every $l \in \{1, \ldots, n+m\}$, ρ_l is an entropy-admissible solution to (6.2) on I_j and, for a.e. $t > 0$, $\rho_l(t, \cdot)$ has a version with finite total variation.
2. For every $l \in \{1, \ldots, n+m\}$, $\rho_l(0, x) = \rho_{0,l}$ for a.e. $x \in I_l$.
3. For a.e. $t > 0$

$$\begin{aligned} &\mathcal{RS}\left(\rho_1(t,0), \ldots, \rho_{n+m}(t,0), r_{n+1}(t), \ldots, r_{n+m}(t)\right) \\ &= \left(\rho_1(t,0), \ldots, \rho_{n+m}(t,0), f^{\mathrm{in}}_{n+1}(t), \ldots, f^{\mathrm{in}}_{n+m}(t)\right) \end{aligned}$$

and

$$r_j(t) = r_{j,0} + \int_0^t \left(f^{\mathrm{in}}_j(s) - f(\bar{\rho}_j(t,0))\right) ds$$

for every $j \in \{n+1, \ldots, n+m\}$.

6.6 Comparison with Other Models with Buffers

The idea of using roads, or in general arcs, with buffers in front was used by various authors. Here we compare our model with those used in recent literature.

In Göttlich et al. (2006), a coupled ODE-PDE model for supply chains (and networks) was proposed. It consists of arcs with dynamics described by a conservation law for part density and ODEs for buffers in front of each arc. More precisely the conservation law is of the type:

$$\rho_t + (\min\{v\rho, \mu\})_x = 0,$$

where ρ is the part density, v the constant velocity, and μ the processing rate. Notice that such model produces waves having only positive speed. Thus, the evolution of part density affects the network only in the forward direction.

The ODEs for buffers are of the type:

$$\dot{r} = f^{\text{in}} - f^{\text{out}} \tag{6.15}$$

where f^{in} is the flux entering the buffer from the previous arc and f^{out} is the flux exiting the buffer to the next arc. f^{in} depends only on the density of the previous arc and is independent from the buffer status, thus the buffer is necessarily with infinite size. f^{out} is defined similarly to our case, namely it is equal to f^{in}, if the buffer is empty (and f^{in} is below the processing capacity of next arc), otherwise it is equal to the processing capacity of next arc.

Summarizing the main differences with our model are the following:

- The conservation laws admits only waves with positive velocity, so no backward effect is possible.
- Consequently fluxes entering buffers cannot depend on buffer status, therefore buffers are necessarily of infinite size.

In Herty et al. (2009), authors propose a model for vehicular traffic, which considers junctions with an arbitrary number of incoming and outgoing roads and a buffer in between. Conservation laws are of the same type we considered here. On the other side, the equation for the buffer is of the type (6.15).
The buffer has limited size, thus fluxes from incoming roads will stop when the buffer is full. If the buffer is not full, then fluxes from incoming roads enter the buffers, possibly limited by a maximal processing rate. Finally, the flux exiting the buffer distribute over outgoing roads according to traffic distribution coefficients.
The main differences with our model are the following:

- There is a unique buffer for all outgoing roads, opposed to our multibuffer.
- The traffic from incoming road is stopped only when the common buffer is full, while in our case a single full buffer will stop the traffic.
- The traffic distribution coefficients do not depend on incoming roads. Indeed a traffic distribution matrix cannot be respected as shown by example of Sect. 6.2.

6.7 Conclusions

We have proposed a new way for describing dynamics at intersections in road networks, when car density evolution is governed by the Lighthill–Whitham–Richards model. Due to finite speed of waves in the LWR model, we can focus on a single junction. We supposed that a buffer is attached in front of each outgoing road of the junction, and we completely described the dynamics inside the buffers and between roads and buffers by means of a Riemann solver with multibuffer. Moreover we provided examples and analytical comparisons between our approach and some previously introduced in the literature.

References

Bardos C, le Roux AY, Nédélec J-C. First order quasilinear equations with boundary conditions. Comm Part Differ Equat. 1979;4(9):1017–34.

Chitour Y, Piccoli B. Traffic circles and timing of traffic lights for cars flow. Discrete Continuous Dyn Syst Ser B 2005;5(3):599–630.

Coclite GM, Garavello M, Piccoli B. Traffic flow on a road network. SIAM J Math Anal. 2005;36(6):1862–86 (electronic).

Daganzo CF. The cell transmission model: a dynamic representation of highway traffic consistent with the hydrodynamic theory. Transport Res Part B 1994;28(4):269–87.

D'Apice C, Manzo R. A fluid dynamic model for supply chains. Netw Heterog Media 2006;1(3):379–98 (electronic).

D'apice C, Manzo R, Piccoli B. Packet flow on telecommunication networks. SIAM J Math Anal. 2006;38(3):717–740 (electronic).

D'Apice C, Piccoli B. Vertex flow models for vehicular traffic on networks. Math Models Methods Appl Sci. 2008;18 Suppl:1299–1315.

Garavello M, Piccoli B. Source-destination flow on a road network. Commun Math Sci. 2005;3(3):261–83.

Garavello M, Piccoli B. Traffic flow on networks. Volume 1 of AIMS series on applied mathematics. Springfield, MO: American Institute of Mathematical Sciences (AIMS); 2006 [Conservation laws models].

Garavello M, Piccoli B. Conservation laws on complex networks. Ann H Poincaré 2009;26(5):1925–51.

Garavello M, Piccoli B. Time-varying Riemann solvers for conservation laws on networks. J Differ Equat. 2009;247(2):447–64.

Newell GF. A simplified theory of kinematic waves in highway traffic, part i: general theory. Transport Res Part B 1993;27(4):281–7.

Newell GF. A simplified theory of kinematic waves in highway traffic, part ii: queueing at freeway bottlenecks. Transport Res Part B 1993;27(4):289–303.

Newell GF. A simplified theory of kinematic waves in highway traffic, part iii: multi-destination flows. Transport Res Part B 1993;27(4):305–13.

Godunov SK. A difference method for numerical calculation of discontinuous solutions of the equations of hydrodynamics. Mat Sb (NS) 1959;47(89):271–306.

Göttlich S, Herty M, Klar A. Network models for supply chains. Commun Math Sci. 2005;3(4):545–59.

Göttlich S, Herty M, Klar A. Modelling and optimization of supply chains on complex networks. Commun Math Sci. 2006;4(2):315–30.

Greenshields BD. A study of traffic capacity. Proc Highway Res Board 1935;14(1):448–77.

Helbing D, Siegmeier J, Lämmer S. Self-organized network flows. Netw Heterog Media 2007;2(2):193–210 (electronic).

Herty M, Klar A, Piccoli B. Existence of solutions for supply chain models based on partial differential equations. SIAM J Math Anal. 2007;39(1):160–73.

Herty M, Lebacque J.-P., Moutari S. A novel model for intersections of vehicular traffic flow. Netw Heterog Media 2009;4(4):813–26 (electronic).

Holden H, Risebro NH. A mathematical model of traffic flow on a network of unidirectional roads. SIAM J Math Anal. 1995;26(4):999–1017.

Lebacque J. The godunov scheme and what it means for first order macroscopic traffic flow models. In: Lesort JB. Proceedings of the 13th ISTTT; 1996. p. 647–77.

Lighthill MJ, Whitham GB. On kinematic waves. II. A theory of traffic flow on long crowded roads. Proc R Soc Lond A 1955;229:317–45.

Marigo A, Piccoli B. A fluid dynamic model for T-junctions. SIAM J Math Anal. 2008;39(6):2016–32.

Richards PI. Shock waves on the highway. Oper Res. 1956;4:42–51.

Sun D, Strub IS, Bayen AM. Comparison of the performance of four Eulerian network flow models for strategic air traffic management. Netw Heterog Media 2007;2(4):569–95 (electronic).

Chapter 7
Cell-Based Dynamic Equilibrium Models

W.Y. Szeto

Abstract Cell-based dynamic equilibrium models are one class of dynamic traffic assignment (DTA) models that can capture equilibrium conditions and realistic traffic dynamics, such as queue spillback, queue formulation, and queue dissipation. However, compared with point-queue DTA models or DTA models using whole-link delay models for flow propagation, cell-based equilibrium models are often more computational demanding. This may raise issues for actual applications, in particular, for the implementation for real-time traffic control and route guidance applications, because the solution must be obtained quickly. Moreover, recent cell-based dynamic equilibrium models tend to capture more realistic travel behavior and traffic dynamics but this made the resulting models even more complicated and more difficult to solve for optimal solutions. Hence, this article aims at reviewing the recent development of cell-based dynamic equilibrium models, the formulation approaches, solution methods used, and the components of these models so as to point out the implementation issues of the latest cell-based dynamic equilibrium model with the consideration supply stochasticity for traffic control and route guidance applications as well as some gaps for future research directions.

7.1 Introduction

Traditionally, transportation planning and traffic operations relied on static traffic assignment models for analysis and policy evaluation. To improve the fidelity and accuracy, dynamic traffic assignment (DTA) models were employed for these purposes. Lo and Szeto (2004) use examples to show that the static traffic assignment and DTA models can produce diametrically opposite results. More importantly,

W.Y. Szeto (✉)
Department of Civil Engineering, University of Hong Kong, Hong Kong
e-mail: ceszeto@hku.hk

S.V. Ukkusuri and K. Ozbay (eds.), *Advances in Dynamic Network Modeling in Complex Transportation Systems*, Complex Networks and Dynamic Systems 2,
DOI 10.1007/978-1-4614-6243-9_7, © Springer Science+Business Media New York 2013

the impacts of transport management policies evaluated by static models could be ill-represented. In some cases, the management schemes such determined could actually worsen the congestion problem. These findings illustrate the importance of adopting the DTA models for planning and operation control policy evaluation, despite that DTA models are more complex and computationally more demanding than the static traffic assignment models

DTA models can be developed based on the simulation approach (e.g., Ben-Akiva et al. 2003; Mahmassani and Liu 1997; Ziliaskopoulos and Rao 1999). This approach emphasizes microscopic traffic flow characteristics. However, the computation burden can be quite high.

DTA models can also be developed through an analytical approach. This approach usually has well-defined properties, in terms of optimality conditions, adherence to a dynamic version of Wardrop's first or second principle (1952). Compared with static traffic assignment models, analytical DTA models can capture the time-varying demands and other temporal effects. However, some analytical DTA models cannot capture realistic traffic dynamics and the spatial effect of queues such as queue spillovers (Peeta and Ziliaskopoulos 2001) because these models adopt either dynamic link performance functions (e.g., Ban et al. 2008; Chow 2009; Friesz et al. 1993; Jayakrishnan et al. 1995; Ran and Boyce 1996), the bottleneck model (e.g., Ramadurai et al. 2010), or exit flow models (e.g., Carey 1987; Wie et al. 1994; Lam and Huang 1995) for modeling flow propagation.

Encapsulating the cell transmission model (CTM), a macroscopic simulation traffic flow model proposed by Daganzo (1994, 1995), into the DTA framework opens a new way to model and analyze the problem. The advantage of this approach is that traffic dynamics such as queue spillback and traffic interaction across links can be captured. One line of research direction is on incorporating the CTM, into a system-optimal DTA framework (e.g., Li et al. 1999, 2003; Ziliaskopoulos 2000). The model can be formulated as a linear program but the traffic can be holding back, which is sometimes considered as an undesirable property. Hence, current efforts focus on how to eliminate this undesirable property. For example, Lin and Wang (2004) propose a penalizing method to address this holding-back problem. Shen et al. (2007) suggest an iterative approach to address this problem. Ramadurai (2009) transforms the CTM model into a linear complementarity formulation but the resultant model cannot address the diverging cell. Pavlis and Recker (2009) and Zhang et al. (2010) reformulate the linear constraints of Ziliaskopoulos (2000) to a mixed integer programming model but their resultant models may not solve large-scale networks. Nie (2011) shows that the algorithm of Ho (1980) can solve the holding-back problem for the cell transmission-based Merchant–Nemhauser model. Zheng and Chang (2011) propose a network flow algorithm to handle the holding-back problem. Doan and Ukkusuri (2012) provide an excellent summary on the existing approaches to address the holding back problem and propose a simulation-based system-optimal DTA formulation that can solve a general network of multiple O–D pairs with multiple paths.

Another line of research is developing equilibrium models with the CTM included. The models have well-defined properties, in terms of adherence to a dynamic version of Wardrop's first principle (1952) or its extensions. These models can be formulated by different approaches. For example, Lo (1999) proposes a user equilibrium-based DTA framework that encapsulates the CTM. The framework models the CTM using mixed integer constraints but it is difficult to efficiently solve such framework. Lo and Szeto (2002a,b) reformulate the user equilibrium-based DTA problem as a variational inequality problem and a nonlinear complementarity problem, respectively. Both formulations model travel time as a function of route flows, and the travel time is obtained by the CTM simulation and average travel time extraction procedure. This approach allows a wide range of solution methods to solve the cell-based DTA models. However, compared with point-queue DTA models or DTA models using whole-link models for flow propagation, these cell-based dynamic equilibrium models are more computational demanding since more than one cell is required to model one link in general. Ukkusuri and Waller (2008) formulate the cell-based user-equilibrium problem as the linear programming problem. The nonlinear propagation of flow is approximated by linear inequalities. Their model is computational efficient but it suffers from the "holding back" problem.

The framework of Lo and Szeto (2002a,b) is extended to further capture more realistic travel behavior and traffic dynamics. For example, Szeto and Lo (2004) and Ukkusuri et al. (2012) consider both route and departure time choices. Szeto and Lo (2005) further consider stochastic dynamic user equilibrium. Han et al. (2011) consider user heterogeneity.

The above studies assumed that travelers select routes/departure based on nominal travel times/costs and the models require cell lengths to be uniform as in the CTM. Szeto and Sumalee (2009) adopt the concept of travel time budget proposed by to capture the risk-averse behavior of travelers—extra time for travel is reserved to avoid late arrival. They also extend the CTM to consider stochasticity in supply by using Monte Carlo simulation. Szeto et al. (2011) modify the framework of Szeto and Sumalee (2009) to avoid the overlapping problem in route choice by C-Logit model. Their models also allow cell lengths being non-uniform. However, some implementation issues are needed to address in order to apply their model for traffic management and route guidance in realistic networks.

In general, capturing more realistic travel and traffic behaviors made the resulting cell-based equilibrium models more complicated and more difficult to solve for optimal solutions. This may raise issues in the implementation for real-time traffic control and route guidance applications because computation time required in these applications is short. Hence, this article aims at reviewing the two building blocks of the cell-based dynamic equilibrium models—namely traffic flow component and the travel choice principle—formulation approaches to the models, and the major solution methods used in the literature in order to point out the implementation issues of the cell-based dynamic equilibrium model proposed by Szeto et al. (2011) for the large-scale, real-time applications of route guidance and traffic control and to identify research gaps for further research.

The rest of article is organized as follows. Section 7.2 depicts the travel flow component of the cell-based dynamic equilibrium models, which relies on the CTM or its extensions. Section 7.3 describes the travel choice principles adopted in the existing cell-based dynamic equilibrium models. Section 7.4 reviews the major formulation approaches of the models and the major solution methods. Section 7.5 discusses the implementation issues and presents some related research directions. Finally, Sect. 7.6 gives some concluding remarks.

7.2 Traffic Flow Component

7.2.1 *Basic Concept*

The traffic-flow component depicts how traffic propagates inside a transport network and hence governs the network performance in terms of travel time. This component can be modeled as a set of side constraints (e.g., Lo 1999). However, representing the traffic-flow component as side constraints explicitly is cumbersome and makes the resultant DTA formulation difficult to obtain solutions efficiently (Lo and Szeto 2002a).

Modeling the traffic-flow component as a unique mapping of route flows (e.g., Lo and Szeto 2002a, b; Ramadurai and Ukkusuri 2010; Szeto et al. 2011) opens up a new way to model and analyze DTA problems. The outputs of this mapping are route travel times. Mathematically, the unique mapping can be expressed as:

$$\mathbf{n} = \Phi(\mathbf{f}), \tag{7.1}$$

where $\mathbf{f}$ is the vector of route flows; $\mathbf{n}$ is the vector of route travel times, and $\Phi(\mathbf{f})$ is a unique travel time mapping from route flows. There are two advantages of this approach. First, it can automatically ensure the consistency between link travel times and link exit flows in DTA because link travel times are uniquely derived from exit link flows. Second, this approach allows us to determine the existence and uniqueness of solutions of DTA problems directly by simply checking whether the unique mapping is continuous and strictly monotonic, respectively. The existing cell-based dynamic equilibrium models adopt this approach to model the traffic flow components in which the CTM or its stochastic extensions are used to describe the traffic flow propagation.

7.2.2 *Cell Transmission Model*

The CTM is a convergent numerical approximation scheme to the Lighthill and Whitham (1955) and Richards (1956) (LWR) model, which is a hydrodynamic (or kinematic wave) traffic flow model. It covers the full range of the trapezoidal fundamental diagram as shown in Fig. 7.1 and can capture traffic dynamics such as queue formulation, queue dissipation, and queue spillback. It requires highways to

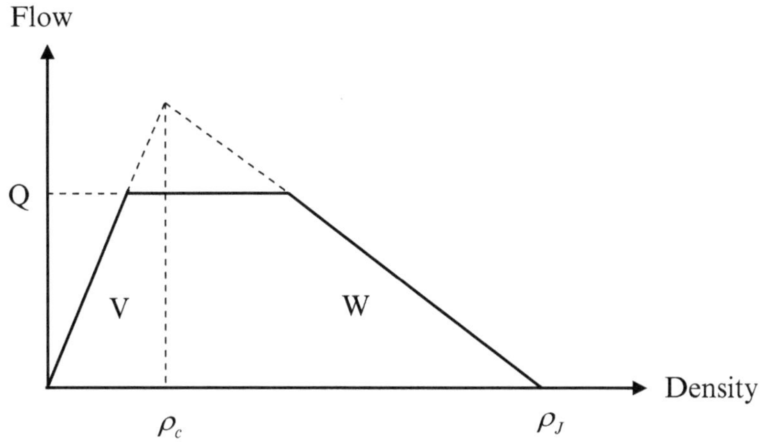

Fig. 7.1 The fundamental diagram used in the CTM

be discretized into many homogenous cells, and time to be discretized into many intervals such that the cell length is equal to the distance traveled by free-flowing traffic in one time interval. Then, the CTM can approximate the LWR results by this set of recursive equations (Daganzo 1994, 1995):

$$n_j(\omega+1) = n_j(\omega) + y_j(\omega) - y_{j+1}(\omega), \tag{7.2}$$

$$y_j(\omega) = \min\{S_{j-1}(\omega), R_j(\omega)\}, \tag{7.3}$$

$$S_{j-1}(\omega) = \min\{n_{j-1}(\omega), \quad Q_{j-1}(\omega)\}, \tag{7.4}$$

$$R_j(\omega) = \min\{Q_j(\omega), \quad (W/V)[N_j(\omega) - n_j(\omega)]\}, \tag{7.5}$$

where the subscript j refers to cell j, and $j+1$ $(j-1)$ represents the cell downstream (upstream) of j. The variables $n_j(\omega)$, $y_j(\omega)$, $N_j(\omega)$ denote the number of vehicles, the actual inflow, and holding capacity – the maximum number of vehicles that can be held in cell j at time ω, respectively. $Q_j(\omega)$ denotes the inflow capacity of cell j or the maximum number of vehicles that can flow into cell j during time ω. W and V are, respectively, the backward wave speed and the free flow speed. $S_{j-1}(\omega)$ is the sending function of cell $j-1$, which defines the number of vehicles leaving the upstream cell $j-1$ during time ω. $R_j(\omega)$ is the receiving function of cell j and determines the number of vehicles entering cell j during time ω.

It is important to differentiate between $Q_j(\omega)$ and $y_j(\omega)$: the former is the inflow capacity while the latter is the actual inflow. Because (7.2)–(7.5) provide a numerical approximation to the LWR equations, all the traffic phenomena demonstrated in the LWR model are replicated in the CTM. The key objective is to determine $y_j(\omega)$ from the minimization (7.3)–(7.5). Once this is accomplished, $n_j(\omega)$ can be determined recursively from the linear equation (7.2). Equations (7.2)–(7.5) provide the basic principle of modeling traffic flow on a series of straight cells.

To apply this principle to a general network with multiple origin–destination (OD) pairs, three extensions are required: (a) modeling merge and diverge junctions; (b) differentiating the OD-specific traffic; (c) maintaining the first-in-first-out (FIFO) property. To maintain the intended routes of traffic and the FIFO property, traffic in each cell is disaggregated by route (p), and waiting time at the cell (τ). The route variable is used to direct traffic along its route. By tracking τ, the FIFO property at the cell level is maintained by ensuring that earlier arrivals (with a larger τ) will leave sooner. The detailed mathematical operations for ensuring these conditions were addressed in Daganzo (1995). The description of which is lengthy; in the interest of space, we do not repeat the analysis here but illustrate it with an example through the origin cell.

For the construction of an origin cell, we create a cell just upstream of an origin cell with an infinite capacity to store the traffic that intends to enter the network. Let's call this cell $r-1$. To load traffic into the network according to the departing demand or the flow $f_p^{rs}(t)$ on route p between OD pair rs departing at time t, we set the inflow into this cell at the time instance $\omega = t$ such that:

$$\begin{aligned} y_{r-1,p}(\omega) &= f_p^{rs}(t), \\ n_{r-1,p}(\omega+1) &= n_{r-1,p}(\omega) + y_{r-1,p}(\omega) - y_{r,p}(\omega). \end{aligned} \tag{7.6}$$

The subscripts r and p extend the definitions of inflow and occupancy for each cell under these conditions: cell location—r, route—p. Note that $y_{r,p}$ is the outflow of $r-1$ but the inflow to the origin cell r, which is subject to the congestion and flow restriction effects at r similar to the expression in (7.3)–(7.5). Once the traffic is loaded to the network, the basic principles of (7.2)–(7.5) and the conditions for merges and diverges (not shown here) hold to ensure proper dynamics of traffic propagation.

In brevity, given a set of time-sequenced inflow $f_p^{rs}(t)$ as in (7.6), based on the traffic propagation conditions [i.e., such as the recursive equations (7.2)–(7.5)], one can obtain a set of unique occupancy counts $n_{j,p}(\omega)$ for traffic in cell j on route p at any time instance ω. Such information on cell occupancy is used to determine the actual route travel times for the entire modeling horizon.

In addition to modeling network traffic, the CTM can be easily modified to handle signalized networks as well as the occurrence of incident or link blockage. Traffic signals and incidents can be modeled by modifying the values of the flow capacity $Q_j(\omega)$ of the affected cells. Specifically, for the case of traffic signals, the inflow capacity of the signalized cells can be altered according to whether the time is in a green or red phase:

$$Q_j(\omega) = s_f \text{ for } \omega \in \text{ green phase and } j \in \text{ a signalized cell,} \tag{7.7}$$

$$Q_j(\omega) = 0 \text{ for } \omega \in \text{ red phase and } j \in \text{ a signalized cell,} \tag{7.8}$$

where s_f is the saturation flow rate. One can model both fixed and dynamic timing plans with the CTM, as demonstrated in Lo (2001). The treatment of incident modeling is similar.

7.2.3 Modified Cell Transmission Model and Switching-mode Model

The modified cell transmission model (MCTM) uses cell densities instead of cell occupancies which permits the CTM to adopt non-uniform cell lengths and leads to greater flexibility in partitioning highways. In the MCTM, the density of cell j evolves according to the conservation of vehicles:

$$\rho_j(\omega+1) = \rho_j(\omega) + \frac{T_s}{l_j}(q_j(\omega) - q_{j+1}(\omega)), \tag{7.9}$$

where $\rho_j(\omega)$ is the vehicle density of cell j at time ω; $q_j(\omega)$ is the total inflows (in vehicles per unit time) entering cell j during the time interval $[kT_s,(k+1)T_s)$, T_s is the sampling duration, k is the time index, and l_j is the length of cell j. The model parameters, including the free-flow speed V, the backward congestion wave speed W, the maximum allowable flow Q, the jam density ρ_J and the critical density ρ_c, are depicted in the trapezoidal fundamental diagram of Fig. 7.1. These parameters can vary from cell to cell over time. Following Daganzo (1994, 1995), $q_j(\omega)$ is determined by the following:

$$q_j(\omega) = \min\{S_{j-1}(\omega), R_j(\omega)\}, \tag{7.10}$$

$$S_{j-1}(\omega) = \min\{V_{j-1}\rho_{j-1}(\omega), Q_{j-1}(\omega)\}, \text{ and} \tag{7.11}$$

$$R_j(\omega) = \min\{Q_j(\omega), W_j[\rho_{J,j} - \rho_j(\omega)]\}. \tag{7.12}$$

Equations (7.9)–(7.12) are the density-based equivalents of (7.2)–(7.5).

Although the MCTM is much simpler than many other higher order hydrodynamics-based partial differential models, the nonlinear nature of the flow-density relationships (7.10)–(7.12) still makes the MCTM difficult to be analyzed and used as a basis for the design of traffic controllers (Munoz et al. 2003). To avoid the nonlinearity, the switching mode model (SMM) is proposed by Munoz et al. (2003). The SMM is a hybrid system (or switched linear system) that switches among different sets of linear difference equations (representing different traffic states of the highway), depending on the mainline boundary data and the congestion status of the cells in a highway segment.

The SMM formulation avoids the nonlinearity of the MCTM at the cost of using the same triangular flow-density relationship for all the cells along the whole freeway segment, and introducing the switching condition based on the at-most-one-wavefront assumption. Based on this assumption, the cell can be in one of the five modes:

1. "Free flow–Free flow (FF)" in which all cells in the freeway segment have free-flow status.
2. "Congestion–Congestion (CC)" in which all cells in the freeway segment have congested status.

3. "Congestion–Free flow (CF)" in which the upstream part of the freeway segment is congested and the downstream part has free-flow status.
4. "Free flow–Congestion 1 (FC1)" in which the upstream part of the freeway segment has free-flow status, the downstream part has congested status, and the boundary (i.e., wave front) separating the two regions is moving downstream.
5. "Free flow–Congestion 2 (FC2)" in which the upstream part of the freeway segment has free-flow status, the downstream part has congested status, and the wave front separating the two regions is moving upstream.

At each time step, the SMM determines its mode based on the measured mainline boundary data and the congestion status of the cells in the freeway segment. If both the measured density at the upstream and downstream of the freeway have free flow status (i.e., both densities are below ρ_c), the FF mode is selected, and if both of these densities are congested (i.e., both densities are at or above ρ_c), the CC mode is selected. If both measured densities are of opposite status, then the SMM performs a search over the ρ_j to determine whether there is a status transition inside the section. This wave front search consists of searching through the cells, in order, looking for the first status transition between adjacent cells.

According to Munoz et al. (2003), the SMM does not fully replicate the CTM merge and diverge laws described in Daganzo (1995). While the on-ramp entering and off-ramp exiting flows are represented in the SMM, the ramps are not modeled by cells; hence, the traffic densities on the ramps are not represented.

7.2.4 Stochastic Cell Transmission Model

The CTM assumes a steady-state speed-density relationship which adopts a number of deterministic parameters (e.g., free-flow speed, jam-density, and capacity) and does not allow fluctuations around the equilibrium (nominal) fundamental flow-density diagram (FD). However, research and empirical studies on the FD have revealed that the FD admits large variations (see Fig. 7.2) due to the variabilities in driving behavior and the characteristics (e.g., acceleration and deceleration abilities) of vehicles, the changing weather conditions, estimation errors, and others (Ngoduy 2011). Therefore, the stochastic cell transmission model (SCTM) is developed by Sumalee et al. (2008) to capture the random evolution of traffic states for freeways.

The SCTM extends the CTM by defining parameters governing sending and receiving functions explicitly as random variables as well as by specifying the dynamics of the basic model parameters of the FD including free flow speed, backward wave speed, saturation flow rate, and jam density in each cell. The SCTM employs a triangular flow-density relation and uses densities as state variables instead of cell occupancies. As in the SMM, the SCTM allows variable cell lengths. As shown in Sumalee et al. (2011), the SCTM is able to give a good estimate on the mean actual traffic flow on freeways. The SCTM has been extended for

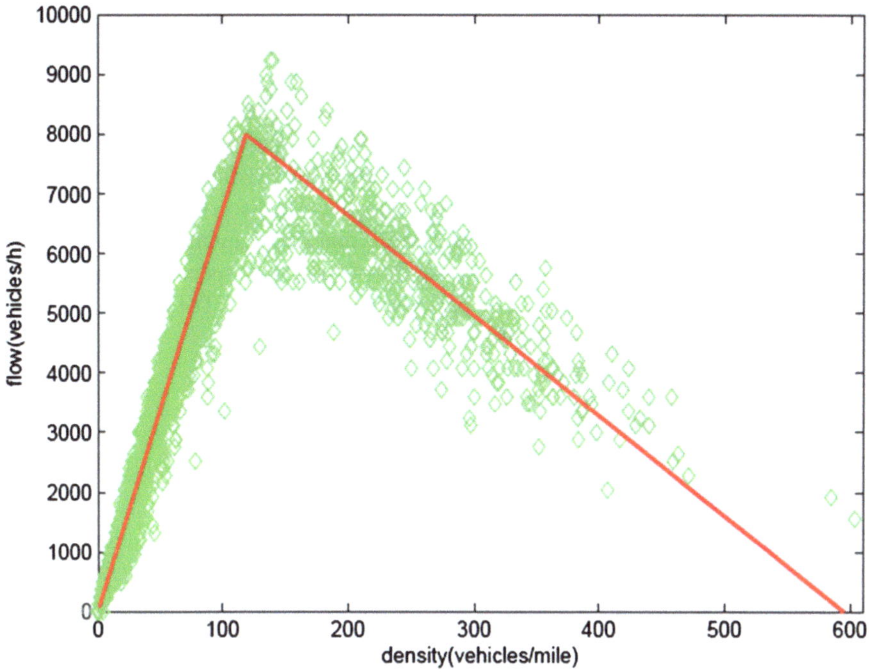

Fig. 7.2 A fundamental flow density diagram of traffic flow with the 24-h traffic flow data of interstate 210 West in Los Angeles collected on April 22, 2008

modeling traffic flow propagation in general networks by Sumalee et al. (2010) and for capturing spatial and temporal correlation by Pan et al. (2010). To have the Markovian property, the SCTM adopts some local white noise assumptions, which restricts the choice of distributions for modeling the parameters.

7.2.5 *Monte-Carlo-based Stochastic Cell Transmission Model*

Szeto and Sumalee (2009) propose the Monte-Carlo-based stochastic cell transmission model (MC-SCTM). The MC-SCTM does not restrict the choice of distributions for modeling the parameters of trapezoidal flow-density relationships at the expense of the increase in computation time. Unlike the SCTM, the MC-SCTM relies on the Monte Carlo simulation to generate traffic states for network loading and collect travel time statistics from each network loading. The network loading is done by the network version of the enhanced lagged CTM (Szeto 2008), which is extended from the lagged CTM (Daganzo 1999) that uses densities as state variables and allows cells having unequal lengths. This network version also requires the cell occupancy information deduced from the cell density for two purposes: (a) to direct traffic at junctions using the merge and diverge concepts proposed by Daganzo (1995) and (b) to deduce route travel time using the averaging scheme proposed by Lo and Szeto (2002a).

Szeto et al. (2011) simplify the formulation of the MC-SCTM by solely using cell occupancies as state variables. The network loading is done by the occupancy-based modified CTM (OM-CTM), which is extended from the existing network version of the CTM (Daganzo 1995). In the OM-CTM, a highway is discretized into many cells and the study horizon is discretized into many equal length intervals. The length of each cell is equal to the product of the *maximum* free flow speed for that cell and the length of each time interval. The basic equations in the CTM still apply to the OM-CTM except the definitions of $S_{j-1}(\omega)$, and $R_j(\omega)$:

$$S_j(\omega) = \min\left\{Q_j(\omega), n_j(\omega), \frac{N_j(\omega)}{d_j/d_0}\right\} \text{ and} \tag{7.13}$$

$$R_j(\omega) = \min\left\{Q_j(\omega), \frac{w_j(\omega)}{v_j(\omega)}\left[N_j(\omega) - n_j(\omega)\right], \frac{N_j(\omega)}{d_j/d_0}\right\}. \tag{7.14}$$

$w_j(\omega)$ and $v_j(\omega)$ are, respectively, the backward wave speed and the free flow speed of cell j during time ω. d_j is the length of cell j. d_0 is the length of a standard cell. Equations (7.13)–(7.14) assume that any cell is made up of many " imaginary" standard cells connected in series. According to (7.13), the sending flow of cell j is constrained by the inflow capacity, the cell occupancy, and the holding capacity of the last "imaginary" standard cell of cell j. Equation (7.14) states that the available capacity of cell j is the minimum of the inflow capacity of cell j, the product of the vacant space $[N_j(\omega) - n_j(\omega)]$ and the factor $w_j(\omega)/v_j(\omega)$ that accounts for the effect of shockwave on the vacant space in cell j, and the holding capacity of the first "imaginary" standard cell embedded in cell j.

To avoid vehicles traveling at a speed higher than the free flow speed, according to the Courant–Friedrichs–Levy condition (Courant et al. 1967), the minimum waiting time $a_j(\omega)$, or equivalently the minimum number of time intervals required by a vehicle to stay in cell j, must be at least equal to the minimum number of time intervals required to leave cell j, $t_j(\omega)$:

$$a_j(\omega) \geq t_j(\omega). \tag{7.15}$$

The latter depends on the length and actual free flow speed of the cell as follows:

$$t_j(\omega) = \frac{d_j}{d_0} \cdot \frac{v_{\max,j}}{v_j(\omega)}, \tag{7.16}$$

where $v_j(\omega)$ and $v_{\max,j}$, respectively, denote the actual and maximum free flow speeds of cell j with $v_j(\omega) \leq v_{\max,j}$, and $v_j(\omega) = v_{\max,j}$ when there is no uncertainty for $v_j(\omega)$. d_j and d_0, respectively, denote the lengths of cell j and the shortest cell. They represent the distance traveled by a vehicle at the maximum free flow speed

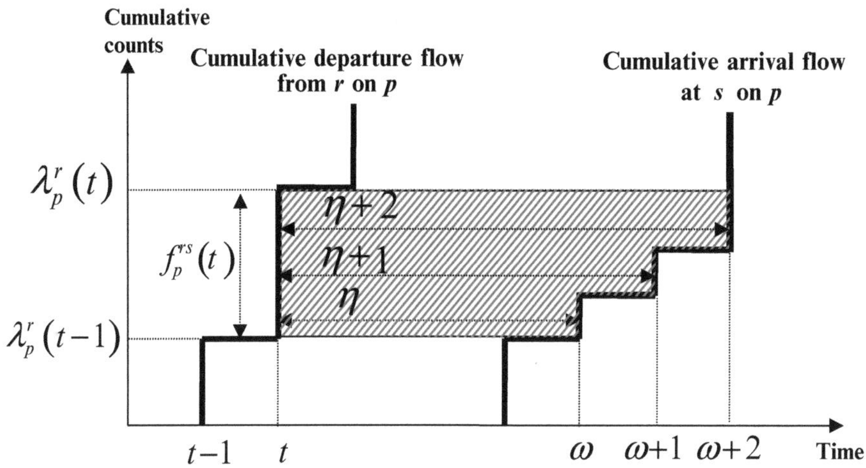

Fig. 7.3 The cumulative vehicle counts of origin cell r and destination cell s

during one time interval. Equation (7.16) states that $t_j(\omega)$ is directly proportional to the length of cell j and inversely proportional to the actual free flow speed $v_j(\omega)$. In the extreme case, when $d_j = d_0$ and $v_j(\omega) = v_{\max,j}, \forall j, \omega$, the OM-CTM is reduced to the CTM.

7.2.6 Determination of Actual Route Travel Time from the CTM Output

Knowing the occupancy of each package of traffic on route p in origin cell r and destination cell s at each time instance, the actual en route travel time $\eta_p^{rs}(t)$ of flow $f_p^{rs}(t)$ can be determined through the use of cumulative counts. Let $\lambda_p^r(t)$ be the cumulative traffic departing from cell r on route p at time t and $\lambda_p^s(\omega)$ be the cumulative traffic arriving at cell s on route p at time ω, defined by:

$$\lambda_p^r(t) = \sum_{t' \le t} n_{r,p}(t'), \text{and} \tag{7.17}$$

$$\lambda_p^s(\omega) = \sum_{\omega' \le \omega} n_{s,p}(\omega'). \tag{7.18}$$

Figure 7.3 shows that the cumulative count curves $\lambda_p^r(t)$ and $\lambda_p^s(\omega)$ on route p between OD pair rs. The actual en route travel time of traffic departing at t (i.e., $f_p^{rs}(t)$) is the horizontal distance between the two cumulative curves as shown in Fig. 7.3. If time is discretized, subject to the en route conditions, there is no guarantee that the entire packet $f_p^{rs}(t)$ will arrive at the destination s in the same discretized time tick ω. As shown in Fig. 7.3, the front part of the packet has an actual travel time of η whereas the latter parts have longer actual travel times. The extent of this difference in en route travel times of the same departing packet depends on the length of the discretized time. In general, higher accuracy can be achieved by using finer discretized time length. However, using finer discretized time length will lead to more cells, more time intervals, and eventually higher computational efforts, because $T_s V = l_s$ has to be held in the CTM.

Lo and Szeto (2002a) propose an *averaging scheme* so that the entire departing traffic $f_p^{rs}(t)$ has one uniquely determined average en route travel time $\eta_p^{rs}(t)$. Mathematically, this scheme can be stated as:

$$\eta_p^{rs}(t) = \frac{\int_{\lambda_p^r(t-1)}^{\lambda_p^r(t)} \left[\lambda_p^{s-1}(\upsilon) - \lambda_p^{r-1}(\upsilon)\right] \mathrm{d}\upsilon}{\lambda_p^r(t) - \lambda_p^r(t-1)}, \tag{7.19}$$

where

$$f_p^{rs}(t) = \lambda_p^r(t) - \lambda_p^r(t-1). \tag{7.20}$$

The numerator on the right-hand side of (7.19) is the area of the shaded region in Fig. 7.3 or the total en route travel time of the entire packet $f_p^{rs}(t)$. The denominator is the packet departing at time t as defined in (7.20). The above averaging scheme is well defined for used routes with positive departure flows. For an unused route with $f_p^{rs}(t) = 0$, we define its actual route travel time to be equal to that of $\lim\limits_{\sigma \to 0^+} f_p^{rs}(t) = \sigma$. For an infinitely small σ, the whole packet shall arrive at the same discretized tick. According to (7.19), $\eta_p^{rs}(t) = \frac{\eta \cdot \sigma}{\sigma} = \eta$ or $\omega - t$.

Lo (1999) adopts the *round off scheme* instead of the averaging scheme. If the cumulative count in r at time t is bounded by the cumulative counts in s between ω and $\omega + 1$, then the path travel time is set to be $\omega - t$ as illustrated in Fig. 7.4 The maximum error in this estimation is one time interval. Mathematically, it can be stated as:

$$\text{If } \lambda_p^s(\omega) \le \lambda_p^r(t) < \lambda_p^s(\omega+1) \text{ then } \eta_p^{rs}(t) = \omega - t. \tag{7.21}$$

Han et al. (2011) propose a scheme based on maximum travel time. To compute the maximum travel time, they define an indicator variable $\tau_{t,\omega}^p$ that indicates whether or not the cumulative departures on path p up to time t are greater than the cumulative arrivals on the same path up to time ω namely,

$$\tau_{t,\omega}^p = \begin{cases} 1 & \text{if } \lambda_p^r(t) > \lambda_p^s(\omega) \\ 0 & \text{if } \lambda_p^r(t) < \lambda_p^s(\omega) \\ [0,1] & \text{if } \lambda_p^r(t) = \lambda_p^s(\omega) \end{cases}. \tag{7.22}$$

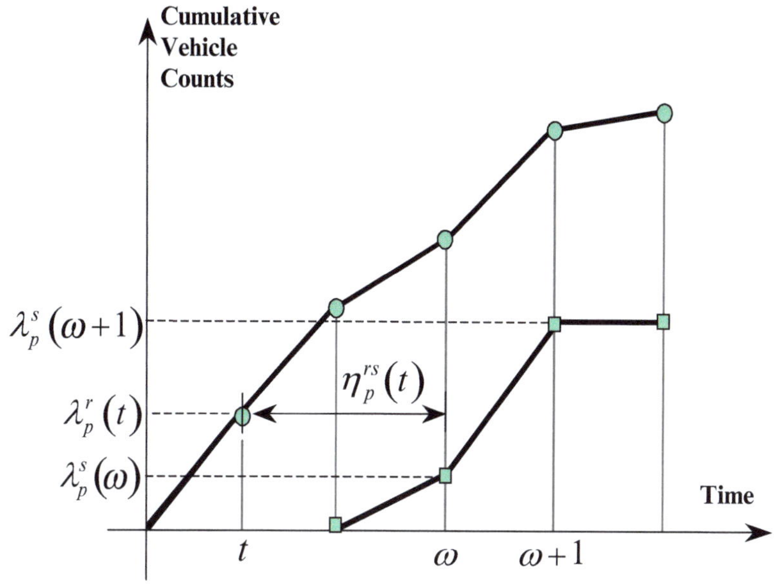

Fig. 7.4 An illustration of the round-off scheme

The travel time of the vehicles departing in time interval t taking path p is given by

$$\eta_p^{rs}(t) = \sum_{\omega} \tau_{t,\omega}^p. \tag{7.23}$$

Notice that when $\lambda_p^r(t) = \lambda_p^s(\omega)$, the indicator can take any value between 0 and 1, and hence is not uniquely defined. This is due to the "convexifaction" method applied to deal with the possible discontinuity when the maximum travel time scheme is adopted.

7.3 Route Choice Component

The travel choice principle models travelers' propensity to travel, and if so, how they select their routes, departure times, modes, or destinations. In making such choices, travel time is one important element of their considerations. The commonly adopted travel choice principles in the cell-based dynamic equilibrium models include

- The dynamic user optimal (DUO) or dynamic user equilibrium route choice principle
- The DUO route/departure time choice principle

- The stochastic dynamic user optimal (SDUO) route choice principle
- The SDUO route/departure time choice principle
- The reliability-based stochastic dynamic user optimal route choice principle

7.3.1 The DUO Route Choice Principle

The DUO route choice principle is the simplest dynamic extension of Wardrop's (1952) first principle, and states that for each origin-destination pair, any routes used by travelers departing at the same time must have equal and minimal travel time. Mathematically, this principle can be expressed as the following conditions:

$$f_p^{rs}(t)\left[\eta_p^{rs}(t)-\pi^{rs}(t)\right]=0,\quad \forall rs,p,t,\text{and} \tag{7.24}$$

$$\eta_p^{rs}(t)-\pi^{rs}(t)\geq 0,\quad \forall rs,p,t, \tag{7.25}$$

where $f_p^{rs}(t)$ and $\eta_p^{rs}(t)$ are, respectively, the flow on route p between OD pair rs departing at time t and its travel time; $\pi^{rs}(t)$ is the lowest travel time between OD pair rs for flows departing at time t. According to (7.24), if route p carries a positive flow at time t (i.e., $f_p^{rs}(t)>0$), then its associated route travel time $\eta_p^{rs}(t)$ must be equal to the lowest travel time $\pi^{rs}(t)$ through the condition $[\eta_p^{rs}(t)-\pi^{rs}(t)]=0$. Equation (7.25) ensures $\pi^{rs}(t)$ to be the lowest travel time among all the possible routes between OD pair rs for flows departing at time t.

7.3.2 The DUO Route/Departure Time Choice Principle

The DUO route/departure time choice principle considers the departure time choice in additional to route choice and considers generalized travel cost instead of travel time. This principle states that for each OD pair, the generalized travel costs incurred by travelers departing at any time are equal and minimal. Mathematically, this principle can be written as follows:

$$f_p^{rs}(t)\left[\phi_p^{rs}(t)-\phi_{\min}^{rs}\right]=0,\quad \forall rs,p,t,\ \text{and} \tag{7.26}$$

$$\phi_p^{rs}(t)-\phi_{\min}^{rs}\geq 0,\quad \forall rs,p,t, \tag{7.27}$$

where $\phi_{\min}^{rs}$ represents the minimal generalized travel cost during the modeling horizon and is independent of t, and $\phi_p^{rs}(t)$ denotes the generalized travel cost incurred by travelers entering route p at time t

The generalized travel cost of each traveler is equal to:

$$\phi_p^{rs}(t)=\alpha_s\eta_p^{rs}(t)+c_s(t)+\tau_p^{rs}(t)+\kappa,\qquad \forall rs,p,t, \tag{7.28}$$

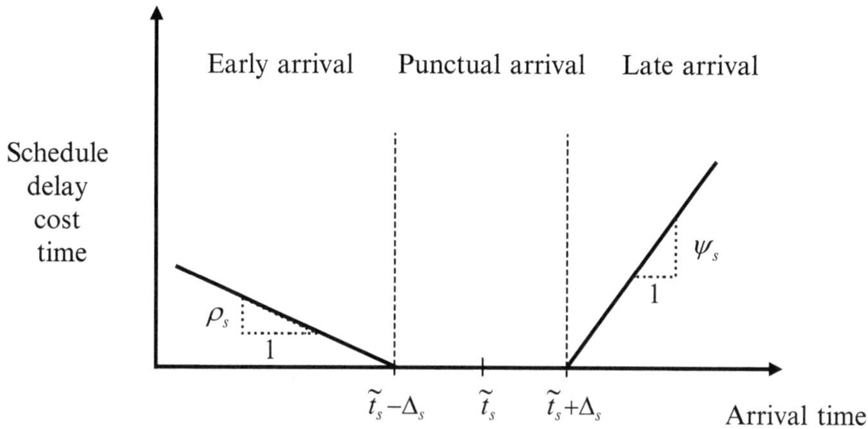

Fig. 7.5 The schedule delay cost as a function of arrival time

where α_s is the cost of unit travel time for travelers heading to destination s. κ and $\tau_p^{rs}(t)$ are, respectively, the fixed and variable out-of-pocket cost, including toll, parking cost, fuel cost, etc. $c_s(t)$ is the schedule delay cost. Travelers acquire no schedule delay cost $c_s(t)$ if they arrive within the desired arrival time window. Otherwise, they incur schedule delay costs for both early and early arrivals outside of this arrival time window. For illustration purposes, the piece-wise linear schedule delay cost function $c_s(t)$ is given below:

$$c_s(t) = \begin{cases} \rho_s\left[(\tilde{t}_s-\Delta_s)-(t+\eta_p^{rs}(t))\right] & \text{if } \tilde{t}_s-\Delta_s > t+\eta_p^{rs}(t) \\ 0 & \text{if } \tilde{t}_s-\Delta_s \le t+\eta_p^{rs}(t) \le \tilde{t}_s+\Delta_s \\ \psi_s\left[(t+\eta_p^{rs}(t))-(\tilde{t}_s+\Delta_s)\right] & \text{if } \tilde{t}_s+\Delta_s < t+\eta_p^{rs}(t) \end{cases}, \quad (7.29)$$

where ρ_s and ψ_s correspond to the unit costs of early and late arrivals for travelers heading to destination s. $\tilde{t}_s$ is the desired arrival time and Δ_s is the interval of arrival time flexibility. $[\tilde{t}_s-\Delta_s, \tilde{t}_s+\Delta_s]$ is therefore the desired arrival time interval. Figure 7.5 shows the schedule delay cost as a function of arrival time.

Two points are worthwhile mentioning. First, α_s, ρ_s, ψ_s, Δ_s, and $\tilde{t}_s$ are independent of origin r but only dependent on destination s, meaning that travelers heading to the same destination have the same desired arrival time window and the same schedule delay cost function. This assumption, as proposed by Yang and Meng (1998), is reasonable for morning commute traffic. Second, the empirical results found in Small (1982) shows that

$$\psi_s > \alpha_s > \rho_s > 0. \quad (7.30)$$

That is, the unit cost of late arrival (ψ_s) is higher than the unit cost of travel time (α_s), which is in turn higher than the unit cost of early arrival (ρ_s).

7.3.3 The SDUO Route Choice Principle

The SDUO route choice principle is the dynamic extension of the Daganzo and Sheffi's (1977) stochastic user equilibrium principle and states that for each origin-destination pair at each instant of time, the actual travel times perceived by travelers departing at the same time are equal and minimal, where the perceived travel time is equal to the sum of actual travel time and the perception error. Mathematically, the SDUO route choice principle can be formulated as

$$f_p^{rs}(t) = w_p^{rs}(t) \cdot q^{rs}(t), \tag{7.31}$$

where $w_p^{rs}(t)$ is the proportion of travelers on route p between OD pair rs departing at time t; $q^{rs}(t)$ is the demand of travelers between OD pair rs at time t. Note that (7.31) only ensures the perceived travel times (not the actual travel times) on all used routes to be the same. The actual travel times on all used routes are not the same because of the perception error.

If the perception errors on each travel time are independently and identically distributed Gumbel variates, the flow proportion can be described by a multinomial logit model. This model has been adopted by Lo and Szeto (2004) for calculating the flow proportion:

$$w_p^{rs}(t) = \frac{\exp\left(-\theta \cdot \eta_p^{rs}(t)\right)}{\sum_k \exp\left(-\theta \cdot \eta_k^{rs}(t)\right)}, \tag{7.32}$$

where the parameter θ represents the perception variations of travelers. A higher value of θ means smaller travel time perception variations, and hence, better information quality. In the limiting case when θ approaches infinity, the corresponding route flow pattern approaches that as modeled by the DUO conditions, in which travelers are assumed to have perfect information about the network status.

7.3.4 The SDUO Route/Departure Time Choice Principle

The SDUO route/departure time choice principle is the generation of the DUO route/departure time choice principle but assumes that the travelers have imperfect information on the network status as in the SDUO route choice principle. This principle states that for each OD pair, the generalized costs $\phi_p^{rs}(t)$ perceived by travelers departing at any time are equal and minimal. Mathematically, this principle can be characterized by the following equation:

$$f_p^{rs}(t) = w_p^{rs}(t) \cdot Q^{rs}, \forall rs, p, t, \tag{7.33}$$

where Q^{rs} is the total demand between OD pair rs in the study horizon.

In Szeto and Lo (2005), a nested logit model is used for determining the proportion:

$$w_p^{rs}(t) = \frac{\exp\left(-\theta_R \cdot \phi_p^{rs}(t)\right)}{\sum_k \exp\left(-\theta_R \cdot \phi_k^{rs}(t)\right)} \times \frac{\exp\left(-\theta_T \cdot \phi_*^{rs}(t)\right)}{\sum_j \exp\left(-\theta_T \cdot \phi_*^{rs}(j)\right)}, \forall rs, p, t, \tag{7.34}$$

$$\text{and } \phi_*^{rs}(t) = -\frac{1}{\theta_R} \ln \sum_p \exp\left(-\theta_R \cdot \phi_p^{rs}(t)\right), 0 < \theta_T \leq \theta_R, \tag{7.35}$$

where θ_T and θ_R are, respectively, parameters representing perception variations on the departure time and route.

The above assumes that first, travelers decide on the departure time, which forms the top-level nest. The probability of choosing a particular departure time t is expressed by the term:

$$\frac{\exp\left(-\theta_T \cdot \phi_*^{rs}(t)\right)}{\sum_j \exp\left(-\theta_T \cdot \phi_*^{rs}(j)\right)}, \forall rs, p, t. \tag{7.36}$$

Note that in deciding the departure time, the performances of all the routes leaving at all the possible departure times are considered. Specifically, the log-sum term $\phi_*^{rs}(j)$ represents the expected maximum utility of all the routes leaving at time j. The expression in (7.36) determines the probability for travelers choosing departure time t after considering all the possible departure times j.

After choosing a particular departure time t, travelers decide on the route, as expressed by the following term in (7.37):

$$\frac{\exp\left(-\theta_R \cdot \phi_p^{rs}(t)\right)}{\sum_k \exp\left(-\theta_R \cdot \phi_k^{rs}(t)\right)}, \forall rs, p. \tag{7.37}$$

Combining (7.36) and (7.37), we obtain the expression in (7.34). This formulation provides the flexibility of modeling different perception variations on the departure time (θ_T) and on the route (θ_R). In the special case, when both parameters on perception variations are equal (i.e., $\theta_T = \theta_R = \theta$), the above nested logit formulation can be simplified to this standard logit model:

$$w_p^{rs}(t) = \frac{\exp\left(-\theta \cdot \phi_p^{rs}(t)\right)}{\sum_k \sum_j \exp\left(-\theta \cdot \phi_k^{rs}(j)\right)}, \forall rs, p, t. \tag{7.38}$$

7.3.5 The RSDUO Route Choice Principle

The RSDUO route choice principle adopted by Szeto and Sumalee (2009) and Szeto et al. (2011) considers the attitude of travelers towards the risk of late arrivals due to uncertain travel time in addition to perception variations of the travel times. Travelers are assumed to select routes based on the dynamic extension of the reliability-based stochastic user equilibrium principle (Shao et al. 2006). This RSDUO principle states that for each travelers departing at any time, they select routes with the minimum perceived effective travel time at the time of departure. The effective route travel time (or travel time budget) is defined as the sum of the mean route travel time and the safety margin:

$$\hat{\eta}_p^{rs}(t) = \eta_p^{rs}(t) + s_p^{rs}(t), \tag{7.39}$$

where $\hat{\eta}_p^{rs}(t)$ and $s_p^{rs}(t)$ denote the effective travel time and safety margin of route p between OD pair rs for travelers departing at t, respectively. $\eta_p^{rs}(t)$ is the mean travel time on route p between OD pair rs.

The safety margin $s_p^{rs}(t)$ in (7.39) is a linear function of the standard deviation (SD) $\sigma_p^{rs}(t)$ of the travel time on route p:

$$s_p^{rs}(t) = Z\sigma_p^{rs}(t), \tag{7.40}$$

where Z is the parameter describing the degree of the risk aversion of travelers. The larger is the value of Z, the greater is the degree of the risk aversion of travelers. In particular, $Z > 0$ if travelers are risk-averse and depart earlier to allow additional time to avoid late arrivals; $Z = 0$ if travelers are risk-neutral and ignore $\sigma_p^{rs}(t)$ when selecting routes. Of course, Z can also be related to the importance of the trip, i.e. by trip purpose (a higher Z value for a more important trip and vice versa).

Szeto and Sumalee (2009) adopt a logit model to determine the flow proportion, where the travel time in (7.32) is replaced by the effective travel time to determine the proportion. To handle path correlation, Szeto et al. (2011) adopt C-logit model for determining the proportion. The modified effective route travel time instead of effective travel time is used to determine the flow proportion and is defined as the sum of the commonality factor and actual effective route travel time:

$$V_p^{rs}(t) = CF_p^{rs} + \hat{\eta}_p^{rs}(t), \tag{7.41}$$

where CF_p^{rs} is the commonality factor of route p between OD pair rs, and $V_p^{rs}(t)$ denotes the modified effective travel time on route p between OD pair rs for travelers departing at time t.

The commonality factor in (7.41) is used to capture the degree of similarity between paths as in Cascetta et al. (1996). It is expressed as follows:

$$CF_p^{rs} = \lambda \cdot \ln \sum_h \left(\frac{L_{hp}}{L_h^{1/2} \cdot L_p^{1/2}} \right)^{\psi}, \tag{7.42}$$

where L_{hp} is the "length" of links or free flow travel time on links common to both paths h and k. L_h and L_p are, respectively, the free flow travel times on paths h and p belonging to the same OD pair. λ and ψ are parameters.

It is not difficult to see that the RSDUO principle includes the SDUO and DUO conditions as special cases. When $Z = 0$, the flow pattern satisfying the RSDUO principle is also the SDUO flow pattern. When θ approaches infinity and $Z = 0$, the corresponding route flow pattern approaches that as modeled by the DUO conditions.

7.4 Formulation and Algorithmic Approaches

The cell-based dynamic equilibrium models can be formulated by at least three approaches:

- Variational Inequality Problem (VIP)
- Nonlinear Complementarity Problem (NCP), and
- Fixed-Point Problem (FPP).

The VIP is to find $\mathbf{f}^* = [f_p^{rs^*}(t)]$ such that

$$(\mathbf{f}-\mathbf{f}^*)^{\mathrm{T}}\mathbf{H}(\mathbf{f}^*) \geq 0, \forall \mathbf{f} \in \Omega, \tag{7.43}$$

where $\mathbf{H}(\mathbf{f})$ represents a general vector function of $\mathbf{f}$, and Ω represents the feasible solution set of the problem. The superscript "$*$" refers to a solution of $\mathbf{f}$ that fulfills the travel choice conditions. The existence of solutions to the VIP (7.43) requires that (i) $\mathbf{H}(\mathbf{f})$ is a continuous function of $\mathbf{f}$ and (ii) Ω is a nonempty compact convex set (Theorem 1.4 in Nagurney 1993). The uniqueness of the solution further requires the mapping function to be strictly monotonic (Theorem 1.8 in Nagurney 1993). When the SDUO or RSDUO route choice principle is adopted in the cell-based dynamic equilibrium model, $\mathbf{H}(\mathbf{f}) = [f_p^{rs}(t) - w_p^{rs}(t) \cdot q^{rs}(t), \forall rs, p, t]$ and Ω is the nonnegative orthant, where $w_p^{rs}(t)$ is a function of route travel time, which in turn is a function of flows $\mathbf{f}$ through the unique mapping (7.1) According to Szeto and Lo (2006), the route travel time may not be a continuous function of route flows, leading to the possibility of the nonexistence of solutions to the cell-based dynamic equilibrium models.

The NCP is indeed a special form of the VIP; their equivalency conditions are discussed in Proposition 1.4 of Nagurney (1993)—the solutions to these problems are equivalent when the feasible solution region is the nonnegative orthant. Based on this proposition, one can express the VIP as an NCP. The NCP is to find an optimal vector $\mathbf{f}^* \geq \mathbf{0}$ such that:

$$\mathbf{f}^{*\mathrm{T}} \cdot \mathbf{H}(\mathbf{f}^*) = 0 \text{ and } \mathbf{H}(\mathbf{f}^*) \geq \mathbf{0}. \tag{7.44}$$

As mentioned before, when the SDUO or RSDUO route choice principle is adopted in the cell-based dynamic equilibrium model, $\mathbf{H}(\mathbf{f}) = [f_p^{rs}(t) - w_p^{rs}(t) \cdot q^{rs}(t), \forall rs, p, t]$. Since $w_p^{rs}(t)$ is always nonnegative, all routes carry flows at optimality (i.e., $\mathbf{f}^* > \mathbf{0}$). It is therefore not difficult to see that $\mathbf{H}(\mathbf{f}^*) = \mathbf{0}$, which is the SDUO route choice conditions.

The FPP is to find $\mathbf{f}^* = [f_p^{rs^*}(t)]$ such that

$$\mathbf{f}^* = \mathbf{Y}(\mathbf{f}^*), \tag{7.45}$$

where $\mathbf{Y}(\mathbf{f})$ represents a general vector function of $\mathbf{f}$. Let the projection operator $P_\Omega(\mathbf{y}) = \arg\min_{\mathbf{z}\in\Omega} \|\mathbf{y} - \mathbf{z}\|$, $\kappa > 0$, and Ω is closed and convex. If $\mathbf{Y}(\mathbf{f}) = P_\Omega(\mathbf{f} - \kappa\mathbf{H}(\mathbf{f}))$, then the FPP (7.45) and the VIP (7.43) have the same set of optimal solutions (Theorem 1.3 in Nagurney 1993). When the SDUO or RSDUO route choice principle is adopted in the cell-based dynamic equilibrium model, Ω is the nonnegative orthant. By putting $\kappa = 1$ into $\mathbf{Y}(\mathbf{f}) = P_\Omega(\mathbf{f} - \kappa\mathbf{H}(\mathbf{f}))$ and simplifying the resultant expression, we can get the SDUO or RSDUO route choice condition (7.31).

The choice of the formulation approach highly depends on the solution method adopted. In Lo and Szeto (2002b) and Szeto and Sumalee (2009), the NCP formulation is adopted to describe the DUO route choice problem and the RSDUO route choice problem respectively. The formulation is transformed into an unconstrained optimization problem via a gap function proposed by Lo and Chen (2000). The optimization problem derived from the DUO route choice problem is then solved by the genetic algorithm (GA), which is used to find a nearly global optimal solution. The optimization problem derived from the RSDUO route choice problem is solved by their stochastic gradient-based solution algorithm that only relies on the statistical estimate of the gap function. Han et al. (2011) formulate the cell-based DUO route/departure time choice problem with elastic demand and user heterogeneity as a complementary problem and use PATH and KNITRO, two well-known state-of-the-art solvers for complementarity problems, for obtaining solutions.

In Lo and Szeto (2002a) and Szeto and Lo (2004, 2005), the VIP formulation is adopted to depict the DUO route choice problem, the DUO route/departure time choice problem, and the SDUO route/departure time choice problem respectively. The first problem is solved by the projection method proposed by Han and Lo (2002) whereas the second and third problems are solved by the projection method proposed by Han and Lo (2004). The convergence of all these methods is guaranteed when the mapping function $\mathbf{G}(\mathbf{f})$ is co-coercive. Ukkusuri et al. (2012) reformulate the cell-based DUO route/departure time choice problem as a VIP and solve it by a projection method.

In Szeto et al. (2011), the fixed point formulation is adopted to describe the RSDUO route choice problem, and is solved by the self-regulated averaging method (SAM) proposed by Liu et al. (2009). This method includes the method of successive averages as a special case.

Indeed, many solution algorithms such as swapping algorithms, the method of successive averages (MSA), the decent direction algorithms, feasible direction algorithm, and other algorithms tailored for solving the VIPs, FPPs, and NCPs are available to solve the cell-based dynamic equilibrium formulations although the convergence can be achieved under some conditions. Nie and Zhang (2010) compare the performance of some of these algorithms for solving the cell-based DUO route/departure time choice problem. They find that introducing line searches provides relatively faster and more stable convergence, compared MSA. When appropriately implemented, the feasible direction algorithms can outperform MSA in terms of computational overhead.

For some special problems such as the DUO route choice problem, cell-based dynamic equilibrium models (e.g., Ukkusuri 2002) can be directly represented as optimization models and solved by combinatorial optimization algorithms proposed by Golani and Waller (2004) for solving the multi-destination case and by Waller and Ziliaskopoulos (2006) for the single-destination case. The algorithm for solving the single-destination case is exact but the algorithm for solving the multi-destination case is a heuristic.

7.5 Implementation Issues and Future Research Directions

Clearly, the RSDUO route choice model that encapsulates the MC-SCTM is more realistic and can be easily extended to consider departure time choice, mode choice (by using nested logit model), activity location choice, activity duration, time of participation (by using the supernetwork representation proposed in Ramadurai and Ukkusuri 2010), and demand elasticity (by using elastic demand functions as in Szeto and Lo 2004). However, the model is associated with the following implementation issues for the applications in dynamic route guidance and traffic control, leading to some future research directions.

7.5.1 Nonexistence and Nonuniqueness of RSDUO Solutions

Due to the effect of physical queues and traffic signals, travel time may not be a continuous function of flows and hence there is a possibility that there is no solution for the cell-based equilibrium model. The relaxation of the DUO condition to the SDUO and RSDUO conditions may not be able to address the problem. Existence of a solution is a fundamental requirement of a model for actual applications. Szeto and Lo (2006) provide some initial thinking on developing a framework to accommodate this. They allow the travel times of all used routes to be unequal but their maximum difference lie within a tolerance or an aspiration level, where the tolerance level is purely a function of the behavior of the network users and can be calibrated through surveys. This relaxation acknowledges the fact that, in reality, the travel times of the used routes between the same origin-destination pair are rarely exactly the same, and that travelers will stop exploring new routes when they perceive no appreciable differences between their current routes and the candidate ones. This relaxation matches the bounded-rationality behavioral notion in Simon (1955) and can be easily generalized to capture departure time, mode, and destination choices, the learning effect of travelers' tolerance over time, imperfect traffic information perceived by travelers, and the risk aversive behavior of travelers. Under this relaxation, Szeto and Lo (2006) show that the existence of a stable solution under the spatial queue consideration depends on the network topology, the demand pattern, and also travelers' behavior on travel time tolerance. Using their proposed day-to-day route swapping algorithm, they further demonstrate that for a small tolerance, the system can keep on evolving without converging to a stable equilibrium. In particular, periodicity in terms of total system travel time and traffic pattern can be observed under varied parameters and initial solution settings. These raise interesting and important questions to be answered in future research on theoretical explorations of the network behavior under non-equilibrium, bounds of the changes in the total system travel time, periodicity of the changes, and their relations to travelers' aggressiveness in route swapping, etc. in the RSDUO context. It is also likely that there are multiple optimal solutions. How to design and manage the network when the solution is nonuniqueness is an important research question.

7.5.2 A Large Time-Dependent Path Set

Actual networks are often large and involve many paths, although most of them are unused. A large path set makes path enumeration impossible. However, to deal with queue spillback properly, we must use path-based DTA models and hence path-based algorithms if alternative formulations for handling queue spillback information do not exist. How to deal with this problem is an important issue for actual model implementation. A simple approach to deal with this problem is to assume that the path set is fixed and can be deduced from GPS data. This path set can be changed over time of day. How to extract a reliable path set from the GPS data deserves further investigation.

Another possible but more complicated approach to deal with the large path set is to generate a small path set in the path-based solution algorithm. Compared with the path-based solution algorithms for static traffic assignment models, an additional effort on the path set generation for DTA models is required because we need to consider time-dependent paths. In particular, the effect of junction blockages can cause the network configuration temporally change, which must be duly handled. This raises questions: Can the path set generation rules used in the static traffic assignment be modified and extended to the dynamic case? Are there any other methods to efficiently deal with the temporally changing network configuration under the effects of physical queues? How does the choice of path set generation rules affect the solution speed and convergence of the algorithm? These questions require further study.

The third approach to handle the large path set is to develop the network loading procedure similar to Dial's algorithm (1971) or the origin-based approach (Bar-Gera 2002) that can avoid the path set information but still can handle the queue spillback information. This again is left for future research.

7.5.3 *Time-Consuming Monte Carlo Simulation*

The MC-SCTM requires the Monte Carlo simulation that can lead to long computation times for large network applications. This is clearly unacceptable for real-time deployment. Szeto et al. (2011) have tested the idea of using smaller sample sizes for early stages of solution processes and using larger ones at the later stages. The results are quite promising in terms of saving computation time. Although the optimal parameter for triggering switching varies case by case, this strategy can be used for reducing the computation time to solve large problems by selecting a nearly optimal switching triggering value.

Another strategy for speeding up the computation process is to use the analytical SCTM (e.g., Pan et al. 2010 and Sumalee et al. 2010, 2011) to approximately solve the problem in the early or some stages of the solution process. This strategy has not been tested and is left for further testing.

The third strategy is to use antithetic sampling technique, Latin hypercube sampling technique, and single point approximation techniques to reduce the sample sizes and hence the computation time while maintaining an acceptable level of solution quality. According to Sharma et al. (2011), using Antithetic Sampling can reduce the sample size 10 times to solve in their problem about the Network of Fort Area, Mumbai while reducing the computation time to 20% of the original time required. Using single point approximations can further lead to a 99% computation time saving. We expect the computation time can be reduced significantly by introducing these techniques in the SAM, but the exact reduction magnitudes may be slightly different. The evaluation of the benefit of introducing these approximations is left for future studies.

The fourth strategy is implementing parallel computing. Given that each M-CTM simulation can be done independently, each simulation can be performed in one computer. The computation time can be reduced by a factor of the sample size. The efficiency of this approach can be tested in the future.

One more strategy is to develop travel time functions to approximate the unique mapping (which involves the MC-SCTM simulation) at some stages to speed up the solution processes. The question is whether this approximation can guarantee convergence and how to develop the travel time function. This question has not been answered yet.

7.5.4 Nonexistence of Efficient and Convergent Solution Methods

The properties of discontinuity and non-monotonic route travel time lead to developing efficient and convergent solution methods for real-time deployment difficult. The SAM proposed by Liu et al. (2009), the projection methods proposed by Han and Lo (2002, 2004) and the projection method used by Ukkusuri et al. (2012) can be treated as heuristics as the mathematical requirements of the convergence may not be satisfied. It seems to imply that the RSDUO problem has to be solved by the method proposed by Lo and Szeto (2002b) in which the problem is reformulated into an unconstrained optimization problem and is solved by some less restrictive global optimization methods. GA is definitely one option to solve the reformulated problem. However, GA has to evaluate the objective values of each trial solution, which is a time-consuming process as the objective function value is obtained through the MC-SCTM simulation. To increase the computation efficiency, a possible approach is implementing parallelized genetic algorithm (e.g., Wong et al. 2001) for solving these models. This approach makes good use of the inherent nature of GA that the evaluation of each trial solution can be done independently, and hence the performance of GA can be greatly improved by means of parallel computing.

The performance of GA highly depends on the parameter setting. However, there is no clue on how to set the parameter value. Moreover, GA may not be the most efficient solution method to solve the RSDUO problem. Other meta-heuristics can also be used to solve the model. It is unclear which meta-heuristic is relatively more efficient to solve the model. Computation tests have been performed in the future to answer these questions.

7.5.5 Ignoring Lane-Changing Behavior, Moving Bottlenecks, and Interaction between Different Types of Vehicles

The MC-SCTM or even SCTM does not consider lane changing traffic behavior and moving bottlenecks. The question is whether it is acceptable to ignore them for the real-time implementation. This can be partially answered by performing validation in the future. If the results show that there is a significant loss of accuracy in predicting the traffic flow pattern, we may need to consider to extend the multiple-pipe and variational theories as well as the modeling theories for lane-changing behavior and moving bottlenecks (see Daganzo and Laval 2005, Daganzo 2006, Leclercq 2007) for the general traffic networks and incorporate them into the MC-SCTM or the SCTM.

Existing cell-based equilibrium models currently adopted either the CTM or the MC-SCTM, but they cannot capture the traffic interaction between different vehicle classes. Enhancing the CTM and SCTM to capture this realistic behavior can be another research direction. One can also validate the resulting model and estimate the gain in modeling accuracy to justify whether to incorporate this behavior in the SCTM or its extension.

7.5.6 Requirement on the Quality of the OD Matrix

The RSDUO route choice model requires a fairly accurate time-dependent OD matrix for actual implementation. If this matrix is not accurate enough, one can consider extending the model to consider departure time choice as well, as the simultaneous route and departure time choice model only requires an OD matrix like the one for the static traffic assignment. The cost is to estimate the desirable arrival time and desirable arrival time interval of travelers. For the peak hour traffic, this information is not difficult to obtain from surveys. Alternatively, one can consider time-dependent OD matrix with a larger discretized time interval and the demand for each larger interval is assumed to be equally split into several smaller time intervals for network loading. In the worse case, we can calibrate a distribution for each OD demand and relax the assumption of deterministic OD demand in the formulation. All these approaches can be studied in the future.

7.5.7 Missing OD Matrix and Travel Time Updating Components for Teal-Time, Large-Scale Applications

One of the applications of RSDUO model is to predict the real-time traffic flow estimate travel time, and update the time dependent OD matrix for route guidance and traffic control based on historical and real-time traffic count information. However, not all junctions can provide traffic counts. Szeto et al. (2009) propose a cell-based travel time prediction method that does not require traffic counts to be obtained at every junctions. This method can be used in the route guidance application. However, this method has not considered the random traffic state and random demand arrivals. On the other hand, Lo (2001) developed a cell-based dynamic signal control formulation. Again, other than equilibrium principles, the random traffic states and random demand arrivals have not been captured. In the future, these two methods can be extended and incorporated into the roll-horizon framework proposed by Ran et al. (2002) for OD matrix updating and into the large-scale implementation method of Ziliaskopoulos et al. (2004) in order to develop a framework for estimating travel time and flow patterns online as well as for managing traffic.

7.5.8 Calibration Issues

The RSDUO route choice model that encapsulates the MC-SCTM is more realistic but at the same time introduces more parameters to calibrate. How to calibrate these parameters becomes an important question and has not been done yet. One off-line method that can be used in the future is to break down the calibration process into three sequential steps. The first step calibrates Z using the method of Jackson and Jucker (1982). The second step calibrates the parameters in the fundamental diagram as in Sumalee et al. (2011). The last step is to calibrate the parameters associated with C-logit model by the maximum likelihood estimation method. When real-time data is available, the parameters in the fundamental diagram can be adjusted in real time. However, an online calibration procedure is also missing at the moment. Moreover, the issues mentioned in Sect. 7.5.4 are still applied here and needed to address.

7.6 Concluding Remarks

There is no doubt that cell-based dynamic equilibrium models have received attention recently. Over time, more advanced cell-based dynamic models are proposed and analyzed. However, the resulting models become even more complicated and harder to solve for exact solutions, and some implementation issues, especially for the cell-based models with the consideration of stochasticity in the fundamental diagram, have not been addressed. It is time for us to review these models and point out the implementation issues for future research. Hence, this article reviews the details of the two building blocks of the existing cell-based dynamic equilibrium models, the major formulation and algorithmic approaches, the development of cell-based equilibrium models, and some implementation issues of the latest version of cell-based equilibrium models that can capture stochasticity in the fundamental diagram. Some initial thinking for handling these issues and some related future research directions are also provided in this article.

Acknowledgements The work described in this paper was fully supported by a grant from the Research Grants Council of the Hong Kong Special Administrative Region, China (HKU 716312E). The author is grateful for the two reviewers for their constructive comments.

References

Ban X, Liu HX, Ferris MC, Ran B. A link-node complementarity model and solution algorithm for dynamic user equilibria with exact flow propagations. Transport Res. 2008;42B:823–842.

Bar-Gera H. Origin-based algorithm for the traffic assignment problem. Transport Sci. 2002;36:398–417.

Ben-Akiva M, Cuneo D, Hasan M, Jha M, Yang Q. Evaluation of freeway control using a microscopic simulation laboratory. Transport Res. 2003;11C:29–50.
Carey M. Optimal time-varying flows on congested networks. Oper Res. 1987;35:58–69.
Cascetta E, Nuzzolo A, Russo F, Vitetta A. A modified logit route choice model overcoming path overlapping problems: specification and some calibration results for interurban networks. In: Lesort JB, editor, Transportation and traffic theory. New York: Pergamon; 1996. pp. 697–711.
Chow AHF. Properties of system optimal traffic assignment with departure time choice and its solution method. Transport Res. 2009;43B:325–344.
Courant R, Friedrichs K, Lewy H. Über die partiellen differenzengleichungen der mathematischen physik. Math Ann. 1967;100:32–74.
Daganzo CF. The cell transmission model: a simple dynamic representation of highway traffic. Transport Res. 1994;28B:269–287.
Daganzo CF. The cell transmission model, part ii: network traffic. Transport Res. 1995;29B:79–93.
Daganzo CF. The lagged cell transmission model. In: Ceder A, editor, Transportation and traffic theory. New York: Pergamon-Elservier; 1999. pp. 81–103.
Daganzo CF. On the variational theory of traffic flow: well-posedness, duality and applications. Network Heterogeneous Media. 2006;1:601–619.
Daganzo CF, Laval JA. On the numerical treatment of moving bottlenecks. Transport Res. 2005;39B:31–46.
Daganzo CF, Sheffi Y. On stochastic models of traffic assignment. Transport Sci. 1977;11:253–274.
Dial RB. A probabilistic multipath traffic assignment model which obviates path enumeration. Transport Res. 1971;5:83–111.
Doan K, Ukkusuri S. On the holding back problem in cell transmission based dynamic traffic assignment models. Transp. Res. Part B. 2012:1218–1238.
Friesz TL, Bernstein DH, Smith TE, Tobin RL, Wie BW. A variational inequality formulation of the dynamic network user equilibrium problem. Oper Res. 1993;41:179–191.
Golani H, Waller ST. Combinatorial approach for multi-destination dynamic traffic assignment. J Transport Res Board. 2004;1882:70–78.
Han D, Lo HK. A new alternating direction method for a class of nonlinear variational inequality problems. J Optim Theor Appl. 2002;112:549–560.
Han D, Lo HK. Solving nonadditive traffic assignment problems: a decent method for co-coercive variational inequalities. Eur J Oper Res. 2004;159:529–544.
Han L, Ukkusuri S, Doan K. Complementarity formulations for the cell transmission model based dynamic user equilibrium with departure time choice, elastic demand and user heterogeneity. Transp Res B. 2011;45:1749–1767.
Ho J. A successive linear optimization approach to the dynamic traffic assignment problem. Transport Sci. 1980;14:295–305.
Jackson WB, Jucker JV. An empirical study of travel time variability and travel choice behavior. Transport Sci. 1982;16:460–475.
Jayakrishnan R, Tsai WK, Chen A. A dynamic traffic assignment model with traffic flow relationship. Transport Res. 1995;3C:51–82.
Lam WHK, Huang HJ. Dynamic user optimal traffic assignment model for many to one travel demand. Transport Res. 1995;29B:243–259.
Leclercq L. Bounded acceleration close to fixed and moving bottlenecks. Transport Res. 2007;41B:309–319.
Li Y, Waller ST, Ziliaskopoulos AK. A decomposition scheme for system optimal dynamic traffic assignment models. Network Spatial Econ. 2003;3:441–455.
Li Y, Ziliaskopoulos AK, Waller ST. Linear programming formulations for system optimum dynamic traffic assignment with arrival time based and departure time based demands. J Transport Res Board. 1999;1667:52–59.
Lighthill MJ, Whitham JB. On kinematic waves. I. Flow movement in long rivers. II. A theory of traffic flow on long crowded road. Proc R Soc. 1955;A229:281–345.
Lin WH, Wang C. An enhanced 0–1 mixed-integer LP formulation for traffic signal control. IEEE Trans Intell Transport Syst. 2004;5:238–245.

Liu HX, He XZ, He BS. Method of Successive Weighted Averages (MSWA) and self-regulated averaging schemes for solving stochastic user equilibrium problem. Network Spatial Econ. 2009;9:485–503.

Lo HK, Luo XW, Siu BWY. Degradable transport network: Travel time budget of travelers with heterogeneous risk aversion. Transp Res Part B: Methodol. 2006;40(9):792–806.

Lo HK. A dynamic traffic assignment formulation that encapsulates the cell-transmission model. In: Ceder A, editor. Transportation and traffic theory. Oxford: Elsevier; 1999. pp. 327–350.

Lo HK. A cell-based traffic control formulation: strategies and benefits of dynamic plans. Transport Sci. 2001;35:148–164.

Lo HK, Chen A. Traffic equilibrium problem with route-specific costs: formulation and algorithms. Transport Res. 2000;34B:493–513.

Lo HK, Szeto WY. A cell-based variational inequality formulation of the dynamic user optimal assignment problem. Transport Res. 2002a;36B:421–443.

Lo HK, Szeto WY. A cell-based dynamic traffic assignment model: formulation and properties. Math Comput Model. 2002b;35:849–865.

Lo HK, Szeto WY. Modeling advanced traveler information services: static versus dynamic paradigms. Transport Res. 2004;38B:495–515.

Mahmassani HS, Liu YH. Models of user pre-trip and en-route switching decisions in response to real-time information. Transportation systems 1997. (TS'97). Proceedings volume from the 8th IFAC/IFIP/IFORS Symposium. 1997;3:1363–1368.

Munoz L, Sun X, Horowitz R, Alvarez L. Traffic density estimation with the cell transmission model, Proceedings of the American Control Conference, Denver, Colorado, June, 2003. 3750–3755.

Nagurney A. Network economics: a variational inequality approach. Norwell, MA: Kluwer Academic; 1993.

Ngoduy D. Multiclass first order model using stochastic fundamental diagrams. Transportmetrica. 2011;7:111–125.

Nie Y. A cell-based merchant–nemhauser model for the system optimum dynamic traffic assignment problem. Transp Res B. 2011;45:329–342.

Nie Y, Zhang HM. Solving the dynamic user optimal assignment problem considering queue spillback. Network Spatial Econ. 2010;10:49–71.

Pan T, Sumalee A, Zhong R, Uno N. The stochastic cell transmission model considering spatial and temporal correlations for traffic states prediction. Proceedings of the third international symposium on dynamic traffic assignment 2010.

Pavlis Y, Recker W. A mathematical logic approach for the transformation of the linear conditional piecewise functions of dispersion-and-store and cell transmission traffic flow models into linear mixed-integer form. Transport Sci. 2009;43:98–116.

Peeta S, Ziliaskopoulos AK. Foundations of dynamic traffic assignment: the past, the present and the future. Network Spatial Econ. 2001;1:233–265.

Ramadurai G, Ukkusuri SV. Dynamic user equilibrium model for combined activity-travel choices using activity-travel supernetwork representation. Network Spatial Econ. 2010;10:273–292.

Ramadurai G. Novel dynamic user equilibrium models: analytical formulations, multi-dimensional choice, and an efficient algorithm. Ph.D. Thesis, Department of civil and environmental engineering. Troy, NY: Rensselaer Polytechnic Institute; 2009.

Ramadurai G, Ukkusuri SV, Zhao S, Pang JS. Linear complementarity formulation for single bottleneck model with heterogeneous commuters. Transp Res B. 2010;44:193–214.

Ran B, Boyce D. Modeling dynamic transportation networks. An intelligent transportation system oriented approach, 2nd revised edn. Heidelberg: Springer; 1996.

Ran B, Lee DH, Shin MSI. Dynamic traffic assignment with rolling horizon implementation. J Transport Eng. 2002;128:314–322.

Richards PI. Shockwaves on the highway. Oper Res. 1956;4:42–51.

Shao H, Lam WHK, Tam ML. A reliability-based stochastic traffic assignment model for network with multiple user classes under uncertainty in demand. Network Spatial Econ. 2006;6:173–204.

Sharma S, Mathew TV, Ukkusuri SV. Approximation techniques for transportation network design problem under demand uncertainty. J Comput Civ Eng. 2011;25:316–329.

Shen W, Nie YM, Zhang H. Dynamic network simplex method for designing emergency evacuation plans. Transport Res Rec. 2007;2022:83–93.

Simon H. A behavioural model of rational choice. Q J Econ. 1955;69:99–118.

Small KA. The scheduling of consumer activities: work trips. Am Econ Rev. 1982;72:467–479.

Sumalee A, Zhong RX, Pan TL, Szeto WY. Stochastic cell transmission model (SCTM): A stochastic dynamic traffic model for traffic state surveillance and assignment. Trans Res B. 2011;45:507–533.

Sumalee A, Zhong RX, Pan TL, Iryo T, Lam WHK. Stochastic cell transmission model for traffic network with demand and supply uncertainties. Transport Res B. 2010;45:507–533.

Sumalee A, Zhong R, Szeto WY, Pan T. Stochastic cell transmission model under demand and supply uncertainties. *Proceedings of the second international symposium on dynamic traffic assignment* 2008.

Szeto WY. The enhanced lagged cell-transmission model for dynamic traffic assignment. Transport Res Rec. 2008;2085:76–85.

Szeto WY, Lo HK. A cell-based simultaneous route and departure time choice model with elastic demand. Transport Res. 2004;38B:593–612.

Szeto WY, Lo HK. The impact of advanced traveler information services on travel time and schedule delay costs. J Intell Transport Syst Tech Plan Oper. 2005;9:47–55.

Szeto WY, Lo HK. Dynamic traffic assignment: properties and extensions. Transportmetrica. 2006;2:31–52.

Szeto WY, Sumalee A. Multi-class reliability-based stochastic-dynamic-user-equilibrium assignment problem with random traffic states. Proceedings of the 88th annual meeting of transportation research board. 2009.

Szeto WY, Ghosh B, Basu B, O'Mahony M. Cell-based short-term traffic flow forecasting using time series modelling. ASCE J Transport Eng. 2009;135:658–667.

Szeto WY, Sumalee A, Jiang Y. A cell-based model for multi-class doubly stochastic dynamic traffic assignment. Comput Aided Civ Infrastruct Eng. 2011;26:595–611.

Ukkusuri S. Linear programs for user optimal dynamic traffic assignment problem. Master's thesis, University of Illinois at Urbana Champaign. 2002.

Ukkusuri S, Han L, Doan K. Dynamic user equilibrium with a path based cell transmission model for general traffic networks. Transp Res B. 2012;46:1657–1684.

Ukkusuri SV, Waller ST. Linear programming models for the user optimal and system optimal network design problem: formulations, comparisons and extensions. Network Spatial Econ. 2008;8:383–406.

Waller ST, Ziliaskopoulos AK. A combinatorial user optimal dynamic traffic assignment algorithm. Ann Oper Res. 2006;144:249–261.

Wardrop J. Some theoretical aspects of road traffic research. Proceedings of the institute of civil engineers, Part II. 1952;325–378.

Wie BW, Tobin RL, Friesz TL. The augmented lagrangian method for solving dynamic network traffic assignment models in discrete-time. Transport Sci. 1994;28:204–220.

Wong SC, Wong CK, Tong CO. A parallelized genetic algorithm for calibration of lowry model. Parallel Comput. 2001;27:1523–1536.

Yang H, Meng Q. Departure time, route choice and congestion toll in a queuing network with elastic demand. Transport Res. 1998;32B:247–260.

Zhang L, Yin Y, Lou Y. Robust signal timing for arterials under day-to-day demand variations. Transp Res Record, 2010;2192:156–166.

Zheng H, Chang YC. A network flow algorithm for the cell-based single-destination system optimal dynamic traffic assignment problem. Transport Sci. 2011;45:121–137.

Ziliaskopoulos AK. A linear programming model for the single destination system optimum dynamic traffic assignment problem. Transport Sci. 2000;34:37–49.
Ziliaskopoulos AK, Rao L. A simultaneous route and departure time choice equilibrium model on dynamic networks. Int Trans Oper Res. 1999;6:21–37.
Ziliaskopoulos AK, Waller ST, Li Y, Byram M. Large-scale dynamic traffic assignment: implementation issues and computational analysis. J Transport Eng ASCE. 2004;130:585–593.

Chapter 8
Information Impacts on Traveler Behavior and Network Performance: State of Knowledge and Future Directions

Ramachandran Balakrishna, Moshe Ben-Akiva, Jon Bottom, and Song Gao

Abstract Advanced Traveler Information Systems (ATIS) have the potential to maximize the operating efficiency of existing transportation infrastructure. Such systems rely on the generation and dissemination of guidance in order to allow drivers to make informed choices about travel mode, route and departure time, etc. The evaluation of the effectiveness of ATIS requires multidimensional study encompassing the analysis of various choice situations arising in the real world, constructing models that explain driver response to information in different contexts, and developing algorithms that can generate traveler information. Since driver confidence in the ATIS is directly related to the accuracy, relevance, and usefulness of the information, a key aspect is the collection of relevant field data that can instruct model development and ATIS evaluation before real-world deployment. This chapter aims to provide a synthesis of both the state of the art and the state of the practice of ATIS modeling and evaluation. We review the literature related to data collection and driver response model development, and classify the same according to the specific choice situations they address. We provide a conceptual discussion of the general framework within which traveler information may be generated, including key ATIS design parameters that may impact the performance

R. Balakrishna (✉)
Caliper Corporation, Newton, Massachusetts, USA
e-mail: rama@caliper.com

M. Ben-Akiva
Massachusetts Institute of Technology, Cambridge, Massachusetts, USA
e-mail: mba@mit.edu

J. Bottom
Steer Davies & Gleave Inc., Boston, Massachusetts, USA
e-mail: Jon.Bottom@sdgworld.net

S. Gao
University of Massachusetts at Amherst
e-mail: sgao@engin.umass.edu

S.V. Ukkusuri and K. Ozbay (eds.), *Advances in Dynamic Network Modeling in Complex Transportation Systems*, Complex Networks and Dynamic Systems 2,
DOI 10.1007/978-1-4614-6243-9_8,

of (and consequently, driver confidence in) the system. We also present brief empirical results from past simulation-based evaluations of ATIS, and conclude with recommendations for future research directions in order to further real-world ATIS deployment.

8.1 Introduction

With steadily growing levels of vehicle ownership and vehicle miles traveled, traveler information has been identified as a potential strategy towards managing travel demand, optimizing transportation networks, and better utilizing available capacity. Towards this goal, Advanced Traveler Information Systems (ATIS) have been conceptually designed with sophisticated travel behavior models and high-fidelity network performance models, made increasingly feasible through the rapid advances in computing power. Crucial components of this problem domain are the modeling of individual drivers' response to traveler information, and the development of algorithms that can generate accurate guidance of relevance to real-world trip-makers. In this chapter, we (a) undertake a detailed literature review to conceptually explore the multifarious dimensions of driver response to information and models of driver response, (b) discuss the key issues in information provision and the factors critical to the successful generation of such information, and (c) provide a flavor for the state of the art of research evaluating the potential benefits of ATIS.

The remainder of this chapter is structured as follows: Sect. 8.2 undertakes a broad discussion of the different types of route guidance as well as key ATIS design parameters that are likely to impact the effectiveness of the guidance system. Section 8.3 is a detailed review of the state of the art and practice dealing with the collection of driver response data, the estimation of mathematical models of said choices, and various real-world situations that might allow drivers the opportunity to respond to guidance. Section 8.4 illustrates the process of evaluating the effects and phenomena associated with the provision of route guidance, mostly with the use of traffic simulation modeling tools. Section 8.5 concludes with a summary of the state of the art, and suggests directions for research on appropriate data collection, driver response modeling, ATIS evaluation and testing.

8.2 Generation of Travel Information

This section discusses the generation of information intended to assist travelers make better travel choices. In other words, the principle underlying the provision of information involves the improvement of the decision making of individual travelers, rather than improvement of network performance overall.

In the discussion below, the term "message" will generally be used to refer to the content disseminated by a travel information provider to its users. However, the terms "information" and "guidance" will also be used interchangeably, and are not meant to imply a particular type (e.g., descriptive, prescriptive or other) of message content. Further, as a majority of the studies in the literature focus on the auto mode, we use the words "traveler" and "driver" interchangeably.

The discussion in this section considers in turn travel information attributes, guidance based on prevailing conditions, and guidance based on predicted conditions. The presentation of these two specific guidance approaches highlights the advantages and disadvantages of each, as well as technology and models typically used in their implementation.

8.2.1 Information Attributes

In this section we briefly describe some of the information attributes that can characterize the messages provided by different travel information systems. This discussion serves to prepare a more detailed discussion below of two specific information generation approaches.

Based on historical, prevailing, predicted conditions: Messages provided to a traveler can be based on typical conditions that have been observed in the past, on measurements and estimates of conditions prevailing at the time the request for information is made, or on predictions of what future conditions will be at the time the traveler is at a particular location.

Descriptive, prescriptive, mixed: The messages provided to a driver can describe network conditions, or recommend a particular action based on the network conditions, or both ("Take route XYZ, severe congestion ahead").

Network coverage: A travel information system might cover only the major transportation facilities in an area, or alternatively might attempt to take into account all facilities.

Information precision: Messages provided by an ATIS will typically provide time and/or delay information intentionally rounded to some level of precision, commensurate with the accuracy of the data sources and processing, to avoid saturating drivers with excessive details.

Information accuracy: Accuracy will be limited by the nature and extent of the data collected and used to generate travel information, and by the ways in which the information is processed.

Information latency: The information disseminated to travelers may not be completely up to date because of lags in data collection, time required for data processing, and compliance with a periodic information update cycle: for example the guidance generator might have a policy of updating messages every 15 min in non-incident conditions.

Other: A variety of other information attributes can be mentioned: push or pull access mode (information that is disseminated automatically to travelers or that must be specifically requested by them); message medium (graphical, text, spoken); the specific message format; etc.

8.2.2 Guidance Based on Prevailing Conditions

As noted above, one approach to guidance generation consists of providing information based on the conditions that prevail over the network at approximately the time that the information is generated (Chen and Mahmassani 1991). This information might simply report on those conditions (descriptive guidance), or might suggest travel options based on the conditions (prescriptive guidance), or both. In any case, the messages sent to travelers reflect the guidance system's estimate of conditions prevailing over the network at the time of guidance generation: there is no attempt to forecast future conditions, although the estimation of prevailing conditions might well combine real-time sensor measurements with historical information on typical traffic conditions.

8.2.2.1 Advantages of This Type of Guidance

With experience, a user of a transportation system is likely to develop a more-or-less accurate mental model that extrapolates from a "snapshot" of prevailing network conditions to an estimate of what the experience is likely to be on alternative travel options. (This model involves both the user's understanding of network dynamics and her interpretation and assessment of the messages received from the information system.) Thus, upon receiving messages about prevailing conditions, the user may be able to transform these into a valid basis for a travel choice.

Advanced traveler information systems that focus on prevailing conditions, while not simple, are arguably less complex than those that attempt to provide guidance messages based on forecast future conditions, as discussed below.

8.2.2.2 Disadvantages of This Type of Guidance

The mental models that users apply to extrapolate from prevailing to experienced conditions are, almost by definition, unlikely to be accurate in many situations of nonrecurrent congestion. (There are of course exceptions to this.) Thus, in these situations providing guidance on prevailing conditions may not be as helpful to travelers as it is under "normal" travel conditions.

8.2.2.3 Technology and Models for Guidance Generation

Guidance generation technology for systems based on prevailing conditions must collect real-time data from a variety of deployed traffic sensors, and process these data into an estimate of the prevailing network state that is sufficient to base the generation of guidance messages on. Depending on the nature of the system and its messages, the estimated network state might be characterized by attributes at the link level (travel times, delays, queue lengths) or route level, with disseminated messages reflecting these attributes (Ben-Akiva et al. 2001).

The sophistication with which the network state estimates are prepared can vary considerably, from local estimates of link speeds based, e.g., on inductive loop detectors, to processing systems that combine multiple types of real-time data (local speeds and volumes from loop detectors, point-to-point volumes and travel times from toll tags, video feeds from traffic management applications, etc.) with historical data in a network-level filtering approach.

8.2.3 *Guidance Based on Predicted Conditions*

An alternative approach to guidance generation involves predicting future network conditions, and basing the guidance messages on these predictions (Kaysi 1992; Papageorgiou et al. 2007). More specifically, predictive guidance attempts to reflect in its messages the conditions that are expected to prevail at network locations at the time the traveler will actually arrive there, rather than those that prevail at the time the guidance is disseminated.

8.2.3.1 Guidance Consistency

Guidance that is based on predicted conditions must confront a fundamental problem: when travelers receive the guidance and react to it, their reactions may invalidate the predictions on which the guidance was based, thus rendering the guidance irrelevant or worse. For example, guidance that is based on predictions of impending congestion on one of two parallel routes may cause travelers to shift to and congest the other route, leaving the original route relatively uncongested and leading to overall higher levels of delay (Ben-Akiva et al. 1991). Guidance is said to be *consistent* when it is based on predictions that are realized (within the limits of modeling accuracy) after the guidance is disseminated and travelers react to it (Bottom 2000). Note that this definition does not necessarily require that travelers comply with prescriptive guidance messages, only that their reactions, whatever they might be, are correctly anticipated when forecasting the conditions that were used to generate the guidance messages.

If relatively few travelers are affected by guidance, then their reactions to it are unlikely to have a significant effect on future network conditions, and consistency may not be an important issue. On the other hand, if many travelers react to guidance, and if their reactions affect network conditions in a significant way, then, as noted, failure to ensure predictive guidance consistency may result in ineffective or counterproductive travel guidance.

8.2.3.2 Advantages of This Type of Guidance

It seems intuitive that, other things being equal, guidance that accurately reflects the conditions at a network location at the time a traveler actually reaches there is likely to be a more relevant basis for decision making than guidance that simply reflects the prevailing or historical conditions at that location at the time the guidance is disseminated. In a dynamic network, such changes can be significant enough that guidance based on conditions prevailing at or around the time the messages are disseminated may be a seriously inaccurate reflection of the actual conditions that would be encountered during a trip on the various travel options available (Bottom et al. 1999; Rathi et al. 2008).

A number of currently active commercial travel information services provide guidance based on predictions of travel conditions (although it does not appear, based on publicly available information from these services, that guidance consistency is a concern). The commercial viability of these offerors suggests that the market recognizes a value in predictive guidance.

8.2.3.3 Disadvantages of This Type of Guidance

Predictive guidance is considerably more complex to generate than other forms of guidance. The discussion below of technologies and models used for this type of guidance explains some of the requirements. As a related matter, the accuracy of network flow and conditions forecasts will almost inevitably be lower than that of estimates of the prevailing network state, although it does not necessarily follow that the effectiveness and usefulness to travelers of guidance based on such forecasts will also be lower than guidance based on state estimates.

Although there is increasing recognition of the potential of predictive guidance, relatively few currently deployed systems provide such information. Those that do appear to be based on models that combine data on prevailing conditions with information on historical travel conditions and patterns, and current and near-term perturbations (weather, special events) in an extrapolation model, but do not consider the effect of the guidance itself on future network flows and conditions—in part, because the penetration of these systems in the market is still too low to produce significant impacts on the network.

This being the case, there is currently no empirical evidence on the effectiveness of consistent predictive guidance in practice. Assessments of such systems, or of alternative designs for them, must currently rely on simulation testing, as discussed in Sect. 8.4 below.

8.2.3.4 Technology and Models for Guidance Generation

Predictive guidance generation is based on estimates of prevailing conditions and so, as noted, requires the deployment of the same sorts of sensor and communications technology that would be required by a system for generating network-level guidance based on prevailing conditions (Kaysi et al. 1995).

Guidance generation systems that are not concerned with consistency would extrapolate future conditions in any of a variety of ways, typically taking account of historical traffic patterns and conditions (using a database that is continually updated with information from the real-time data feed and network state estimation), as well as available information on current and near-term perturbations (weather, work zones, incidents, special events, etc.) There is no explicit attempt to account for traveler reactions to the guidance, or for the impacts of those reactions on network conditions. Guidance messages (descriptive, prescriptive, or mixed) are generated from the predicted conditions.

Systems for generating consistent guidance require data processing capabilities that include (a) forecasting models capable of predicting the evolution over time of future network flows and conditions from their prevailing state following the dissemination of particular guidance messages; and (b) algorithms for generating guidance messages in a way that ensures mutual consistency between them and the condition and flow forecasts that they reflect (Kaysi et al. 1993). It is worth noting that travel forecasting models that are sensitive to the effects of guidance messages on traveler behavior are still the subject of research, and that algorithms for generating consistent guidance can be quite computationally intensive.

8.2.4 Summary

This section summarized attributes of traveler information and discussed the features and technological characteristics of information systems based on prevailing and on predicted travel conditions. We now present a review of literature relevant to drivers' response to various types of information, and under different choice circumstances.

8.3 Driver Response to Travel Information

In this section, we present a synthesis of empirical studies of individual traveler response to ATIS, including both revealed preference (RP) and stated preference (SP) studies.

A traveler makes decisions based on her knowledge of the available alternatives and their attributes, subject to time and cognitive capacity constraints. Such knowledge is obtained through both personal experience and exogenous information, and is usually limited by the inability to explore all available alternatives, the inherent uncertainties of external influences (e.g., bad weather, incidents), and the collective effects of other travelers' decisions in the system. ATIS provides exogenous information that could potentially expand travelers' knowledge of the decision environment, and improve decision quality.

Travel information can be characterized by its attributes (see Sect. 8.2 for a detailed discussion of information typology) along many dimensions. We focus on real-time (or dynamic) information that reflects travel conditions at or close to the decision time, in contrast to static information that usually describes the average decision environment over a relatively long period. However, as both dynamic and static information contribute to the formation of the traveler's knowledge and are usually not separable, we also have some discussions of static information when appropriate. Note that the time scale is by nature continuous and thus the notion of dynamic and static information is also relative.

In the remainder of this section, we start with a discussion of data collection methods for the study of traveler response to ATIS. Next we discuss travelers' valuation of and willingness to pay for real-time information. We then present evidence of information impacts on travelers' learning process. Next we turn to how drivers respond to ATIS, both in the short run and in the long run. Note that this is not intended as a comprehensive literature review; rather we focus on empirical studies since 2000, unless the last major development of a research area was before 2000. For pre-2000 studies on driver response to ATIS, the reader is referred to Lappin and Bottom (2001).

8.3.1 Data Collection Methods

Due to the relatively limited deployment of ATIS on a large enough scale, most collected data suitable for modeling come from SP surveys or laboratory experiments. RP questions are usually simple qualitative ones, and used for summary statistics rather than modeling. SP and RP data are combined in some situations to improve the model estimation efficiency (e.g., Polydoropoulou et al. 1996). Some researchers have made an effort to validate SP data against RP data (e.g., Khattak et al. 1994; Bonsall et al. 1997).

Notable data collection efforts and studies that have used the data include:

- RP and SP surveys from TravInfo field operation test (San Francisco): Khattak et al. (1999), Mehndiratta et al. (2000a), Yim et al. (2002), Khattak et al. (2003), and Wolinetz and Khattak (2004).
- RP and SP surveys from Puget Sound Regional Council's panel travel diary study: Mehndiratta et al. (2000b), Tsirimpa et al. (2007).
- Route choice simulator experiments developed by Bonsall and collaborators: Bonsall et al. (1997).

- Laboratory experiments in a three-route highway network developed by Mahmassani and collaborators: Mahmassani and Liu (1999), Srinivasan and Mahmassani (2003).
- Laboratory experiments in a relatively more realistic network developed by Abdel-Aty and collaborators: Abdel-Aty and Abdalla (2004, 2006).

8.3.2 Valuation of and Willingness to Pay for Information

Some common questions about ATIS are: What kind of information do users want? How much are users willing to pay for the information? These questions are related. A traveler responds to information as the information changes the decision environment. For example, it could suggest route alternatives unknown to the traveler or indicate a work zone downstream. Changed decisions suggest that the traveler perceived benefits from the change, otherwise she would not have made the change. In addition, information might provide benefits even though no changes are made, such as reduced anxiety or the validation of a choice already made. The valuation of and willingness to pay (WTP) for information are conceivably directly related to the perceived benefits of information, which in turn depend on a variety of information attributes. It is thus probably more informative to talk about the WTP for a certain information attribute.

For example, Mehndiratta et al. (2000a) estimated the WTP for information attributes such as update frequency, coverage, and customization using the 1998 survey carried out among a small sample of people who had called TravInfo in April 1997. More frequent updates were found to be the highest priority among the range of possible information enhancements explored in the survey, followed by an extension of coverage to include major arterials in addition to freeways. The authors also cautioned that the absolute values of WTP should not be used to inform pricing decisions due to the nature of the data. Molin and Timmermans (2006) collected SP data on a number of categories of attributes for transit information, including ticket pricing, transfers, real-time information, private transport, walking route, destination, on-board comfort and service, and planning options. Attitudinal questions about the importance of the attributes were asked and the WTP for each attribute category was estimated from trade-off questions. The rankings of information attributes from the two approaches were the same, with real-time information being the first and additional planning options the second.

Broadly speaking, three approaches have been used to estimate the WTP for information, almost all based on hypothetical situations (with the advent of for-fee ATIS markets, more RP studies could become available). A direct question was asked in Aultman-Hall et al. (2000) about the most a user would pay per call before she would stop using the traffic advisory telephone service (TATS) in the Greater Cincinnati and northern Kentucky area. Most WTP studies ask the user to make choices involving tradeoffs among systematically varied combinations of price and information attribute levels, and estimate a utility maximization model based on the data. The WTP can then be derived as the negative ratio of the marginal utilities of a certain information attribute and price. Khattak et al. (2003) used a slightly different approach with a similar idea, where the stated frequency of using TATS was regressed over combinations of information price and attributes, among others. The authors concluded that respondents were willing to accept a small per-call fee for customized travel information services based on the positive coefficient of the indicator variable of customized travel information priced at 25 cents/call. The third is the experimental economics approach, where human subjects participate in laboratory experiments and are granted an initial endowment to buy information, as is done in Denant-Boemont and Petiot (2003). All three methods potentially suffer from the deficiency of SP data, which is that due to the lack of actual commitment subjects tend to overstate their WTP. The third approach potentially has less of a problem as real monetary transactions are involved, though still in a hypothetical situation.

8.3.3 Information Impacts on Day-to-Day Learning

Exogenous information directly influences a traveler's perception of her travel decision environment; combined with personal experience it could alter the traveler's habits and knowledge of the transportation system. Polydoropoulou and Ben-Akiva (1999) described successive stages in the learning process of ATIS users including: awareness, consideration set formation, choice set formation, trial use, repeat use, and travel response. Conventional travel behavior analysis has been mainly done in a static framework, most notably using random utility models estimated on cross-sectional data. In this work, the level of knowledge is not explicitly considered and simplified assumptions of perfect information are made. In the study of traveler response to ATIS, however, information has to be explicitly considered. Thus it is only natural that we abandon the perfect information assumption and model the process of forming habit and establishing knowledge and how it is affected by different types of information. Mahmassani and Chang (1986), Iida et al. (1992), and Mahmassani and Liu (1999) are among the earlier efforts in studying empirically the day-to-day learning process. In the past decade there has been an increasing level of interest in this problem, from both transportation professionals and researchers in psychology and economics, e.g., Barron and Erev (2003), Srinivasan and Mahmassani (2003), Avineri and Prashker (2005), Avineri and Prashker (2006), Selten et al. (2007), Ben-Elia et al. (2008), Ziegelmeyer et al. (2008), Bogers (2009), Lu et al. (2011).

To the best of our knowledge, all empirical studies of day-to-day learning in the literature are based on laboratory experiments, which is not surprising given the difficulty in collecting long-term longitudinal disaggregate data. We also distinguish between two types of experimental settings: competitive and noncompetitive. In a noncompetitive experiment, the alternative attribute values (e.g., route travel times) are generated through an underlying sampling process by the experimenter, and are usually unknown to the subject and not affected by her decisions. In a competitive (game-like) experiment, many subjects make decisions simultaneously and the alternative attributes are determined collectively by a group of subjects' decisions. For example, in a route choice experiment, underlying link volume-delay functions are specified by the experimenter, but the actual travel times are determined by applying the number of subjects choosing a particular route as an input to the delay functions.

Some of the major research questions to be answered from a day-to-day learning study include: How fast do users learn (usually indicated by the declining number of switches among alternatives)? Do users eventually behave "rationally" as speculated in most static choice behavior or traffic equilibrium assignment models? All studies provide users with immediate feedback on chosen alternatives (usually routes in a route choice context) to emulate the reality that a traveler always knows her own travel time after the trip is finished. In addition, various types of pre-trip, post-trip, and en-route information were provided.

A subject switches among alternatives mainly for two purposes: exploration and exploitation. Intuitively more information in addition to personal experience (pre-trip, en-route, or post-trip) should reduce the extent of exploration as it helps the subject learn about the decision environment faster, as was shown in, e.g., Avineri and Prashker (2006), Ben-Elia et al. (2008), Lu et al. (2011), and Iida et al. (1992). However, post-trip information on unchosen alternatives could also generate regret and actually increase the switch propensity, as shown in, e.g., Srinivasan and Mahmassani (2003) and Lu et al. (2011). As for pre-trip or en-route real-time information on prevailing traffic conditions, it generally increases the switches at the point of information provision, as a subject is provided the opportunity to exploit better alternatives revealed by the information, especially in the case of an accident [see, e.g., Mahmassani and Liu (1999), Srinivasan and Mahmassani (2003), Bogers (2009), Lu et al. (2011)]. However, route switches upstream of the information provision node might actually decrease since the information reduces the uncertainty level of the network in general, as shown in Lu et al. (2011).

Information other than personal experience could potentially give the subject a better picture of the decision environment and make the "perfect information" assumption usually needed for utility-maximization choice models or equilibrium traffic assignment models more valid. However in reality our decision environment can never be completely deterministic, and risk attitude is an important factor in decision making. The impacts of information on subjects' revealed risk attitudes vary considerably, and do not necessarily result in a trend towards an "optimal choice" as predicted in a "perfect information" based model. In a noncompetitive environment, for example, Avineri and Prashker (2006) found that pre-trip static information on expected route travel times in a noncompetitive environment did not result in more subjects minimizing expected travel times; rather it increased the heterogeneity in risk attitudes, and actually decreased the propensity to choose the route with the less expected travel time, while Bogers (2009) found that the most elaborate information scenario, with both post-trip information on unchosen routes (often dubbed foregone payoff, or FP information) and en-route real-time information, produced the highest travel time savings. In a competitive environment, Ziegelmeyer et al. (2008) found that FP information did not significantly increase the subjects' ability to find optimal departure times, and Lu et al. (2011) found that FP information either increased or decreased the average travel time depending on whether en-route real-time information was provided, and suggested that information impacts in a competitive route choice environment depended to a large extent on the network structure, and more information could make things worse off as in a conventional Braess paradox.

8.3.4 Traveler Response to Real-Time Information

Individuals make travel and travel-related decisions at various time scales. Travel demand is derived from the need to participate in social and economic activities, such as going to work and visiting friends. The locations of these activities thus determine the origins and destinations of trips, and in particular residential and employment location choices that are usually made in the long run. The scheduling of daily activities includes travel as an integral part, and real-time travel information potentially will affect activity schedules. Mode choice is a relatively shorter-term decision, while departure time and route decisions are made in an even shorter term. Further down along the time scale, real-time adjustments of some of these decisions are made possible by ATIS, for example, route diversion to avoid an incident. These decisions are inter-related, and individual traveler response to ATIS potentially spans across all of them. We first give an overview of the likely responses in the short term, namely, temporary deviations from habitual travel and schedule decisions, and then discuss how ATIS could potentially influence long-term decisions.

8.3.4.1 Short-Term Response

Real-time traveler information provided by ATIS potentially reduces the level of uncertainty in the decision environment of a traveler, and could prompt a traveler to change her previous choices that were made without the updated information. Since the decisions have to be made in real-time, the traveler has to rely on options at her disposal at that time and her existing knowledge of the environment, if no exogenous guidance is available. Therefore the likelihood of a choice change depends on, among others, the availability of alternatives and traveler's familiarity with them.

Departure time choice. Many RP surveys show that departure time and route changes are among the most frequent responses to ATIS, see, for example, Aultman-Hall et al. (2000), Yim et al. (2002), and Martin et al. (2005). Furthermore, a Mitretek study (Shah et al. 2001) provides evidence from simulated yoked driver experiments involving the Washington DC area that pre-trip ATIS is more likely to produce departure time changes than route choice changes. These findings are not surprising, given that alternative departure times are always available, and a traveler is most likely familiar with the consequence of choosing an alternative departure time if she sticks to the same mode/route.

One of the major benefits of real-time information is to avoid schedule delay (defined as the difference between the preferred arrival time and the actual arrival time for a given commute) at the destination, especially for trips with a rigid arrival time requirement. A traveler usually has to reserve enough "buffer" time by leaving home early to account for the unexpected trip delay to ensure arriving on time. With pre-trip real-time information, the departure time can be adaptive to actual traffic conditions, and as a result, the traveler might be able to depart just in time in any situation to arrive on time. This flexibility in departure time enables better use of the previous "buffer" time at home, and/or reduce expected schedule delay (Shah et al. 2001), which is a major explanatory variable in departure time choice studies (Mahmassani and Liu 1999; Jou 2001).

Route choice. It is straightforward to change routes, if the traffic network is dense enough to provide viable alternative routes and the traveler is familiar with them. However the level of network knowledge potentially varies significantly among the traveler population (Ramming 2002), and it is conceivable that in some situations, a traveler does not feel like deviating from her familiar route(s) and stepping into the unknown.

Many surveys have the respondents' reported route choice changes as one of the major responses to ATIS, see, e.g., Aultman-Hall et al. (2000), Yim et al. (2002), Dai (2002), and Martin et al. (2005). Route choice is arguably the most researched area of traveler response to ATIS, with a particularly large body of research in binary route switching decision-making in response to VMS or radio information in real life, or more advanced hypothetical ATIS in SP surveys. However as we mentioned in the introductory paragraphs, there is a lack of generalizable model or method to predict that x% of travelers receiving information with an array of attributes a, b, c, ..., d will switch to route y.

Some researchers (Razo and Gao 2010, 2011; Tian et al. 2011) study the response even before the information is received for travelers with look-ahead abilities. A traveler does not need to commit to a particular route, but can decide later at a switching point based on then revealed traffic conditions, and pick the route with a lower travel time for the remaining trip. The option value of downstream real-time information thus could potentially make a collection of alternatives that share a common starting link more attractive than other links out of the same decision node. Therefore the travelers respond to the information upstream of the actual point where it is received. Empirical studies of the look-ahead behavior have been carried out with SP data only. On the other hand, there have been a large number of algorithmic studies, which generate optimal routing decisions depending on traffic conditions revealed by real-time information in a stochastic network (Hall 1986; Polychronopoulos and Tsitsiklis 1996; Pretolani 2000; Miller-Hooks and Mahmassani 2000; Waller and Ziliaskopoulos 2002; Gao and Chabini 2006; Gao and Huang 2011), and a recent and comprehensive literature review can be found in

Gao and Chabini (2006). These algorithmic studies are potentially useful for choice set generation in RP studies of such look-ahead behavior in real networks.

Such deficiency is largely due to the lack of detailed field observations of route choice changes. Earlier field observations focused on switchings at VMS locations, see, e.g., Emmerink et al. (1995b), Chatterjee and McDonald (2004). Richards and McDonald (2007) showed that it was difficult to capture a meaningful sample size of respondents passing an "active" VMS in a real-life incident scenario. Less than 1% of the commuter sample stated that they had diverted to an alternative route during the travel diary week as a result of VMS information, although this did correspond to 53% of those 45 drivers originally intending to travel past the incident location. More detailed tracking of individual route choices has been made possible by the advent of Global Positioning System (GPS) technologies. For example, Papinski et al. (2009) compared travelers' planned and actually taken routes (observed by GPS) and found that 20% of surveyed travelers switch routes for various reasons (one of them was ATIS). There have been a large number of GPS data collection efforts throughout the world, especially with the ever increasing popularity of GPS-enabled smart phones, and it is expected that better empirical evidence of route choice response to ATIS can result from these efforts and more quantitative conclusions can be drawn.

Mode choice. Generally real-time information is not found as the major drive of mode switch in many surveys, e.g., Henk and Kuhn (2000), Yim et al. (2002). Aultman-Hall et al. (2000) found that mode change happened only 5.7% of the time after the callers received information from a TATS, lowest among all inquired trip-making changes. There results are intuitive, as mode switch on the spot requires the immediate availability of an alternative mode and enough familiarity with both modes. An en-route traveler virtually cannot have these conditions met, and only pre-trip changes are possible. Conceivably a regular commuter with spare vehicles in an urban area with well-developed transit system might be able to change mode after receiving real-time information; however the frequency of such changes must be much lower than that of departure time or route changes. Dziekan and Kottenhoff (2007) studied the benefits of displaying real-time information at stops of public transport, and found that although passengers appreciated the benefits, empirical evidence of actual mode switching was still lacking.

There are a number of experimental studies of mode choice changes in response to ATIS, e.g., Abdel-Aty and Abdalla (2006) and Chorus et al. (2011). However, it is questionable how realistic the settings were, given that mode choice is mostly habitual and significantly affected by the context (e.g., the availability of a spare vehicle and/or a viable transit route).

Destination choice and trip cancellation. The change of destination and cancellation of the trip could happen only when the trip is discretionary, since generally there are no alternative work locations and not going to work is not an option, see, e.g., Henk and Kuhn (2000), Aultman-Hall et al. (2000), Yim et al. (2002), Martin et al. (2005).

For discretionary trips, the effects of ATIS on shopping trip destination choice was investigated in a set of Internet-based stated preference surveys by Kraan et al. (2000) and Mahmassani et al. (2003). In the survey, respondents were asked to make a (simulated) shopping trip from a central location in Austin, Texas to a major suburban mall. It was found that respondents who were less familiar with the Austin area were more likely to switch destination, but not route. We postulate that this result might be explained by the fact that individuals are more likely to have landmark knowledge than route knowledge, as defined in Freundschuh (1992) and reviewed in Ramming (2002).

Yim et al. (2002) summarized empirical evidence from several behavioral surveys conducted in the San Francisco Bay Area between 1995 and 1999, and found that noncommuting drivers changed their travel choices more than commuting drivers, including the occasional cancellation of trips, perhaps reflecting the flexibility inherent in nonwork trips.

Daily activity schedule adjustment. As travel is part of a daily activity schedule, the impact of real-time information can potentially extend beyond the trip-making itself and reach related activities.

With a more precise estimate of trip time, travelers could make minor changes to activities at both ends of the trip and derive more utility from the given amount of time. For example, if a bus rider knows that the bus is delayed from a real-time bus location information system, she could stay at office and make good use of the time (say, finish up a report) rather than wait at the bus stop (especially when the weather is bad). Similarly, if a driver learns of a severe incident downstream on her way to an appointment, she could notify relevant people to make other arrangements. Empirical evidence of minor activity adjustments at trip ends is generally connected to departure time choice change, see, e.g., Shah et al. (2001), Dziekan and Kottenhoff (2007).

A major change of activity schedules might be related to trip cancellation, destination change, and/or a major change of departure time (e.g., moving the trip from AM to PM, rather than from 7:00 am to 7:30 am). People schedule the activities that they need to accomplish in a day based in part on the time taken by each activity and the time required to travel between activities in different locations, thus a major change of the timing and/or locations of a trip usually accompanies rearrangements of other activities. Although conceivable, such empirical evidence is not as common, with only two studies in the literature to the best of our knowledge. Sun (2006) conducted a laboratory experiment, where activity rescheduling is explicitly explored as one of the responses to ATIS. Subjects were found to be willing to reschedule activities as well as change travel choices when notified of abnormal traffic conditions. Tsirimpa (2010) developed a model system that includes two models for travelers' response to ATIS estimated on RP and SP data from the Athens metropolitan area. The first model incorporates decisions (including modifying activity daily schedule) before or during the first primary tour, after acquiring traffic information from current traffic information systems. The second model concerns en route traffic information acquisition from ATIS and incorporates decisions, such as add activity, delete activity, and change the sequence of activities. The model estimation results indicate that travelers are willing to change their daily activity pattern, provided that they would be confident of the reliability of the information content.

Stress and anxiety reduction. Many surveys have found that trip-makers appreciate having travel information available even if they do not or cannot modify their trip-making behavior in any way because of it, see, e.g., Yim et al. (1999), Petrella and Lappin (2004). Lee (2000) has attempted to make the notion of travel stress relief more precise by arguing that the value of time spent in travel includes at least two distinct components: the opportunity cost of the activities foregone by traveling, and the disutility of the travel experience itself. Reduced stress and anxiety improve the travel experience, and likely reduce the value of time.

8.3.4.2 Long-Term Response

Habitual trip-making behavior adjustments. The responses discussed in the previous section are temporary deviations from regular travel choices and/or activity schedules. If real-time information is available for a long time period, a traveler's habitual choices might change as well. Real-time information provides a traveler with more flexibility in travel choices and activity schedules, which will result in more efficient use of time. The option value of information in route choice as discussed in the previous section on short-term responses could potentially be realized in the context of all travel choices and activity schedules, provided that alternative travel options or schedules exist and it is feasible to switch in real-time. In addition, even if a change is impossible due to the lack of alternatives, real-time information could have psychological benefits (e.g., reduced anxiety), which over time might also encourage a habitual preference to the alternatives with real-time information.

The study of habitual behavior requires longitudinal behavioral data at the individual level and is generally lacking. Among the limited RP evidence, Uchida et al. (1994) surveyed commuters in a three-route corridor in Osaka, Japan, following the installation of a variable message sign (VMS) network that provided predicted travel time information, and found that drivers showed a reluctance to switch away from their habitual route, but over time, roughly 40% of respondents reported that they had changed their habitual route as a result of the ATIS.

Historical information might be useful in influencing habitual choices. Bogers (2009) conducted a series of laboratory experiments on binary route choice where subjects were provided with post-trip information of the travel time on unchosen routes. Such information might affect habitual choice, as travelers were provided the opportunity to evaluate both their chosen and unchosen alternatives. Kenyon and Lyons (2003) worked with a number of focus groups in the UK on mode choice and found that mode choice was mostly automatic and habitual, based on subconscious perceptions of the viability and desirability of travel by modes other than the dominant mode. Their results suggested that presentation of a number of modal options for a journey in response to a single enquiry could challenge previous perceptions of the utility of non-car modes, overcoming habitual and psychological barriers to consideration of alternative modes.

Residence and/or employment location choice. The variety of changes brought about by ATIS in the trip-making context could lead people to reconsider their decisions regarding residential and/or employment location. As one example, if more predictable travel times became available from an ATIS, households could move farther away from job locations while still maintaining the same average commute time. Again, rearrangements in daily activity schedules brought about by ATIS could allow more time for outdoor activities, and incite households to take advantage of this by moving. Through these kinds of effect, ATIS could ultimately have an impact on urban form and structure. Boyce (1988), in an early paper, evoked this possibility.

However there is little empirical evidence, largely due to the limited deployment of ATIS. Argiolu (2008) (see also Argiolu et al. (2008)) studied how three ITS-related concepts affect office location choice, namely, automatic car lane, automatic bus lane, and people mover from park and ride, which potentially increased the accessibility of office locations. ATIS however was not explicitly studied. Rodriguez et al. (2011) found that in a laboratory experiment, providing multimodal accessibility information to people who were relocating enhanced the attractiveness of locations that support multiple travel modes.

8.3.5 Summary of Driver Response Literature

Our review of the driver response literature indicates that the available information tends to be highly specific to particular situations. In the data analysis and modeling efforts, an indicator variable is usually used to represent a specific information system, leading to results that are difficult to generalize. As mentioned before and also discussed in Sect. 8.2, there are a large number of information types, and presumably a general model should include an array of ATIS attributes that characterize the information type. To calibrate such a model, however, will require more deployments, more experience with deployed systems, and more research and analysis.

8.4 Evaluating the Effectiveness of ATIS

Owing to their complexity ATIS should be evaluated thoroughly before being implemented. ATIS evaluation can take place either through field tests or in a laboratory. Well-conducted field tests are more likely to reveal the ground truth reliably, since they are based on real-world conditions and actual drivers. However, such tests are generally very expensive and potentially disruptive, restricting the range and scope of strategies that might be evaluated. Public officials and administrators may also be wary of the potential for adverse effects resulting from some or all of the proposed strategies. Laboratory evaluation thus has many advantages, since it can take place in a controlled environment and identify issues before the ATIS is actually deployed. Traffic simulation tools provide unique opportunities for laboratory testing and evaluation, owing to their rich and flexible modeling capabilities that can capture time-varying travel demand, detailed driver behavior, demand–supply interactions, and network performance. Such flexibility nevertheless comes with the need to ensure that the models are accurate reflections of observed processes and phenomena. Simulation-based ATIS evaluation is therefore conditional on the availability of well-calibrated choice models that faithfully capture individual drivers' response to the range of situations observed in the field. This topic is naturally rather complex, relying on specialized data collection and model estimation methodologies. The reader is referred to Polydoropoulou (1997) and Polydoropoulou and Ben-Akiva (1999) for an in-depth review and treatment of this subject, one that continues to generate significant research interest.

8.4.1 Evaluating the Economic Benefits of ATIS

The economic benefits that an ATIS user derives from ATIS services are very closely tied to the user's response to ATIS: they are both aspects of the same internal evaluation and decision-making process. The discussion in Sect. 8.3 covered many aspects of ATIS user response, ranging from relatively simple behavioral responses like route switching to complex responses such as rearranging the daily activity schedule. This range exceeds the gamut of responses conventionally considered in transportation benefit evaluation exercises, which tend to focus on travel cost and time savings, and indicates that considerable care must be taken in thinking about and quantifying ATIS benefits.

For example, travelers' rearrangement of their daily activity schedules may lead to more rather than less time being spent in travel, as they are able to carry out more activities because of more precise trip planning information. If one were to ask such travelers if they were better off because of ATIS, they would reply affirmatively, even though they spend more time traveling: the benefits that they derive from the additional things they do more than offset the disutility of the time spent traveling.

Because understanding of individual traveler response to ATIS is still relatively primitive, research into the impacts of ATIS tends to focus on network-level effects, using simulation tools that incorporate assumptions about traveler response. The remainder of this section discusses current research in this area.

8.4.2 Evaluation Parameters

ATIS are characterized by a host of models, modeling assumptions, input parameters, and design criteria. ATIS evaluation should therefore consider the effects of the associated modeling, design, and operational choices.

The surveillance system is characterized by the location, type, and number of sensors in the field. Sensors can also have measurement errors. The detection and communication of incidents and their severity will also impact the effectiveness of the ATIS. Typically, there may be delays in obtaining and transmitting information about an incident, and uncertainty about its duration. The importance of such design characteristics of the surveillance system can thus be evaluated. The data and guidance communication interfaces between different ATIS entities can be modeled and their performance assessed. Some important aspects include latency in information transmission and errors in the information.

ATIS design parameters such as the estimation and prediction horizon lengths, the frequency of information updating, and the time resolution of the provided guidance influence ATIS effectiveness. The computational time required to generate predictive guidance depends on the size of the network and the available computational resources (this time also determines the minimum feasible time between information updates). The information generation module may also use a number of models to simulate demand aspects of the transportation system (such as route choice and departure times) and network performance (such as queue formation and dissipation). These models can be imperfect reflections of the ground truth, and the associated modeling errors must be studied and their impact on the effectiveness of the generated guidance assessed.

8.4.3 Evaluation Frameworks

Different frameworks for the evaluation of ATIS are described in Balakrishna et al. (2005). ATIS can be evaluated either in a (simulation) laboratory or in the real world. A real-world test uses real-time traffic data directly from the surveillance system. In contrast, a laboratory test replaces online communications with either an archived data set of past surveillance observations, or a simulator of the real world surveillance system.

The evaluation can be either without guidance dissemination (open loop) or closed loop. When guidance is not disseminated to any drivers, the evaluation compares estimated and predicted network performance measures (which form the basis for the generation of guidance) with corresponding real-world traffic measurements. When some or all drivers have access to the information generated by the ATIS, the evaluation provides a feedback mechanism for informing equipped drivers of the current guidance strategy and for evaluating the resulting driver response to the disseminated information.

8.4.3.1 Experimental Design

The previous classification results in four possible evaluation approaches:

1. **Laboratory evaluation without ATIS.** Archived surveillance data are used to estimate current network state. State predictions for future time periods are evaluated through comparisons with the corresponding observations in the archived data set. A simulation laboratory (such as a microscopic traffic simulator) may also be used to compare the predicted network state with the ground truth to establish the benefits of prediction. This step is critical in assessing the appropriateness of the models within the guidance module and could provide valuable feedback for model refinement. Such validation tests with the DynaMIT system have been conducted on networks in Irvine, California (Balakrishna et al. 2004), lower Westchester County, New York (Antoniou et al. 2004), and Hampton Roads, Virginia. Similar studies with DYNASMART have focused on networks in Fort Worth, Texas (Huynh et al. 2003) and Knoxville, Tennessee. Several papers have also addressed the impacts and benefits of variable message signs, high occupancy tolls, and hybrid DTA approaches on network performance (Chiu and Mahmassani 2002; Doan et al. 1999; Murray et al. 2001).
2. **Laboratory evaluation with ATIS.** A simulation laboratory replaces the real world. Traffic data from the simulated surveillance system are transmitted in real time to the guidance generation system. Prediction-based guidance is delivered to equipped drivers in the microscopic simulator, and network performance measures are computed to ascertain the effectiveness of the guidance. This test helps validate the models that capture route choice and response to information but requires that the simulation laboratory first be calibrated against real data. Laboratory evaluations with the dissemination of guidance are limited. Yang et al. (2000) report on MITSIMLab for the evaluation of ATIS. Mahmassani and Jayakrishnan (1991), Stephanedes et al. (1989), and Jayakrishnan et al. (1994) study simulation-based evaluations of route diversion. Jayakrishnan et al. (2001) provide a brief qualitative discussion on the possibility of coupling the mesoscopic simulator DYNASMART with the microscopic program PARAMICS. In their framework, DYNASMART uses a simplified network abstracted from the PARAMICS model to generate paths. These paths are subsequently input to PARAMICS. Guidance about these paths may be used by PARAMICS drivers, thus allowing for the testing of ATIS.
3. **Real-world evaluations without ATIS.** Traffic data received in real-time from the actual surveillance system in the field are used to perform state estimation, state prediction, and guidance generation. The predicted network state is compared with actual sensor measurements as they become available, to validate the congestion reduction capability of the guidance generation system. This step also involves the testing of the communication interfaces between the surveillance system and the guidance module. Real-world evaluations of DTA-based prediction models, without actually disseminating guidance, have been documented on large-scale networks: DynaMIT has been applied in

lower Westchester County, New York; downtown Los Angeles, California; and Hampton Roads, Virginia. DYNASMART has been tried in Houston, Texas. These tests are steps towards real-world, closed-loop ATIS operations in a traffic management center (discussed next).

4. **Real-world evaluations with ATIS.** The feedback loop is closed with real equipped drivers on the network receiving the generated guidance. Limited field tests with guidance dissemination have been reported. Smith and Perez (1992) present the evaluation of INFORM, a traffic management system designed for Long Island, New York. Similar evaluations of ATIS include TravInfo (Miller 1998) and ADVANCE (Saricks et al. 1997).

Few papers discuss detailed simulation tests designed to objectively assess the quality of route guidance. Some such papers are reviewed next, and their primary results and conclusions summarized.

8.4.3.2 Select Results

Kaysi et al. (1993; 1995) considered ATIS design and identified the importance to system effectiveness not only of accurate real-time traffic information, but also of a predictive capability able to forecast the effects on traffic conditions of drivers' reactions to ATIS messages that they receive. Simulation methods were used to investigate the impact on congestion reduction of system design parameters such as message update frequency and routing strategy. They recognized the phenomenon of overreaction, in which a significant number of motorists respond similarly to ATIS messages, and so exacerbate or displace congestion, and showed that it could be reduced through frequent updates or strategies that explicitly attempt to spread traffic over multiple routes.

Emmerink et al. (1995a) provide an early attempt to quantify the network impacts of ATIS under nonrecurrent congestion. The paper provides a qualitative discussion on the factors and issues, including expectations about drivers' psychological behavior that might determine the effectiveness of ATIS. Using a network with a single OD pair and several routes, the authors empirically analyze the effects of ATIS market penetration and information update frequency. The findings illustrate the concept of over-reaction: as more drivers begin to use the ATIS, network benefits increase before tapering off or even worsening (Fig. 8.1).

Figure 8.1 further indicates that over-reaction at higher ATIS usage levels may be minimized or even eliminated by the providing more frequent information updates, so that drivers may adapt their route choice decisions more closely with the evolving network conditions. A maximum travel time improvement of about 25% is reported.

Levinson (1999) reviews several studies that assess the impact of ATIS on network travel times. His discussion reveals that travel time savings have been found to vary over a wide range, between 2.7% and 55%. Wunderlich et al. (2001) analyze a survey for the Washington DC area, concluding that ATIS improved on-time reliability without a significant reduction of in-vehicle travel time.

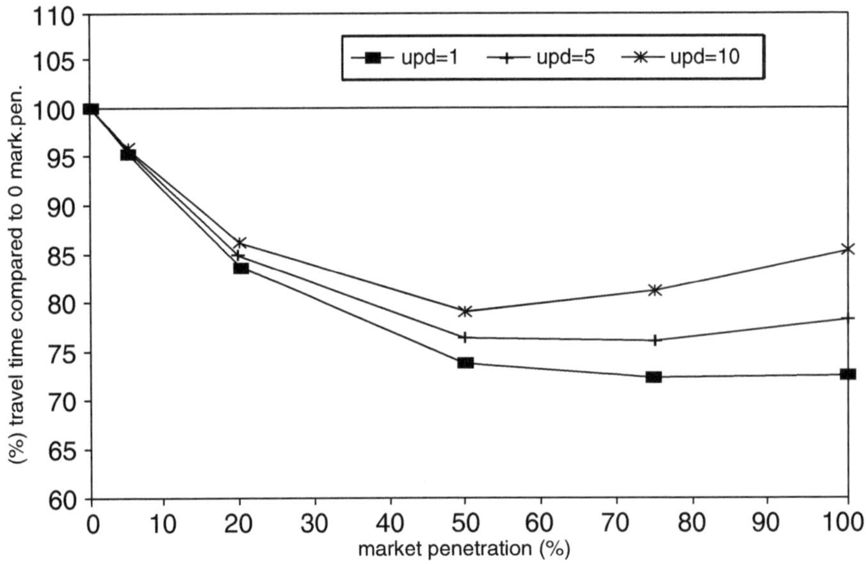

Fig. 8.1 ATIS impacts by usage level and update frequency (Emmerink et al. 1995a)

Bottom et al. (1999) discuss consistent, anticipatory route guidance in the framework of fixed-point problems and present a detailed analysis of different solution algorithms to tackle the same. The paper evaluates various guidance generation parameters by using a microscopic traffic simulator to reflect real-world drivers and their reactions to guidance generated in a rolling horizon. The study confirmed that shorter guidance recomputation intervals (the time between two successive horizons) led to higher speeds and lower travel times. A similar effect was observed when more frequent updates were initiated within each guidance recomputation interval. It was found, however, that the overall benefits of guidance dissemination reduced on either side of a horizon length of 30 min. The authors hypothesize that very short prediction horizons are unable to account for demand patterns that change in the future, while poorer quality forecasts further in the future also somehow reduce the overall quality of the generated guidance.

Balakrishna et al. (2005) analyze the impact of several key ATIS parameters by using an evaluation framework similar to the one in Bottom et al. (1999). The paper focuses on three aspects: the frequency with which new guidance is disseminated to drivers, the market penetration of ATIS services, and the extent of errors in predicting future demand. DynaMIT, a simulator designed to generate consistent, anticipatory route guidance, was applied to an incident scenario on a freeway network from Boston, Massachusetts. A greater update frequency (or a shorter update interval) was found to significantly decrease travel times (Fig. 8.2(a)). The results also indicate that the marginal benefit of a very high update frequency may not justify the additional computational resources required to support it. The choice

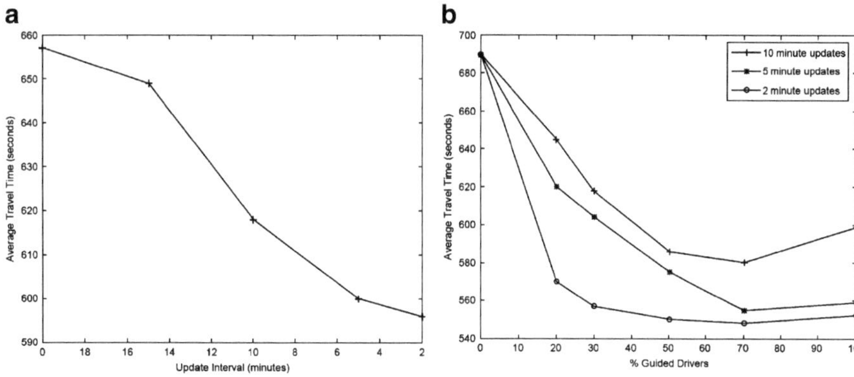

Fig. 8.2 (a) Impact of update frequency (b) Impact of guidance penetration

of update frequency may thus depend on other factors, such as the computational power available or the desired level of network performance.

Figure 8.2(b) illustrates the effect of market penetration rates on the effectiveness of ATIS. While the average travel time generally decreases with greater ATIS usage, there is evidence of slight over-reaction with high market penetration. However, over-reaction may be reduced or even eliminated through more frequent guidance dissemination. Differences between these results and those from previous studies may reflect the impact of the network, demand characteristics, assumptions regarding ATIS design, and the overall structure of the evaluation methodology (which underlines the importance of establishing a simulation-based laboratory for detailed testing).

Prediction-based ATIS must assess the short-term evolution of future demand in order to predict the evolution of network congestion. Demand prediction models adjust historical demand patterns according to recent sensor measurements, usually a two-step process involving state estimation (to match immediately previous time intervals to real-time sensor data) followed by demand prediction that extrapolates these deviations into the future. Balakrishna et al. (2005) study the impact of demand prediction errors on guidance quality. Figure 8.3 shows that systematically over- or underpredicting network demand results in a deterioration in the benefit of ATIS. Further, it is better to overpredict the demand rather than estimate fewer vehicles for the short term future.

Florian et al. (2006) evaluate the impact of ATIS using an incident scenario on a simple network with two route alternatives. The simulation experiments focused on varying levels of guidance penetration. The paper reports that the mean travel times improve only until about 50% guidance penetration. Mean travel times are mostly flat beyond this point, with only a slight degradation in performance. An analysis of the distribution of mean travel times between guided and unguided drivers (Fig. 8.4) indicates that the mean travel time for guided drivers approaches that for unguided drivers at higher levels of ATIS market penetration. Further, DynaMIT guidance is

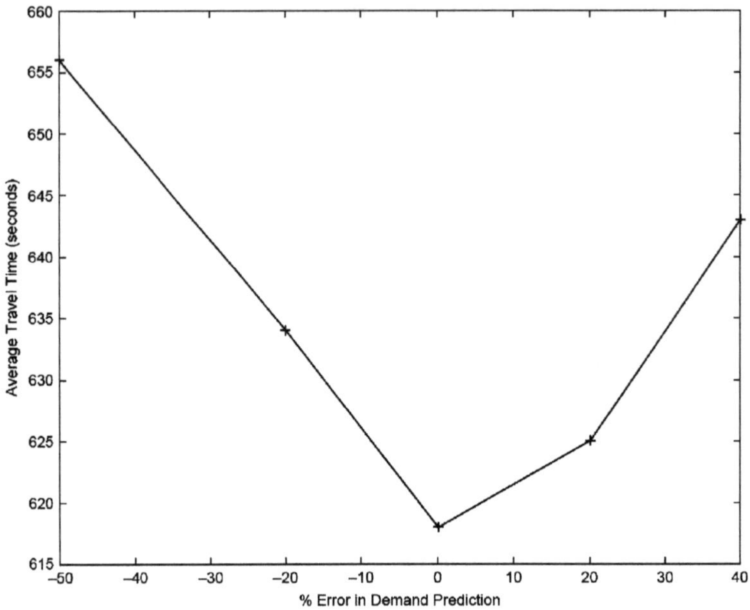

Fig. 8.3 Impact of demand prediction error

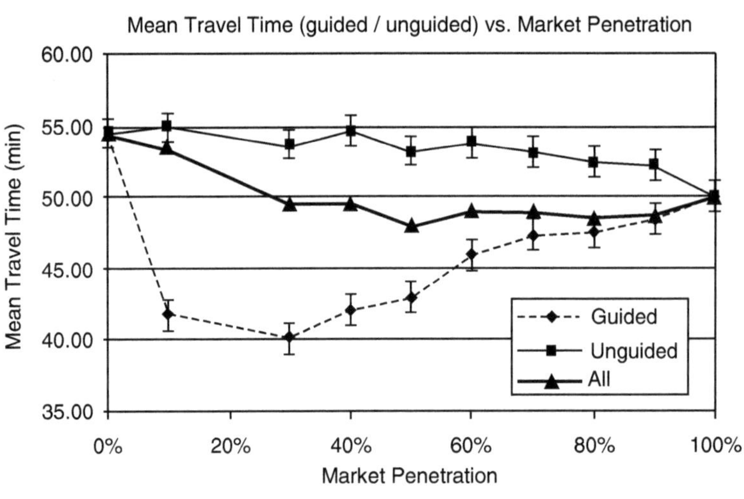

Fig. 8.4 Guided vs. unguided mean travel time

beneficial to both guided and unguided drivers. However, as more guided drivers switch to the alternative path, they impose increasing congestion costs on each other, and must therefore share the benefits. Travel time reliability for guided drivers remained higher (with lower standard deviation) than for unguided drivers.

Fig. 8.5 Lower Westchester County network and incidents [source: Google]

In recent work, Paz and Peeta (2009a; 2009c) propose a route guidance generation method that integrates both the information provider's network control objectives and its real-time estimation of driver response behavior. The model of traveler response to guidance has a fuzzy multinomial logit structure, where the systematic utility component is obtained using aggregate if–then rules. In Paz and Peeta (2009b), the authors describe an online approach to calibrate and refine the behavior model based on discrepancies between the dynamic actual and estimated system states. They use simulation experiments of the Borman expressway network in Indiana to test the effectiveness of the proposed approach.

A recent application (Rathi et al. 2008) involves the simulation-based evaluation of predictive route guidance disseminated through variable message signs (VMS) on the Lower Westchester County network in New York state. The paper attempts to reduce network travel times when incidents disrupt highly congested commuter traffic traveling north to south on freeways and parkways. Figure 8.5 shows the study network and the locations of the incidents that were considered. The two 1-hour incidents were modeled as separate scenarios to isolate the impacts on the surrounding OD pairs. VMS locations (shown as triangles in the figure) were selected to provide drivers with route switching options upstream of the incidents.

The evaluation was closed-loop—DynaMIT's travel time guidance generated with a prediction horizon of 30 min was disseminated to those drivers in

Table 8.1 Lower Westchester County: incident scenarios and statistics

Scenario	Mean TT (sec)	Vehicle Hrs	Vehicle Hrs Saved
No Incident	956.4	53,533	–
Incident	1532.6	85,852	–
VMS - 1 Pred Iter	1261.8	70,673	15,179
VMS - 2 Pred Iters	1266.2	70,920	14,932
VMS - 3 Pred Iters	1239.4	69,419	16,433
VMS - 4 Pred Iters	1246.9	69,841	16,011
VMS - 5 Pred Iters	1239.9	69,451	16,401

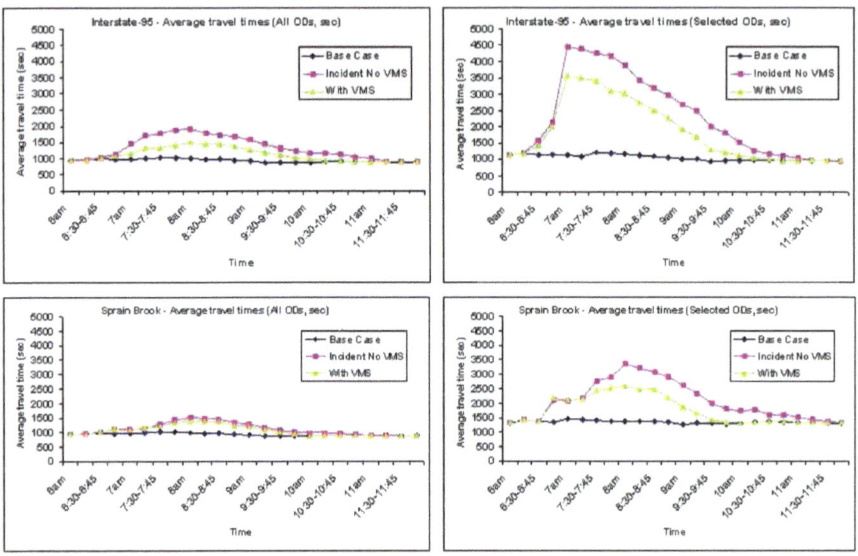

Fig. 8.6 Lower Westchester County: results by departure time

MITSIMLab who had access to en-route information. The use of VMS to disseminate guidance had the expected effect of reducing the impact of the incident, indicated by Table 8.1. Performing three prediction iterations was found to yield the maximum savings in travel time and vehicle hours traveled.

A comparison of trip frequencies by travel time indicated that a significant number of vehicles with lower travel times had shifted to a higher travel time range due to the incident. VMS guidance caused a substantial drop in the number of trips with large travel times. Predictive route guidance significantly decreased the average travel times for those departing in time intervals most affected by the incident (Fig. 8.6). The Lower Westchester County case study thus reinforces the congestion mitigating potential of route guidance on congested, real networks.

8.4.4 Summary of Results

Given that real-world ATIS implementations are bound to be expensive, it is not surprising that the laboratory evaluation of ATIS has drawn considerable research interest. While several papers have focused on determining the network impacts of ATIS, the quantitative benefits are hard to generalize. Contributing factors include case study situations that vary in the ATIS market penetration, network size and topology and demand characteristics, and the type of guidance disseminated. However, the results largely seem to agree on the qualitative aspects of the said benefits, and advocate for predictive information at a suitably high update frequency. Given the speculative nature of the studies thus far encountered, a natural direction for the future is in the rigorous and objective testing of ATIS on real-world networks with a strong focus on collecting detailed behavioral response data to support the various hypotheses, rather than relying on often uncalibrated simulation tools to fill this current gap.

8.5 Conclusion

This chapter undertook a detailed review of three key aspects in the modeling, design, evaluation, and deployment of ATIS: (a) models of drivers' behavioral response to ATIS under diverse contexts, (b) ATIS design concepts and parameters, and (c) the mostly simulation-based evaluation of ATIS properties and network performance. Our review illustrates that ATIS research methodologies are largely based on limited field data supported by an assortment of simulation models and outputs. Key limitations of the current state of the art are the lack of substantial RP data about drivers' real-world responses to ATIS, and the rudimentary nature of the choice models embedded in most simulation tools. These two drawbacks in effect reinforce each other, since more complex behavioral models require RP data for calibration and validation.

The future, however, looks promising. Technological advances are lowering the traditional barriers to collecting large volumes of relevant information at the level of individual users of the transportation infrastructure. Rapid adoption of instrumented devices such as GPS navigation units, bluetooth devices, and smart phones can help boost sample sizes as well as the accuracy with which driver choices are reported back to the modelers and planners. It is hoped that such rich RP datasets will facilitate the move from simplistic, hypothetical driver response models to realistic, well-calibrated, and validated models that may then be integrated into network simulation tools for practical analyses that reflect the ground reality. Such models may include an array of ATIS attributes, each with multiple levels, introducing unprecedented sensitivity and fidelity into the driver response models. A comparison of subsequent results similar to those reviewed in this chapter should be highly interesting and instructive.

References

Abdel-Aty M, Abdalla MF. Modeling drivers' diversion from normal routes under ATIS using generalized estimating equations and binomial probit link function. Transportation 2004;31:327–348.

Abdel-Aty M, Abdalla MF. Examination of multiple mode/route-choice paradigms under ATIS. IEEE Trans Intell Transport Syst. 2006;7(3):332–348.

Antoniou C, Koutsopoulos HN, Ben-Akiva M, Chauhan AS, Cusack M. Development and evaluation of traffic diversion strategies. In: Working paper. 2004.

Argiolu R. Office location choice behaviour and Intelligent Transport Systems. PhD thesis, Radboud University Nijmegen, 2008.

Argiolu R, van der Heijden R, Bos I. Intelligent transport systems and preferences for office locations. Environ Plann. 2008;40:1744–1759.

Aultman-Hall L, Bowling S, Asher JC. ARTIMIS telephone travel information service: Current use patterns and user satisfaction. Transport Res Rec. 2000;1739;9–14.

Avineri E, Prashker JN. Sensitivity to travel time variability: Travelers' learning perspective. Transport Res C 2005;13:157–183.

Avineri E, Prashker JN. The impact of travel time information on travelers' learning under uncertainty. Transportation 2006;33:393–408.

Balakrishna R, Koutsopoulos HN, Ben-Akiva M. Evaluation of the estimation and prediction capability of a dynamic traffic assignment system. In: Mahmassani HS, editor. 16th international symposium on transportation and traffic theory. London: Elsevier; 2004.

Balakrishna R, Koutsopoulos HN, Ben-Akiva M, Fernandez-Ruiz BM, Mehta M. Simulation-based evaluation of advanced traveler information systems. Transport Res Rec. 2005;1910:90–98.

Barron G, Erev I. Small feedback-based decisions and their limited correspondence to description-based decisions. J Behav Dec Making 2003;16:215–233.

Ben-Akiva M, De Palma A, Kaysi I. Dynamic network models and driver information systems. Transport Res A 1991;25A(5):251–266.

Ben-Akiva M, Bottom J, Ramming MS. Route guidance and information systems. J Syst Contr Eng. 2001;215(14):317–324.

Ben-Elia E, Erev I, Shiftan Y. The combined effect of information and experience on drivers' route-choice behavior. Transportation 2008;35:165–177.

Bogers EAI. Traffic information and learning in day-to-day route choice. PhD thesis, Delft University of Technology, 2009.

Bonsall P, Firmin P, Anderson M, Palmer I, Balmforth P. Validating the results of a route choice simulator. Transport Res C 1997;5:371–387.

Bottom J. Consistent anticipatory route guidance. PhD thesis, Department of Civil and Environmental Engineering, Massachusetts Institute of Technology, September 2000.

Bottom J, Ben-Akiva M, Bierlaire M, Chabini I, Koutsopoulos HN, Yang Q. Investigation of route guidance generation issues by simulation with DynaMIT. In: Avishai Ceder, editor. 14th international symposium on transportation and traffic theory, pp. 577–600. Oxford: Pergamon; 1999.

Boyce DE. Route guidance systems for improving urban travel and location choices. Transport Res A 1988;22:275–281.

Chatterjee K, McDonald M. Effectiveness of using variable message signs to disseminate dynamic traffic information: Evidence from field trials in European cities. Transport Rev. 2004;24(5):559–585.

Chen PS-T, Mahmassani HS. Reliability of real-time information systems for route choice decisions in a congested traffic network: Some simulation experiments. In: Proceedings of the vehicle navigation and information systems conference, pages 849–856, 1991.

Chiu Y-C, Mahmassani HS. Hybrid real-time dynamic traffic assignment approach for robust network performance. Transport Res Rec. 2002;89–97.

Chorus CG, Walker JL, Ben-Akiva ME. The value of travel information: A search-theoretic approach. J Intell Transport Syst. 2011;14:154–165.
Dai H. An agent-based approach to modelling driver route choice behaviour under the influence of real-time information. Transport Res C 2002;10:331–349.
Denant-Boemont L, Petiot R. Information value and sequential decision-making in a transport settting: An experimental study. Transport Res B 2003;37:365–386.
Doan DL, Ziliaskopoulos A, Mahmassani H. On-line monitoring system for real-time traffic management applications. Transport Res Rec. 1999;1678:142–149.
Dziekan K, Kottenhoff K. Dynamic at-stop real-time information displays for public transport: Effects on customers. Transport Res A 2007;41:489–501.
Emmerink RHM, Axhausen KW, Nijkamp P, Rietveld P. The potential of information provision in a simulated road transport network with non-recurrent congestion. Transport Res C 1995a;3(5):293–309.
Emmerink RHM, Nijkamp P, Rietveld P, van Ommeren JN. Variable message signs and radio traffic information: An integrated empirical analysis of drivers' route choice behavior. Transport Res A 1995b;30(2):135–153.
Florian D, Balakrishna R, Ben-Akiva M, Wen Y. Evaluation of on-line dynamic traffic assignment using micro-simulation. In: Proc. second international symposium of transport simulation, 2006.
Freundschuh SM. Is there a relationship between spatial cognition and environmental patterns?, vol 639 of *Lecture Notes in Computer Science*. New York: Springer; 1992.
Gao S, Chabini I. Optimal routing policy problems in stochastic time-dependent networks. Transport Res B 2006;40(2):93–122.
Gao S, Huang H. Real-time traveler information for optimal adaptive routing in stochastic time-dependent networks. Transport Res C 2011. doi:10.1016/j.trc.2011.09.007.
Hall RW. The fastest path through a network with random time-dependent travel times. Transport Sci. 1986;20(3):91–101.
Henk RH, Kuhn BT. Assessing the effectiveness of advanced traveler information on older driver travel behavior and mode choice. Technical report, Texas Transportation Institute, 2000.
Huynh N, Chiu Y-C, Mahmassani HS. Finding near-optimal locations for variable message signs for real-time network traffic management. Transport Res Rec. 2003;1856:34–53.
Iida Y, Akiyama T, Uchida T. Experimental analysis of dynamic route choice behavior. Transport Res B 1992;26(1):17–32.
Jayakrishnan R, Mahmassani HS, Hu Y. An evaluation tool for advanced traffic information and management systems in urban networks. Transport Res C 1994;2(3):129–147.
Jayakrishnan R, Oh J-S, Sahraoui A-E-K. Calibration and path dynamics issues in microscopic simulation for advanced traffic management and information systems. Transport Res Rec. 2001;1771:9–17.
Jou R-C. Modeling the impact of pre-trip information on commuter departure time and route choice. Transport Res B 2001;35:887–902.
Kaysi I, Ben-Akiva M, Koutsopoulos HN. Integrated approach to vehicle routing and congestion prediction for real-time driver guidance. Transport Res Rec. 1993;1408:66–74.
Kaysi I, Ben-Akiva M, de Palma A. Design aspects of advanced traveler information systems. In: Gartner NH, Improta G, editors. Urban traffic networks: dynamic flow modeling and control. New York: Springer; 1995. pp. 59–81.
Kaysi IA. Framework and models for the provision of real-time driver information. PhD thesis, Department of Civil Engineering, Massachusetts Institute of Technology, February 1992.
Kenyon S, Lyons G. The value of integrated multimodal traveller information and its potential contribution to modal change. Transport Res F 2003;6:1–21.
Khattak A, Yim Y, Stalker L. Does travel information influence commuter and noncommuter behavior? Results from the San Francisco Bay Area TravInfo project. Transport Res Rec. 1999;1694:48–58.
Khattak A, Yim Y, Prokopy LS. Willingness to pay for travel information. Transport Res C 2003;11:137–139.

Khattak AJ, Kanafani A, Le Colletter E. Stated and reported route diversion behavior: implications of benefits of advanced traveler information systems. Transport Res Rec. 1994;1464:28–35.

Kraan M, Mahmassani HS, Huynh N. Interactive survey approach to study traveler responses to ATIS for shopping trips. In: 79th TRB Annual Meeting CD-ROM, 2000.

Lappin J, Bottom J. Understanding and predicting traveler response to information: A literature review. Technical report, Prepared for USDOT, FHWA, 2001.

Lee DB. Benefit-cost evaluation of traveler information: Seattle's Washington State Department of Transportation website. Transport Res Rec. 2000;1739:25–34.

Levinson D. The value of advanced traveler information systems for route choice. Transport Res C 1999;11.

Lu, X, Gao, S and Ben-Elia, E. Information Impacts on Route Choice and Learning Behavior in a Congested Network: An Experimental Approach. Transport Res Rec. 2011; 2243:89–98.

Mahmassani H, Chang GL. Experiments with departure time choice dynamics of urban commuters. Transport Res B 1986;20(4):297–320.

Mahmassani H, Liu YH. Dynamics of commuting decision behaviour under advanced traveller information systems. Transport Res C 1999;7:91–107.

Mahmassani HS, Jayakrishnan R. System performance and user response under real-time information in a congested traffic corridor. Transport Res A 1991;25(5):293–307.

Mahmassani HS, Huynh NN, Srinivasan K, Kraan M. Tripmaker choice behavior for shopping trips under real-time information: Model formulation and results of stated-preference internet-based interactive experiments. J Retailing Consum Serv. 2003;10:311–321.

Martin PT, Lahon D, Cook K, Stevanovic A. Traveler information systems: Evaluation of UDOT's ATIS techologies. Technical report, University of Utah, 2005.

Mehndiratta SR, Kemp M, Pierce S, Lappin J. Users of a regional telephone-based traveler information system - a study of TravInfo users in the San Francisco Bay Area. Transportation 2000a;27:391–417.

Mehndiratta SR, Kemp MA, Lappin JE, Nierenberg E. Likely users of advanced traveler information systems: Evidence from the Seattle region. Transport Res Rec. 2000b;1739:15–24.

Miller M. TravInfo evaluation: Traveler information center study. Technical Report UCB-ITS-PWP-98-21, University of California, Berkeley, 1998.

Miller-Hooks ED, Mahmassani HS. Least expected time paths in stochastic, time-varying transportation networks. Transport Sci. 2000;34(2):198–215.

Molin EJE, Timmermans HJP. Traveler expectations and willingness-to-pay for web-enabled public transport information services. Transport Res C 2006;14:57–67.

Murray PM, Mahmassani HS, Abdelghany KF. Methodology for assessing high-occupancy toll-lane usage and network performance. Transport Res Rec. 2001;1765:8–15.

Papageorgiou M, Ben-Akiva M, Bottom J, Bovy PHL, Hoogendoorn SP, Hounsell NB, Kotsialos A, McDonald M. Its and traffic management. In: Barnhart C, Laporte G, editors. Transportation, vol 14 of Handbooks in operations research and management science. London: Elsevier; 2007. pp. 715–774.

Papinski D, Scott DM, Doherty ST. Exploring the route choice decision-making process: A comparison of planned and observed routes obtained using person-based GPS. Transport Res F 2009;12:347–358.

Paz A, Peeta S. Information-based network control strategies consistent with estimated driver behavior. Transport Res B 2009a;43(1):73–96.

Paz A, Peeta S. On-line calibration of behavior parameters for behavior-consistent route guidance. Transport Res B 2009b;43(4):403–421.

Paz A, Peeta S. Behavior-consistent real-time traffic routing under information provision. Transport Res C 2009c;17(6):642–661.

Petrella M, Lappin J. Comparative analysis of customer response to online traffic information in two cities: Los Angeles, California and Seattle, Washington. Transport Res Rec. 2004;1886:10–17.

Polychronopoulos G, Tsitsiklis JN. Stochastic shortest path problems with recourse. Networks 1996;27:133–143.

Polydoropoulou A. Modeling user response to advanced traveler information systems (ATIS). PhD thesis, Massachusetts Institute of Technology, Cambridge, MA, 1997.
Polydoropoulou A, Ben-Akiva M. The effect of advanced traveller information systems (ATIS) on travellers' behaviour, pages 317–352. Ashgate, 1999.
Polydoropoulou A, Ben-Akiva M, Khattak A, Lauprete G. Modeling revealed and stated en-route travel response to advanced traveler information systems. Transport Res Rec. 1996;1537:38–45.
Pretolani D. A directed hyperpath model for random time dependent shortest paths. Eur J Oper Res. 2000;123:315–324.
Ramming MS. Network knowledge and route choice. PhD thesis, Massachusetts Institute of Technology, 2002.
Rathi V, Antoniou C, Wen Y, Ben-Akiva M, Cusack M. Assessment of the impact of dynamic prediction-based route guidance using a simulation-based, closed-loop framework. In: 87th annual meeting of the Transportation Research Board, 2008.
Razo M, Gao S. Strategic thinking and risk attitudes in route choice: A stated preference approach. Transport Res Rec. 2010;2085:136–143.
Razo M, Gao S. A rank-dependent expected utility model for strategic route choice with stated preference data. Transport Res C 2011. doi:10.1016/j.trc.2011.08.009.
Richards A, McDonald M. Questionnaire surveys to evaluate user response to variable message signs in an urban network. IET Intell Transport Syst. 2007;1(3):177–185.
Rodriguez DA, Levine J, Agrawal AW, Song J. Can information promote transportation-friendly location decisions? A simulation experiment. J Transport Geogr. 2011;19:304–312.
Saricks CL, Schofer JL, Soot S, Belella PA. Evaluating effectiveness of real-time advanced traveler information systems using a small test vehicle fleet. Transport Res Rec. 1997;1588:41–48.
Selten R, Chmura T, Pitz T, Kube S, Schreckenberg M. Commuters route choice behaviour. Games Econ Behav. 2007;58(2):394–406.
Shah VP, Wunderlich K, Larkin J. Time management impacts of pretrip advanced traveler information systems: Findings from a Washington DC, case study. Transport Res Rec. 2001;1774:36–43.
Smith SA, Perez C. Evaluation of INFORM: Lessons learned and application to other systems. Transport Res Rec. 1992;1360:62–65.
Srinivasan KK, Mahmassani H. Analyzing heterogeneity and unobserved structural effects in route-switching behavior under ATIS: A dynamic kernel logit formulation. Transport Res B 2003;37:793–814.
Stephanedes YJ, Kwon E, Michalopoulos P. Demand diversion for vehicle guidance, simulation, and control in freeway corridors. Transport Res Rec. 1989;1220:12–20.
Sun Z. Travel information impact on activity-travel patterns. PhD thesis, Eindhoven University of Technology, 2006.
Tian H, Gao S, Fisher DL, Post B. Route choice behavior in a driving simulator with real-time information. In: 90th TRB Annual Meeting DVD, 2011.
Tsirimpa A. Development of a simulation model of individuals activity travel patterns in an information rich environment. PhD thesis, University of Aegean, 2010.
Tsirimpa A, Polydoropoulou A, Antoniou C. Development of a mixed multi-normal logit model to capture the impact of information systems on travelers' switching behavior. J Intell Transport Syst. 2007;11(2):79–89.
Uchida T, Iida Y, Nakahara M. Panel survey on drivers' route choice behavior under travel time information. In: IEEE vehicle navigation and information systems conference proceedings, pp. 383–388, 1994.
Waller ST, Ziliaskopoulos AK. On the online shortest path problem with limited arc cost dependencies. Networks 2002;40(4):216–227.
Wolinetz LD, Khattak A. Why will some individuals pay for travel information when it can be free? Analysis of a Bay Area traveler survey. Transport Res Rec. 2004;1759:9–18.
Wunderlich K, Hardy M, Larkin J, Shah V. On-time reliability impacts of advanced traveler information services (ATIS). Technical report, Mitretek Systems, 2001.

Yang Q, Koutsopoulos HN, Ben-Akiva ME. Simulation laboratory for evaluating dynamic traffic management systems. Transport Res Rec. 2000;1710:122–130.
Yim Y, Hall R, Koo R, Miller M. TravInfo 817-1717 caller study. In: 78th TRB Annual Meeting CD-ROM, 1999.
Yim Y, Khattak A, Raw J. Traveler response to new dynamic information sources: Analyzing corridor and areawide behavioral surveys. Transport Res Rec. 2002;1803:66–75.
Ziegelmeyer A, Koessler F, My KB, Denant-Boemont L. Road traffic congestion and public information: An experimental investigation. J Transport Econ Pol. 2008;42:43–82.

Chapter 9
Modeling Within-Day Activity Rescheduling Decisions under Time-Varying Network Conditions

Yunemi Jang, Yi-Chang Chiu, and Hong Zheng

Abstract The within-day activity rescheduling decision process is an integral part of the travel choices when a traveler fulfills his/her daily travel activities. The within-day activity rescheduling decision takes place when the currently executed activity schedule is being interrupted and time pressure or time surplus is being created by traffic condition and/or activity attribute changes. Adjustment may also be triggered by changes that reduce time pressure and create time surplus.

It is postulated in this research that a traveler aims to maximize his/her utility while rescheduling the remaining activities. A utility maximization activity rescheduling model is proposed to depict this decision process. Moreover, time-varying travel times between activity locations are explicitly incorporated in the proposed activity adjustment model and solution algorithm, establishing consistency between the adjusted activities, schedules and the time-varying traffic conditions. Numerical studies demonstrate the solution properties of the proposed activity rescheduling model.

9.1 Introduction

It is well understood that the traveling public makes trips to fulfill daily activities in order to satisfy certain desires and needs. The travel decisions include both long-term and short-term choices. Those choices are often made by considering the interrelationship between the start time and the duration of the activities as well as the travel needs from one activity location to the next. The traditional trip-based travel-demand modeling approach addresses the interrelationship between travel agenda and network conditions via a sequential iterative procedure. However,

Y. Jang • Y.-C. Chiu (✉) • H. Zheng
The University of Arizona, Department of Civil Engineering and Engineering Mechanics, 1209 E 2nd St. Room 206, Tucson, AZ 85721, USA
e-mail: yunemi@email.arizona.edu; chiu@email.arizona.edu; hzheng@email.arizona.edu

S.V. Ukkusuri and K. Ozbay (eds.), *Advances in Dynamic Network Modeling in Complex Transportation Systems*, Complex Networks and Dynamic Systems 2,
DOI 10.1007/978-1-4614-6243-9_9,

several drawbacks of applying the sequential procedure have been pointed out by various studies (Bowman and Ben-Akiva 1997; Bhat and Koppelman 1999; Lam and Yin 2001; Lin et al. 2008). The primary limitation of the trip-based approach is the lack of a cohesive linkage among activities composing an activity schedule. As a result, temporal and spatial interdependency that bounds the feasibility of a traveler's daily activities is likely to be violated. The activity-based modeling approach, an emerging modeling technique, uses an individual's activity as a basic modeling entity, arguably possessing a sound modeling concept and behavioral basis describing a traveler's sequenced travel behavior.

Activity-based models (ABM) have been extensively studied in the last three decades (Hägerstrand 1970; Clarke 1986; Recker et al. 1986a, 1986b; Gärling et al. 1989; Kitamura et al. 1996; Miller and Roorda 2003; Bhat et al. 2004). Based on the concept of a space–time prism proposed by Hägerstrand (1970), several activity-based travel studies have been conducted, including CARLA (Combinatorial Algorithm for Rescheduling List of Activities) (Clarke 1986), STARCHILD (Simulation of Travel/Activity Responses to Complex Household Interactive Logistic Decisions) (Recker et al. 1986a, 1986b), SCHEDULER (Gärling et al. 1989), AMOS (The Activity-Mobility Simulator) (Kitamura et al. 1996), CEMDEP (A Comprehensive Econometric Micro-simulator for Daily Activity-Travel Patterns) (Bhat et al. 2004), and TASHA(Travel Activity Scheduler for Household Agents) (Miller and Roorda 2003). Nowadays, several transportation agencies have implemented or are in the process of promoting activity-based models in practical applications, including Portland (Bradley 1998), San Francisco (Jonnalagadda et al. 2001), Denver (Sabina and Rossi 2007), Sacramento Area (Bradley and Bowman 2010), and Columbus (Vovsha et al. 2003).

Most of the prior ABM research has focused on pre-planned schedules, in which activities' attributes and sequences are determined in compliance with certain space–time constraints for assigning activities to each individual (Wigan and Morris 1981; Gärling et al. 1989; Recker 1995; Bowman and Ben-Akiva 1997; Kitamura et al. 1998; Arentze et al. 2000; Miller and Roorda 2003). In contrast, activity adjusting/rescheduling behavior triggered by an unexpected event—which is a common situation in a within-day travel decision process—has not been well investigated. In this domain, noteworthy work appears until recently; examples include mathematical and analytical models (Timmermans et al. 2001, Joh et al. 2002; Gan and Recker 2008), parameter calibration (Joh et al. 2004), and data collection and interview design (Joh et al. 2005; Ruiz and Timmermans 2006; Roorda and Andre 2007).

Timmermans et al. (2001) embodied the time pressure concept suggested by Gärling et al. (1999) in a rescheduling decision model framework. Further, they suggested an S-shape utility function and proposed the generic form to model a positive correlation between utility and activity duration. Joh et al. (2002) proposed the AURORA (Agent for Utility-driven Rescheduling of Routinized Activities) model, in which a tree structure contains a set of rescheduling decisions including duration change, activity insertion, resequencing, etc., applying the utility maximization rule to model the rescheduling behavior. The parameters of AURORA are estimated using activity-travel diaries in the subsequent research (Joh et al. 2004). Recker et al. (1986a, 1986b; 1995) proposed a rule-based simulation model, called STARCHILD to model daily activity schedules. Gan and Recker (2008) extended Recker's framework to investigate activity rescheduling process behavior, based on a resistance assumption that people always prefer to reschedule activities in a way that maintains the similarity to the original preplanned schedule. Miller and Roorda (2003) used a simulation model to schedule within-a-day activities by sequentially inserting activity "episode" for each individual member of a household.

Meanwhile, some researchers have conducted surveys attempting to understand how the actual activity rescheduling decision mechanism is triggered. In this regard, Joh et al. (2005) conducted an empirical survey to investigate the incentive of occurrence and types of adjustment associated with the triggered rescheduling behavior. Ruiz and Timmermans (2006), by an Internet-based survey, analyzed the feasibility of activity scheduling conflict when a traveler inserts an activity between two consecutive ones. They tested several plausible duration distribution functions and showed some findings related to rescheduling behavioral tendencies. Roorda and Andre (2007) performed a computer-aided telephone survey of a hypothetical scenario of a 1-h-delay and tried to correlate decision strategies to the scenario. Their results validate that sudden network congestion is likely to activate the subsequent activity adjustment/rescheduling decision.

While above studies shed lights on understanding the rationale of the rescheduling process, they omitted the time-varying traffic/network dynamics component as well as its influence on the rescheduling behavior mechanism. The relevance of time-varying travel time to the activity rescheduling problem is that, when one makes a decision to re-timing and/or resequencing activities, the time-dependent travel time from the preceding activity to its next needs to be valid subject to the time–space constraint imposed by the adjusted timing.

The emphasis of our research is to explicitly account for time-varying network travel time in the within-day activity rescheduling decision. The network/traffic conditions vary over time, due to recurrent/nonrecurrent traffic congestions, and so do the travel time between activity locations. When activity attributes are modified or activities are resequenced, the travel time between activities will need to be accounted for to obtain consistent decisions, meaning that the travel time used for decisions shall be the same as those experienced in the actual travel. Time-varying nature is often considered in the dynamic traffic assignment context. These models assign/simulate entities of individual travelers and their route choice decisions along a given origin–destination-departure time (O–D-T) journey. Some prior studies performed dynamic traffic equilibrium analysis in the context where O–D journeys are expressed as given activity chains or tours (Abdelghany et al. 2001; Lam and Yin 2001; Abdelghany and Mahmassani 2003; Kim et al. 2006; Maruyama and Harata 2005,2006; Lin et al. 2008) but none of them studied activity resequencing. In a more recent study, an innovative activity travel network (ATN) concept and approach was proposed. In this ATN representation, virtual links were created to represent additional activity choices dimensions and each route in the augmented network represents a set of travel and activity arcs. Therefore, choosing a route is analogous to choosing an activity location, duration, time of participation, and travel route (Ramadurai and Ukkusuri 2010). This approach allows one to establish a dynamic user equilibrium solution combining the activity and travel decisions in a preplanning context.

Overall, it is apparent that less attention has been given to the decision process in adjusting the remaining activity schedule due to time pressure or time surplus, caused by either the unexpected changes in network condition or occasionally modified activity agenda. To model activity adjustment decisions resulted from network condition changes, one needs to explicitly account for time-varying travel time in the decision process.

This paper first presents a utility-based within-day activity rescheduling model to capture the activity adjustment decision process. The suggested utility maximization formulation simultaneously determines the new activity sequence, and the reoptimized start time and end time for each activity. Second, a solution algorithm incorporating a branch-and-cut technique is proposed with the goal to maintain computational efficiency in a simulation environment. Finally, the activity decision model and algorithm are demonstrated with the unique consideration of accounting for time-varying travel cost in the case studies.

The paper is structured as follows: The rescheduling decision framework is presented in Sect. 9.2, including model assumptions and decision context. Section 9.3 develops the model formulation of rescheduling decision process and the solution algorithm. In Sect. 9.4, the capability of the proposed rescheduling model is numerically verified in two simplified case studies. Section 9.5 concludes this paper.

9.2 Rescheduling Decision modeling Framework

9.2.1 Modeling Considerations

When rescheduling the remaining daily activities, the possible decisions may involve start time, duration, and precedence relationship between activities. With regard to rescheduling decision modeling, several considerations are discussed herein in order to clarify the modeling assumptions and scopes.

Three activity schedules are defined: executed schedules, preplanned schedule, and updated schedule. Given a preplanned schedule, the executed schedule refers to the portion of the preplanned schedule that has been executed until the instance at which the rescheduling decision is to commence. The updated schedule refers to the newly generated schedule replacing the preplanning schedule from the decision time instance onward. Thus, within-day temporal adjustment of a preplanned schedule stemming from exogenously introduced events is the primary concern, which necessitates incorporating the notion of time budget constraint into the rescheduling problem. The time budget constraint is imposed under both time pressure and time surplus situations. We consider time pressure occurs when (1) a traveler needs to insert additional activities into his/her existing schedule or (2) prolonged travel time causes arrival delay to the subsequent activity. In these cases, one may need to adjust the subsequent activities' start time and/or duration or cancel one or several preplanned activities in order to accommodate the unexpected event. In contrast, time surplus arises when the exogenous event calls for the cancelation of one or several existing activities or shortening the duration of remaining activities in the preplanned schedule.

Different from prior research, this research does not consider randomly generating new activity decisions in the rescheduling process. The underlying assumption of the proposed decision mechanism is to retime and re-sequence the existing activity schedule in order to retain the original preplanned activities as much as possible. The activity rescheduling process follows a utility maximization framework subject to the time budget constraints. The utility maximization concept applied in our daily activity rescheduling model was also adopted in several studies (Becker 1965; Ashiru et al. 2004; Kim et al. 2006; Ye et al. 2009). However, our model is aimed at optimizing the new timing and sequencing simultaneously to maximize total utility driven by participating activities. A quadratic marginal utility function is incorporated into the utility maximization model in order to achieve the computational tractability in the practical problems.

Moreover, the proposed model focuses on linking activity-based decisions and daily traffic dynamics in a consistent manner. Several prior rescheduling studies considered the feasibility of travel time in the time–space prism, but with a simplified hypothesis of constant travel time. For example, AURORA adopted a static travel time to reschedule activities under time pressure (Joh et al. 2002). The static travel time was also referenced in the reschedule model with a formulation of pick-up and delivery problem with time-window constraints proposed by Gan and Recker (2008). Although those models addressed the importance of travel cost in a rescheduling decision model, they did not consider time-varying travel cost.

Presenting a time-dependent dynamic traffic condition in the activity rescheduling model is not trivial. When considering fine-grained network/link travel times, the model formulation and solution algorithm need to explicitly search a rescheduling solution in which the travel time among activity pairs is consistent with the time-dependent travel times recognized/anticipated by the traveler. This consistency becomes important when linking the within-day rescheduling decisions with simulation-based dynamic network models in a high-fidelity temporal resolution. This feature distinguishes our study from most of the prior research.

9.2.2 *Decision Context*

An individual's rescheduling behavior is triggered by an unexpected incident due to either a change of the existing schedule (e.g., last minute canceling of a preplanned meeting) or a change in the network condition (e.g., accident).

The decision process of the proposed model is depicted in Fig. 9.1. At a given time instance t, each traveler would check whether he/she needs to reschedule

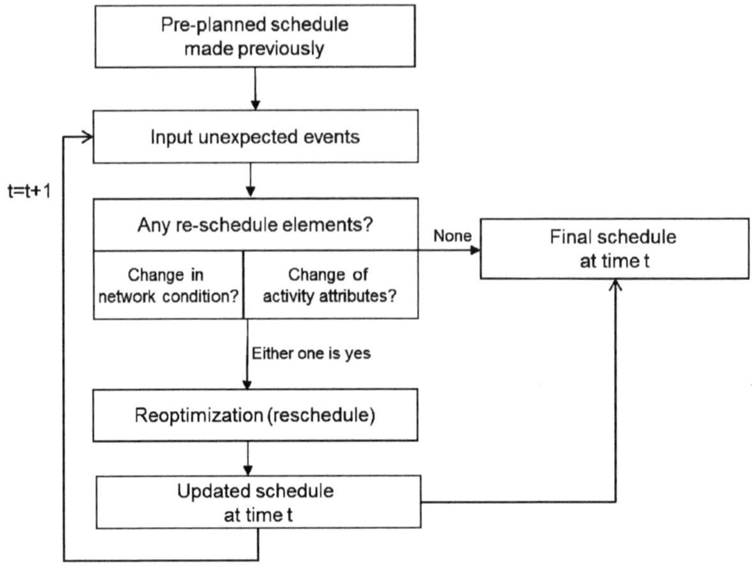

Fig. 9.1 Framework of within-day schedule modification

the activities. If rescheduling is not needed, the individual continues executing the preplanned schedule without adjustment; otherwise, the rescheduling model will generate a different schedule according to the nature of the event. Two events are considered in the proposed model, i.e., individual's activity agenda may change suddenly or the traffic condition in the assignment framework changes. The same process is repeated when the decision time instance is advanced from time t to $t+1$. One may note that herein the status of an individual's activity agenda could change passively, rather than actively. For example, let event E be a meeting between two individuals A and B in a preplanned schedule, if A's endogenous decision is to drop E in his/her rescheduling process due to a time pressure, then this event becomes an exogenous event to B's schedule and B's activity agenda is changed suddenly. In the following discussions we refer to this exogenous change of an activity agenda as *activity agenda change*.

9.3 Model and Solution Methodology

9.3.1 Rescheduling Decision Process

As discussed above, two decision contexts are considered in the proposed rescheduling framework, including *network condition change* and *activity agenda change*. In the *network condition change* situation, as shown in Fig. 9.2, the traveler

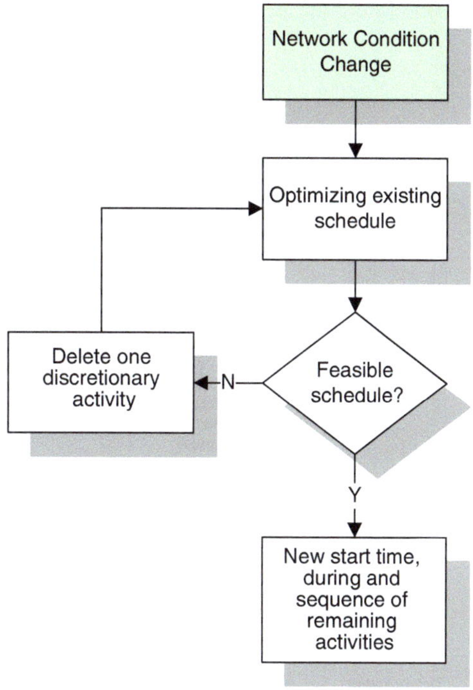

Fig. 9.2 Rescheduling decision process due to network condition change

receives the network condition change information, and first assesses if the network condition change leads to a time surplus or time pressure. If time pressure arises, the traveler will need to retime and re-sequence the remaining activities to obtain a new optimal schedule with the updated activity location-to-activity location travel time. If no feasible solution could be found—this is likely if travel delay is too large to commence all remaining activities within the total time budget—then one discretionary activity (with more flexibility in terms of trip purpose compared with anchor activities) is removed from the activity list. Note that in this rescheduling process, it is assumed that the traveler aims to keep the same preplanned activities unless doing so becomes infeasible. In such a case, the individual drops a discretionary activity and searches for a new schedule. This process is repeated until an optimal solution is found.

Activity attribute change may involve several cases, including insertion of an activity, change of duration, or deletion of an activity resulted from external factors. If the activity attribute change situation creates a time surplus in which more time is permitted, one desired discretionary activity may be added. The expanded activity set is then re-optimized to see if a feasible schedule can be obtained. If so, the revised schedule is obtained with the new activity included; otherwise, the contemplated activity is removed, and the original activity set is re-optimized. On the other hand, when a time pressure situation is created, the traveler first tries to re-optimize the current schedule according to the requirements imposed by the

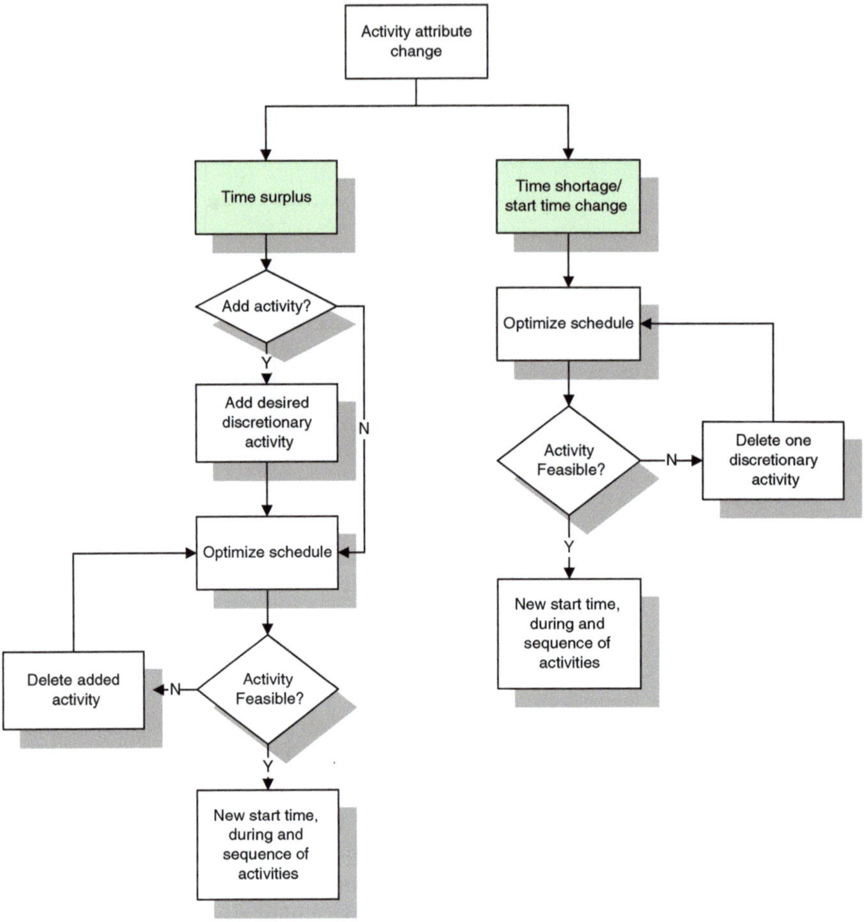

Fig. 9.3 Rescheduling decision process due to activity attribute change

updated activity attribute. If keeping the current schedule becomes infeasible, one discretionary activity has to be removed and all activity schedules are re-optimized. This activity-removal and re-optimization process is repeated until an optimal schedule is obtained. This procedure is illustrated in Fig. 9.3.

9.3.2 Mathematical Model for Rescheduling Decision

The proposed within-day rescheduling model adopts a utility maximization framework with a goal to maximize the total utility associated with the participated activities. This model yields a solution with optimized start time, duration of all remaining activities as well as the activity sequence.

The within-day rescheduling problem is formulated as follows:

$$MAX\ Z = \sum_{a\in A'(i)} \left[F_a\left(t_a^s + d_a\right) - F_a(t_a^s)\right] + \sum_{g\in A'(i)} \sum_{a\in A'(i)} \theta_{g,a} \cdot w_{g,a}(t_g^s + d_g) \cdot y_{g,a} \quad (9.1)$$

Subject to:

$$t_a^s - t_h^s + My_{a,h} + d_a \le M - w_{a,h}\left(t_a^s + d_a\right), \forall a \in A'\left(i\right), h \in A'\left(i\right), a \ne h \quad (9.2)$$

$$y_{a,h} + y_{h,a} = 1, \forall a \in A'\left(i\right), h \in A'\left(i\right), a \ne h \quad (9.3)$$

$$t_a^{s\min} \le t_a^s \le t_a^{s\max}, \forall a \in A'\left(i\right) \quad (9.4)$$

$$d_a^{\min} \le d_a \le d_a^{\max}, \forall a \in A'\left(i\right) \quad (9.5)$$

$$t_a^{e\min} \le t_a^s + d_a \le t_a^{e\max}, \forall a \in A'\left(i\right) \quad (9.6)$$

where

$A(i)$ = Set of all activities in a schedule for traveler i
$A'(i)$ = Set of remaining activities for traveler i
t_a^s = Start time of activity $a \in A'(i)$
d_a = Duration of activity $a \in A'(i)$
$y_{g,a}$ = Binary sequence variable $\forall g \in A'\left(i\right), \forall a \in A'\left(i\right), y_{g,a} = 1$, if activity g precedes activity a
$F_a\left(t\right)$ = Integral of marginal utility (total utility) for activity $a \in A'(i)$ when departing at time t
$w_{g,a}\left(t\right)$ = Travel time from activity $g \in A'(i)$ to activity $a \in A'\left(i\right)$ when departing at time $t, w_{g,a}\left(t\right) < 0$
$\theta_{g,a}$ = Weight of journey from activity $g \in A'\left(i\right)$ to $a \in A'(i)$
$t_a^{s\min}$ = Earliest start time for activity $a \in A'(i)$
$t_a^{s\max}$ = Latest start time for activity $a \in A'(i)$
$t_a^{e\min}$ = Earliest end time for activity $a \in A'(i)$
$t_a^{e\max}$ = Latest end time for activity $a \in A'(i)$
$d_a^{\min}$ = Shortest duration length for activity $a \subset A'(i)$
$d_a^{\max}$ = Longest duration length for activity $a \in A'(i)$
M = Penalty value, $M > 0$

The objective function (9.1) computes the total utility including two utility terms: one is the positive utility for commencing each activity and the other is the disutility associated with travel. The first term in (9.1) expresses the sum of the integral of marginal utility (total utility) for participating in an activity a starting from t_a^s and lasting d_a, i.e., the area underneath the marginal utility function. The value of this term increases along with increased duration of the participated activities. The second term stands for the disutility of the associated time-varying travel time/cost from one activity to the next. The second term influences the re-sequencing decision in that the traveler would wish to minimize the total travel time utility while maximizing activity participation utility. Note that the weight of this term takes negative sign to represent the disutility of travel.

It is assumed that the marginal utility function included in the objective function follows a quadratic form which depends on the duration of an activity. The total marginal utility value is obtained by using the timing information of the earliest start time, the latest end time and the maximum utility value of each activity. The reason for using a quadratic marginal utility function is to obtain an analytically closed and tractable objective function form that proves to have a unique optimal solution, solvable by efficient solution algorithm. Also, the proposed marginal utility function is a comparable with the bell shape function proposed in prior study (Joh et al. 2002).

The integral of the quadratic marginal utility function can be further rewritten into Eq. (9.8). Constraints (9.2) and (9.3) maintain a valid sequence among activities. Constraint (9.2) indicates that, if activity a precedes activity h, then the start time of activity h cannot be earlier than the end of activity a plus the necessary travel time from a to h. Constraint (9.3) enforces the logical equivalence that only one of the two possible preceding relationships between any pair of activities would hold. In other words, only $y_{a,h}$ or $y_{h,a}$ can take value 1 and the other has to take value 0. If $y_{a,h} = 1$, it means that activity h follows a (not necessary immediately follow). Equations (9.4), (9.5) and (9.6) restrict the extent of start time, duration, and end time for valid activities in a feasible schedule.

The total utility $F_a(t)$ takes the form $F_a(t) = \int MU_a(t)\,\mathrm{d}t$,
where,

$$MU_a(t) = a_0 t^2 + b_0 t + c_0, \text{ where } a_0 < 0, b_0, c_0 \text{ are } \quad \text{parameters } a_0 < 0, b_0, c_0 \tag{9.7}$$

As a result, the following is established:

$$\begin{aligned} F_a(t_a^s + d_a) - F_a(t_a^s) &= \int_{t_a^s}^{t_a^s + d_a} MU_a(t)\,\mathrm{d}t \\ &= 1/3 \left\{ \frac{-U_a^{\max}}{\left(\frac{t_a^{e\text{-}\max} - t_a^{e\text{-}\min}}{2}\right)^2} \right\} \left(t - \frac{t_a^{e\text{-}\max} + t_a^{s\text{-}\min}}{2}\right)^3 + U_a^{\max}\left(t - \frac{t_a^{e\text{-}\max} + t_a^{s\text{-}\min}}{2}\right) \end{aligned} \tag{9.8}$$

Where $U_a^{\max}$ = maximum marginal utility value for activity $a \in A'(i)$.

9.3.3 Solution Algorithm

The solution algorithm is designed with two goals in mind. First, it needs to solve for the binary sequence variables and non-integer start time and duration variables with a nonlinear objective function. Second, the solution needs to be consistent with the given time-varying travel time/cost from one activity to another.

The algorithm consists of three steps: the *Relaxation* step, the *Branching* step, and the *Consistency Search* step. The *Relaxation* step solves the rescheduling problem as a relaxed nonlinear optimization in which binary sequence variables are relaxed to be real-valued variables. Meanwhile, a static average travel time instead of the time-varying travel time is used in this step.

The next step, specified as the *Branching* step, implements a branch-and-cut algorithm that constructs a binary solution tree to find the best k feasible sequences. The initial solution obtained by *Branching* is specified as the tree root; then one activity pair is selected of which the feasibility is tested. For example, for the relationship branch in which activity 1 precedes activity 2, two new constraints are constructed (i.e., $y_{12} \geq 1$ and $y_{21} \leq 0$). Constraints $y_{12} \geq 1$ and $y_{21} \leq 0$ are included for the activity 1 following activity 2 relationship branch. Each branch also needs to include constraint $y_{1,2} + y_{2,1} = 1$ to enforce that only one of the two variables takes a value of one. This equality constraint is critical to ensure that no conflicting schedule is produced in the solution.

Next, each branch is solved and tested for feasibility. If a solution branch is infeasible, then the entire branch is cut from the solution tree; no need to evaluate all subsequent branches. The cut process takes the advantage of the fact that in order for the entire solution to be feasible all activity-pair relationships need to be at least feasible. One infeasible solution would nullify the validity of subsequent branches, eliminating the need to further evaluate other solutions along that branch and significantly reduces the algorithm complexity. The next-level branch is created by selecting another activity pair and by specifying their precedence relationship. The branching process continues until all activity pairs are selected. In the end, a tree with a maximum of C_2^n depths is created, where n is the number of remaining activities.

There are a total of C_2^n depths generated from a base node in a tree. Without the cut-process, the complexity of branch algorithm in the worst case is $2^{C_2^n}$. With a branch algorithm that includes the cut-process, when only one feasible solution exists, the number of computations is $2 \cdot C_2^n$. With two feasible solutions, the complexity in the worst case is determined by $4 \cdot C_2^n - 2$.

Figure 9.4 illustrates an example of a solution tree for three activities. Node 1 is the initial solution from the *Relaxation* step. At the first depth, Node 2 institutes the precedence relationship in which activity 2 precedes activity 1 (e.g., $y_{12} \leq 1$ and $y_{21} \geq 0$). However, given this condition, no feasible solution can be found; therefore, the algorithm discontinues on the Node 2 branch. On the other hand, Node 3 is feasible, so the algorithm continues on the next level from this point on and solves for the solution based on the activities 1 and 3 precedence relationships. It turns out that Node 6 is infeasible, so the algorithm continues to the next level from Node 7. Finally Node 14 is found to be the only feasible and optimal solution.

Although the above example demonstrates the special case in which only one feasible solution remains at algorithm termination, it is likely that, in a general case, multiple feasible solutions may exist. It should also be noted that at this point all the feasible solutions are solved for based on time-invariant, average, activity-to-activity travel time. The next *Consistency Search* step incorporates the time-varying

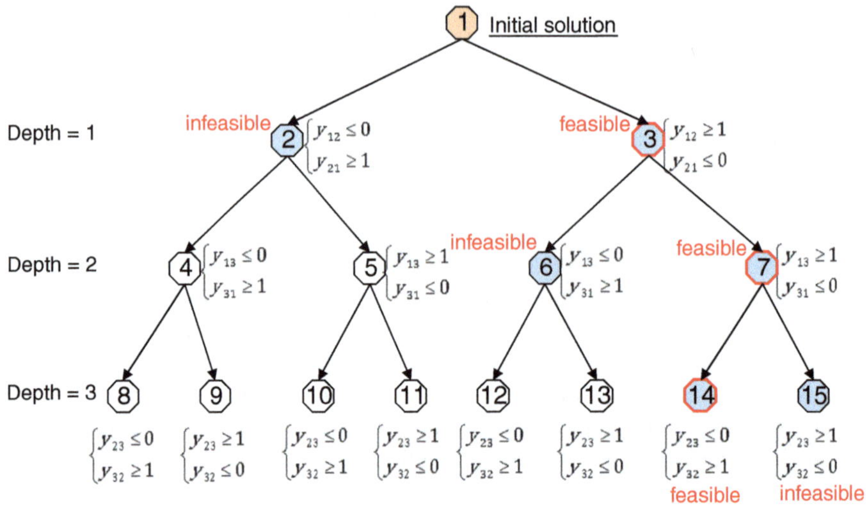

Fig. 9.4 Example of a solution tree for three activities

travel time into these k feasible solutions and finally arrives at an optimal and consistent solution. In other words, for each one of the k feasible solutions, all binary precedence variables y values are fixed, but the optimal departure time and duration for each activity are then resolved with the travel time $w_{g,a}(t)$ in Eq. (9.2) depending on the departure time t of the preceding activity g.

A convergence criterion is defined as follows.

$$\sum_{a \in A'} \left| t_a^{s\_new} - t_a^{s\_previous} \right| + \sum_{a \in A'} \left| d_a^{new} - d_a^{previous} \right| \le \varepsilon \tag{9.9}$$

Where,

$t_a^{s\_new}$ = obtained start time of activity $\boldsymbol{a} \in \boldsymbol{A}'(\boldsymbol{i})$ at current iteration
$t_a^{s\_previous}$ = obtained start time of activity $\boldsymbol{a} \in \boldsymbol{A}'(\boldsymbol{i})$ at previous iteration
d_a^{new} = obtained duration of activity $\boldsymbol{a} \in \boldsymbol{A}'(\boldsymbol{i})$ at current iteration
$d_a^{previous}$ = obtained duration of activity $\boldsymbol{a} \in \boldsymbol{A}'(\boldsymbol{i})$ at previous iteration
ϵ = Stopping threshold value

The solution algorithm is summarized as follows.

/ find initial solution having non–integer values for sequence variables* /*

Call Relaxation

If feasible solution is found from Relaxation

*/*applying Branch algorithm including a cut – process to do Branching* /*

```
Run Branch algorithm
  If amount of feasible solution > 0
   Find the best k sequences in the final leaves of a tree.
   /* applying iterative process in Consistency Search*/
   For each k sequences
     Do Find time-dependent travel time associated with the end time
        of each activity
        Calculate a scheduling problem
     While satisfying a convergence criterion
   Endfor
   Find the best sequences among alternatives
   Stop /* the end of Consistency Search*/
  Endif
  /* the end of Branching*/
Endif
Stop /* the end of Relaxation*/
```

9.4 Experiments

The rescheduling model and algorithm are coded in C language with a nonlinear optimization function linked with the IMSL Numerical Library optimization solver, so that the compiled executable of the entire algorithm is portable without needs of external solvers like CPLEX. This feature is designed to plan for future run-time integration/communication with the dynamic traffic assignment models.

The capability of the developed rescheduling model is tested in two case studies. In the case studies, a schedule diary of one hypothetical traveler is supposed given. Starting from the preplanned schedule, we observe how the preplanned schedule is adjusted by the rescheduling model triggered by an unanticipated event.

In the first case study, a time shortage situation is created. A preplanned schedule is given in Table 9.1 as an initial schedule. Three activities are prescheduled to be executed such as home, work and home. At 12 PM, the traveler is asked to mail out a document at a post office before the end of business hours, causing a schedule conflict between this event and the preplanned schedule in Table 9.1. According to a preplanned schedule, the traveler needs to work until 5 PM. The duration of

Table 9.1 Description of activities in schedule set

ID	1	2	3	4
Activity	Home	Work	Home	Post office
Min duration	6 h	5 h	30 min	20 min
Max duration	8 h	10 h	10 h	60 min
Earliest start time	00:00 AM	7:30 AM	1:00 PM	12:00 PM
Latest start time	00:00 AM	10:00 AM	11:30 PM	4:40 PM
Earliest end time	6:00 AM	1:00 PM	12:00 AM	8:20 AM
Latest end time	8:00 AM	7:00PM	12:00 AM	5:00 PM
Max utility	40	120	40	20
Start time	12:00 AM	7:50 AM	5:40 PM	–
Duration	7 h 20 min	9 h 10 min	6 h 20 min	–
End time	7:20 AM	5:00 PM	12:00 AM	–

Table 9.2 Updated schedule by solution algorithm

Relaxation	ID	Activity	Start time		Duration	End time
	2	Work	12:00 PM		6 h 10 min	6:10 PM
	3	Home	2:00 PM		10 h	12:00 AM
	4	Post office visit	12:00 PM		20 min	12:20 PM
	y_{23}	y_{32}	y_{24}	y_{42}	y_{34}	y_{43}
	0.5	0.5	0.5	0.5	0.5	0.5
Branching	ID	Activity	Start time		Duration	End time
	2	Work	12:00 PM		4 h 20 min	4:20 PM
	3	Home	5:30 PM		6 h 30 min	12:00 AM
	4	Post office visit	4:40 PM		20 min	5:00 PM
	y_{23}	y_{32}	y_{24}	y_{42}	y_{34}	y_{43}
	1	0	1	0	0	1
Consistency Search	ID	Activity	Start time		Duration	End time
	2	Work	12:00 PM		4 h 20 min	4:20 PM
	4	Post office visit	4:40 PM		20 min	5:00 PM
	3	Home	5:30 PM		6 h 30 min	12:00 AM

this post office visit event is at least 20 min, and it must be finished no later than 5 PM. The traveler faces a time pressure situation to insert the new event so that a rescheduling decision is hence initiated.

The solution algorithm is tested in the first case study of time shortage situation. As shown in the *Relaxation* section in Table 9.2, the start time of work activity and the post office visit are still conflicting because the binary precedence variables are relaxed at this step and all take value of 0.5.

The *Branching* step is designed to resolve any possible schedule conflicts found in the solution from the *Relaxation* step. The *Branching* step enforces that the binary precedence variables y take a 0–1 binary value. Further, the Branching step includes the cut-process, producing a solution tree of the best k solutions. Figure 9.5 shows how an example tree is generated to seek 0–1 variables associated with the sequence connections.

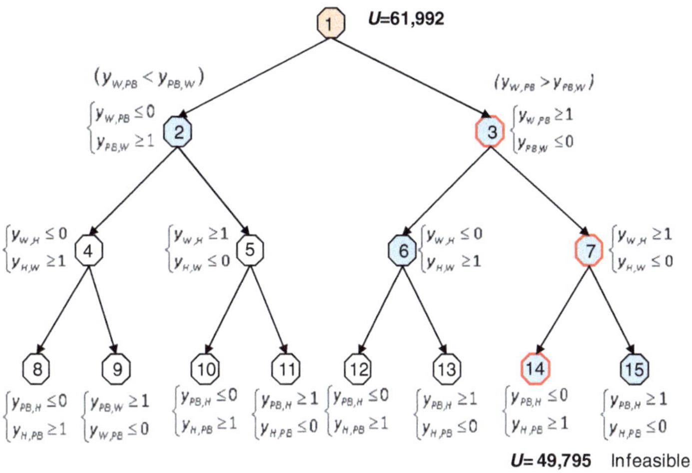

Fig. 9.5 Solution tree in branching

The first node is obtained from the *Relaxation* step; the branching algorithm then generates a solution tree. An optimized schedule having a valid sequence among activities is calculated at Node 14 in this example. The valid sequence from the Node 14 is, for example, work, post office visit, and home. It is apparent that the schedule conflict in a schedule at the *Relaxation* stage is resolved by the *Branching* algorithm, and the total utility is reduced to 49,795, compared to 61,992 units in the initial solution given by the *Relaxation* step.

The *Consistency Search* step applies the time-dependent travel time to the currently found feasible solutions. The time-varying travel times used in this case study are illustrated in Fig. 9.6. Examining the final optimal solution (*Consistency Search* section of Table 9.2), one can find that the traveler adjusts the schedule to leave work earlier at 4:20 PM instead of 6:10 PM in the pre-planned schedule, and arrive at the post office at 4:40 PM, stay 20 min, and then reach home at 5:30 PM, instead of 5:40 PM in the original schedule.

The second case study is aimed at examining the within-day activity rescheduling decision process when the traffic condition is changed. The scenario, based on the same preplanned schedule applied in the first case study, assumes that the traveler receives incident information at 2:00 PM, just before he/she leaves the office as scheduled. It is assumed that the traveler is able to estimate the delay time based on perceived traffic information and experience. Following the preplanned schedule, the traveler is supposed to leave office at 4:20 PM, and arrive at a post office at 4:40 PM. Because a significant delay occurred when enroute to the post office, the traveler needs to adjust the schedule with the updated time-varying travel time as shown in Table 9.3.

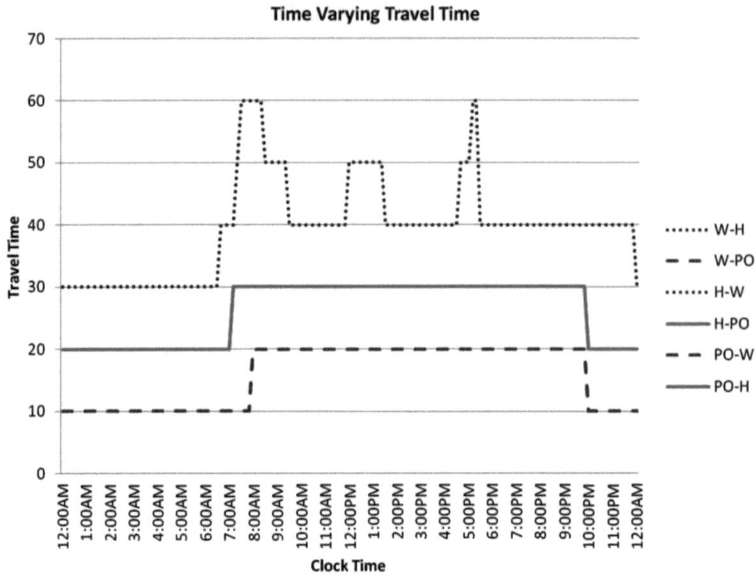

Fig. 9.6 Time-varying travel time

Table 9.3 Travel time information

	Original travel time information (min)				Updated travel time information (min)			
	W->PB	PB->W	PB->H	H->PB	W->PB	PB->W	PB->H	H->PB
4:00 PM	20	20	30	30	70	20	30	30
4:10 PM	20	20	30	30	70	20	30	30
4:20 PM	20	20	30	30	70	20	30	30
4:30 PM	20	20	30	30	80	60	70	80
4:40 PM	20	20	30	30	80	60	70	80
4:50 PM	20	20	30	30	70	50	60	70

Figure 9.7 shows the comparison of the preplanned schedule and the adjusted schedule. The new schedule reduces the working time so that the traveler leaves the office earlier with the same activities sequence to accommodate the extended travel time.

9.5 Conclusion

This research develops a daily activity rescheduling behavior model integrating both activity decision choice and time-varying traffic information. The proposed model investigates the activity rescheduling decision process involved in an unexpected event created by either network traffic condition changes or activity attributes

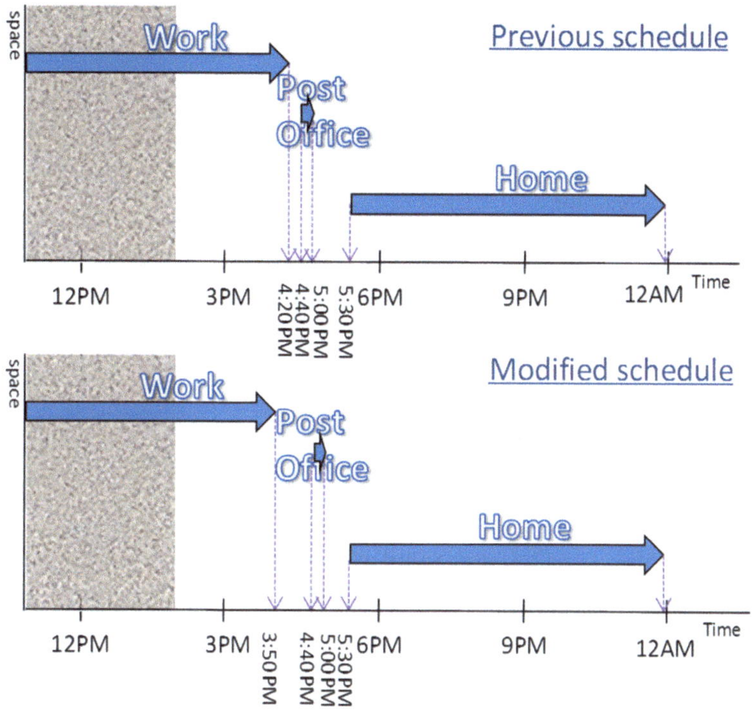

Fig. 9.7 New modified schedule due to travel time change

change. The model is formulated as a utility maximizing rescheduling problem, designed to solve for start time, duration, and activity sequence consistent with time-varying traffic dynamics. An algorithm is proposed to solve the optimal solution in a computationally effective manner. The numerical case studies demonstrate the characteristics of the proposed model behaved under both time shortage and traffic condition changes situations. These case studies show how preplanned schedule is adjusted in response to unforeseen events through a rescheduling decision process.

The further research underway is to conduct the empirical survey to calibrate the parameters of the proposed utility model; this is important to validate the proposed methodology before any real-life application. The survey should emphasize on understanding the perceived preference of the maximum utility value and the influence on the activity adjustment procedure, as well as the real dairy data about timing and sequencing choices travelers intend to execute in the preplanned schedule.

Recently, several researchers have attempted to identify new travel survey technology suitable for the activity-based models (Wolf et al. 2001; Asakura and Hato 2006; Frignani et al. 2010; Moiseeva et al. 2010). Global Positioning System (GPS) technology appears to be promising in collecting individual diaries with high response rate and satisfactory quality. Use of the mobile phone has also been practiced in collecting activity-based travel information. It seems promising to use those well-developed technologies in current practice to calibrate and validate the proposed model.

Another potential research effort is to integrate the proposed model with a dynamic traffic assignment framework to verify the performance of proposed rescheduling methodology in the contexts of interacting activity decisions with dynamic traffic conditions.

Acknowlegments The authors acknowledge the partial financial support from FHWA EARP Project: DTFH61–07-R-00117: Modeling the Urban Continuum in an Integrated Framework: Location Choice, Activity-Travel Behavior, and Dynamic Traffic Patterns.

References

Abdelghany AF, Mahmassani HS. Temporal-spatial microassignment and sequencing of travel demand with activity-trip chains. Transport Res Rec J Transport Res Board. 2003;1831:89–97.

Abdelghany AF, Mahmassani HS, et al. Spatial microassignment of travel demand with activity trip chains. Transport Res Rec J Transport Res Board. 2001;1745:36–46.

Arentze T, Hofman F, et al. ALBATROSS: multiagent, rule-based model of activity pattern decisions. Transport Res Rec J Transport Res Board. 2000;1706:136–144.

Asakura Y, Hato E. Tracking in individual behaviour using mobile phones: recent technological development. The 11th International Conference on Travel Behaviour Research, Kyoto, Japan. 2006.

Ashiru O, Polak JW, et al. The utility of schedules: theoretical model of departure-time choice and activity-time allocation with application to individual activity schedules. Transport Res Rec J Transport Res Board. 2004;1894:84–98.

Becker GS. A theory of the allocation of time. Econ J. 1965;75:493–517.

Bhat CR, Guo JY, et al. Comprehensive economic microsimulator for daily activity-travel patterns. Transport Res Rec J Transport Res Board. 2004;1894:57–66.

Bhat CR, Koppelman FS. Activity-based modeling for travel demand. Handbook of transportation science. R. W. Hall: Springer. 1999.

Bowman JL, Ben-Akiva M. Activity-based travel forecasting. Activity-based travel forecasting conference, Texas Transportation Institute, 1997.

Bradley M. A system of activity-based models for Portland, Oregon. Report prepared for the Federal Highway Administration Travel Model Improvement Program. Washington, DC: FHWA; 1998.

Bradley M, Bowman JL. SACSIM: an applied activity-based model system with fine-level spatial and temporal resolution. J Choice Model. 2010;3(1):5–31.

Clarke M. Activity modelling-a research tool or a practical planning technique? Behavioral research for transport policy. Utrecht, The Netherlands: VNU Science Pres; 1986. p.3–15.

Frignani MZ, Auld J, et al. Urban Travel Route and Activity Choice Survey (UTRACS): An internet-based prompted recall activity travel survey using GPS data. Proceedings of the 89th Annual Meeting of the Transportation Research Board, Washington, DC. 2010.

Gan LP, Recker W. A mathematical programming formulation of the household activity rescheduling problem. Transport Res B Methodological. 2008;42(6):571–606.

Gärling T, Brännäs K, et al. Household activity scheduling. Transport policy, management & technology towards 2001: Selected proceedings of the fifth world conference on transport research. Ventura, CA, Western Periodicals Co. IV. 1989;231–248.

Gärling T, Gillholm R, Montgomery W. The role of anticipated time pressure in activity scheduling. Transportation. 1999;26(2):173–191.
Hägerstrand T. What about people in regional science? Reg Sci. 1970;24(1):6–21.
Joh CH, Arentze T, et al. Modeling individuals' activity-travel rescheduling heuristics: theory and numerical experiments. Transport Res Rec J Transport Res Board. 2002;1907:16–26.
Joh CH, Doherty ST, et al. Analysis of factors affecting the frequency and type of activity schedule modification. Transport Res Rec J Transport Res Board. 2005;1926:19–25.
Joh CH, Arentze TA, et al. Activity-travel scheduling and rescheduling decision processes: empirical estimation of Aurora model. Transport Res Rec J Transport Res Board. 2004;1898:10–18.
Jonnalagadda N, Freedman J, et al. Development of microsimulation activity-based model for San Francisco: destination and mode choice models. Transport Res Rec J Transport Res Board. 2001;**1777**:25–35.
Kim H, Oh JS, et al. Application of activity chaining model incorporating a time use problem to network demand analysis. Transport Res Rec J Transport Res Board. 2006;1977:214–224.
Kitamura R, Chen C, et al. Traveler destination choice behavior: effects of time of day, activity duration, and home location. Transport Res Rec J Transport Res Board. 1998;**1645**:76–81.
Kitamura R, Pas EI, et al. The Sequenced Activity Mobility Simulator (SAMS): an integrated approach to modeling transportation, land use and air quality. Transportation. 1996;23(3):267–291.
Lam W, Yin Y. An activity-based time-dependent traffic assignment model. Transport Res B Methodological. 2001;35(6):549–574.
Lin DY, Eluru N, et al. Integration of activity-based modeling and dynamic traffic assignment. Transportation Research Board 87th Annual Meeting. Washington, DC: Transportation Research Board; 2008.
Maruyama T, Harata N. Incorporating trip-chaining behavior into network equilibrium analysis. Transport Res Rec J Transport Res Board. 2005;1921:11–18.
Maruyama T, Harata N. Difference between area-based and cordon-based congestion pricing: investigation by trip-chain-based network equilibrium model with nonadditive path costs. Transport Res Rec J Transport Res Board. 2006;1964:1–8.
Miller EJ, Roorda MJ. Prototype model of household activity-travel scheduling. Transport Res Rec J Transport Res Board. 2003;1831:114–121.
Moiseeva A, Jessurun J, et al. Semi-automatic imputation of activity-travel diaries using GPS traces, prompted recall and context-sensitive learning algorithms. Transport Res Rec J Transport Res Board. 2010;2183:60–68.
Ramadurai G, Ukkusuri S. Dynamic user equilibrium model for combined activity-travel choices using activity-travel supernetwork representation. Network Spatial Econ. 2010;10:273–292.
Recker WW. The household activity pattern problem: general formulation and solution. Transport Res B Methodological. 1995;29(1):61–77.
Recker WW, McNally MG, et al. "A model of complex travel behavior: part 1 - theoretical development. Transport Res A Gen. 1986a;20(4):307–318.
Recker WW, McNally MG, et al. A model of complex travel bepavior: Part 2-operational model. Transport Res A Gen. 1986b;20(4):319–330.
Roorda MJ, Andre BK. Stated adaptation survey of activity rescheduling: empirical and preliminary model results. Transport Res Rec J Transport Res Board. 2007;2021:45–54.
Ruiz T, Timmermans H. Changing the timing of activities in resolving scheduling conflicts. Transport Plann Pol Res Pract. 2006;33(5):429–445.
Sabina E, Rossi T. Denver's activity-based model project: status report, lessons learned so far, and advice on how you can do it better than we have. 11th TRB National Transportation Planning Applications Conference, Florida, 2007.
Timmermans H, Arentze T, et al. Modeling effects of anticipated time pressure on execution of activity programs. Transport Res Rec J Transport Res Board. 2001;1752:8–15.
Vovsha P, Petersen E, et al. Explicit modeling of joint travel by household members: statistical evidence and applied approach. Transport Res Rec J Transport Res Board. 2003;1831:1–10.

Wigan MR, Morris JM. The transport implications of activity and time budget constraints. Transport Res A Gen._1981;15:63–86.

Wolf J, Guensler R, et al. Elimination of the travel diary: an experiment to derive trip purpose from GPS travel data. Transport Res Rec J Transport Res Board. 2001;1768:125–134.

Ye X, Konduri K, et al. Formulation of activity-based utility measure of time use: application to understanding influence of constraints. Transport Res Rec J Transport Res Board. 2009;2135:60–68.

Chapter 10
Dynamic Navigation in Direction-Dependent Environments

Irina S. Dolinskaya

Abstract This chapter examines optimal path-finding problems with direction-, location- and time-dependent environments. The dependence of the cost function and path constraints on the location of the mobile agent and time creates the need for a dynamic navigation algorithm, capable of adjusting the path in real time as more information about the environment becomes available. In addition, the direction-dependent nature of the environment results in an asymmetric cost function, which is not a metric and prohibits the use of more traditional and established approaches to solving optimal path-finding problems. Moreover, the triangle inequality is often violated for the direction-dependent cost functions, further preventing the use of analysis and results developed for Euclidian shortest path problems. To add another dimension of reality to our model, we integrate the system dynamics and constrain feasible paths by maximum sharpness of a turn that a mobile agent can make.

The presented work delivers a more realistic optimal path-finding model while reducing the computational time required to find such a path. This is particularly important since real-time implementation is essential for our applications. In addition, many analytical results derived here provide insights into the structure of the problem, its objective function, and the optimal solution. These insights provide a closed-form solution to a large subset of problems where additional assumptions are applicable. For such problems, we easily construct the analytical solutions instead of implementing more involved, and often approximate, methods presented in the literature.

We describe the Optimum Vessel Performance in Evolving Nonlinear Wavefields Project that motivated our work and deliver computational results to demonstrate the applicability and performance of our path-finding methods.

I.S. Dolinskaya (✉)
Department of Industrial Engineering and Management Sciences,
Northwestern University, Evanston, IL 60208, USA
e-mail: dolira@northwestern.edu

S.V. Ukkusuri and K. Ozbay (eds.), *Advances in Dynamic Network Modeling in Complex Transportation Systems*, Complex Networks and Dynamic Systems 2,
DOI 10.1007/978-1-4614-6243-9_10,

10.1 Introduction

Over the past few decades, researchers in a wide range of disciplines have been studying optimal path-finding problems within a variety of applications. These approaches find an optimal way to traverse a complex medium or network under a diverse set of constraints and outside factors. For example, researchers in computational geometry and geographical information systems analyze the shortest paths defined by Euclidean distance and other metrics, often with the presence of polygonal obstacles and weighted homogeneous regions. Optimal robot routing problems incorporate the system's physical properties and constraints with the objective of finding the fastest or minimum energy-consumption paths over various terrain. In naval vessel path-finding and navigation, researchers integrate the vessel's hull structure and forces exerted by waves and wind to minimize travel time to a destination. Each aforementioned application adds complexity to the optimal path-finding problem, while integrating a number of assumptions in each scenario to make the problem more tractable. In this chapter, we relax a set of restrictive assumptions to broaden application of optimal path-finding results to direction-dependent environments and to create an accurate and tractable models suitable for real-life implementation.

We study optimal path-finding problems in a direction-, location- and time-dependent environment. Since the objective of a problem depends on the actual application, we do not restrict our analysis to a specific objective function whenever possible. Throughout this chapter we discuss the problems of minimizing travel time, fuel consumption, and motions, as well as more general objective functions. The dependence of the cost function and path constraints on the location of the mobile agent, as well as time, creates the need for a dynamic navigation algorithm, capable of adjusting the path in real time as more information about the environment becomes available. In addition, the direction-dependent nature of the environment results in an asymmetric cost function, which is not a metric and prohibits the use of more traditional and established approaches to solving optimal path finding problems. Moreover, the triangle inequality is often violated for the direction-dependent cost functions, further preventing the use of analysis and results developed for Euclidian shortest path problems. To add another dimension of reality to our model, we integrate the system dynamics and constrain feasible paths by maximum sharpness of a turn that a mobile agent can make.

The presented work delivers a more realistic optimal path-finding model while reducing the computational time required to find such a path. This is particularly important since real-time implementation is essential for our applications. In addition, many analytical results derived here provide insights into the structure of the problem, its objective function, and the optimal solution. These insights provide a closed-form solution to a large subset of problems where additional assumptions are applicable. For such problems, we easily construct the analytical solutions instead of implementing more involved, and often approximate, methods presented in the literature.

This chapter is an overview of a number of the author's prior publications (some co-authored with colleagues) on the said subject. It brings together our work within the various research fields into a single comprehensive dynamic navigation system for direction-dependent environments. For more detailed discussion of the presented work see, Dolinskaya (2009), Dolinskaya (2012), Dolinskaya et al. (2009), Dolinskaya and Maggiar (2012), and Dolinskaya and Smith (2012).

10.1.1 Motivation: Optimum Vessel Performance in Evolving Nonlinear Wavefields Project

Our work was motivated by an optimal vessel routing project entitled "Optimum Vessel Performance in Evolving Nonlinear Wavefields." This 5-year project funded by the Office of Naval Research (ONR) Multidisciplinary University Research Initiative (MURI) grant is a collaboration with the Department of Naval Architecture and Marine Engineering and the Department of Industrial and Operations Engineering at the University of Michigan, the Applied Physics Laboratory at the University of Washington, and the Department of Electrical and Computer Engineering at The Ohio State University. Here, we provide a brief overview of the project and the research tasks of the teams involved. Throughout the chapter, we continually revisit this project to illustrate the real-life application of the developed methodology and results.

The goal of this project was to develop a system that can, in real-time, control the behavior of a vessel, based on real-time measurements and forecasts of the wavefield surrounding the vessel. Four major groups divided the project into the following parts based on the areas of expertise:

1. *Real-Time Measurement of Ocean Wavefields.* The first group of researchers develops and tests a coherent (Doppler) X-band radar for measurement of the ocean wavefield surrounding a moving or stationary vessel in real-time.
2. *Short-Term Forecasts of Evolving Nonlinear Wavefields.* The second team uses data collected by the radar to forecast the time-dependent evolution of the wavefield.
3. *Time-Domain Computation of Nonlinear Ship Motions.* Based on the forecast of the evolving wavefield, this group develops a numerical model to predict nonlinear ship motions in the multidirectional wavefield.
4. *Dynamic Real-Time Path Optimization and Vessel Control.* As part of the fourth team, we use the developed motion prediction model to evaluate the vessel speed, motions and other operability criteria conditional on a path chosen to traverse the forecasted wavefield. We then integrate this information into our optimal path-finding algorithm to determine the most favorable path. An adaptive control system developed by our colleagues guides the vessel along the found optimal path as closely as possible.

This project considers a wide range of problems, and the objective varies depending on the specific application. Wavefield forecast can be used to predict time periods and areas of calm seas to ensure a safe landing onto an aircraft carrier or a successful launch-and-recovery operation in rough weather. Finding a path that minimizes ship motion is important for improving safety and comfort of the passengers on board. Minimizing travel time is crucial in emergency rescue missions and improves efficiency of the ship-to-ship or port-to-ship cargo transfer operations. Alternatively, finding a path that minimizes fuel consumption instead of the vessel's travel time is favorable for some naval transportation problems. Consequently, in our work we predominantly study a very general set of path-finding problems, such that any one of the aforementioned applications can be addressed with our models.

To summarize, our objective, as part of this project, is to develop computationally efficient and numerically robust algorithms to solve path optimization problems in time-varying media. We are given information about the environment surrounding the vessel up to the radar visibility horizon and the dynamic restriction of the vessel: operability constraints such as probability of wet deck and maximum root mean squared roll, and minimum turning radius constraining curvature of a feasible path. We incorporate this information to find an optimal path to a specified desired destination. It is important to note that the wave forecasting model developed as part of this project is precise, and the path-finding problem is considered to be deterministic if the initial condition (i.e., the observed wavefield) is accurate.

10.1.2 Literature Overview

Existing literature details a wide variety of optimal path-finding problems. While some work analyzes path finding in a location, and possibly, time dependent medium, others look at the scenarios of anisotropic (i.e., direction-dependent) environment. However, no previous research studies a generalized model that includes all aforementioned aspects of the environment into a single analysis. In this section, we present an overview of various areas of studies and applications that look at the optimal path finding problems as they relate to our work.

Geometric shortest path finding is a fundamental problem extensively studied in computational geometry. Mitchell's survey (Mitchell 2000) gives a comprehensive overview of the current work conducted in this field. Most computational geometry research is restricted to finding an optimal path defined by Euclidean distance or other metrics, such as L_1-metric (the *Manhattan distance*) (Mitchell 1992) and *C-oriented* paths (Widmayer et al. 1987). Asymmetric direction-dependence is occasionally considered in the literature (Chew and Drysdale 1985; Reif and Sun 2004); however, the introduced anisotropy makes a strong assumption of distance function convexity which we relax in our analysis. The path-finding problems in a location-dependent environment examine the presence of polygonal obstacles (Alt and Welzl 1988; Kapoor et al. 1997; Lozano-Pérez and Wesley 1979; Mitchell 1991a, 1996) and uniform-weighted regions (Cheng et al. 2008;

Mitchell and Papadimitriou 1991b; Sun and Reif 2006). On the other hand, all the problems studied in the field of computational geometry are predominantly static, and time-dependence is not considered in these settings. It is important to note that Geographic Information Systems (GIS) is one of the primary application areas for computational geometry, and a number of papers published in GIS journals (Collischonn and Pilar 2000; de Floriani et al. 2000; Stefanakis and Kavouras 1995; Yu et al. 2003) also discuss shortest path finding problems.

Optimal path-finding research extends to other applications, such as robot, vessel, airplane and unmanned aerial vehicle routing. In each of these areas, researchers create the models specific to said application; unfortunately their analysis and results cannot be easily transferred to other problems. For example, the problem of computing an optimal path for a mobile robot considers friction and gravity forces for various regions of terrain, and then uses this direction- and location-dependent cost function to find a path that minimizes the total energy consumption of the robot (Lanthier et al. 1999; Rowe 1997; Rowe and Ross 1990; Sun and Reif 2005). Since surface contour does not change over time, this set of problems only considers path finding in a static environment.

Optimal vessel routing evaluates how waves and wind affect vessel speed and dynamics in finding an optimal path. For example, Philpott et al. (1993) apply mathematical programming methods to create a yacht velocity prediction program that computes the vessel speed for a specified range of wind speeds and yacht headings. The resulting velocity prediction data is used in stochastic dynamic programming models to find the yacht's fastest path in uncertain weather (Allsopp et al. 2000; Philpott 2005; Philpott and Mason 2001).

A significant amount of work assumes that the vessel speed function can be written analytically. This assumption allows researchers to invoke various methodologies from calculus of variations and optimal control theory to characterize an optimal path (Faulkner 1963a,b; Marks et al. 1968; Papadakis and Perakis 1990; Perakis and Papadakis 1989; Zermelo 1931). However, researchers typically use a simplified form of the speed function in order to make the analysis more manageable. Our colleagues working on the Optimum Vessel Performance in Evolving Nonlinear Wavefields project have developed more accurate and involved models to evaluate vessel dynamics and wave evolution. From our experience of working on this project, it is clear that analytical functions cannot accurately describe the vessel movement through the waves, thus obliging us to look for alternative methods to solve the problem.

Airline industry researchers analyze how weather affects airplane path planning and air traffic management. For example, Nilim and his colleagues (2004; 2001) model the weather as Markov chains where storms have a certain probability of becoming the obstacles, thus preventing the airplanes from passing through those regions. Then, a path-finding model identifies a path minimizing the expected travel time and dynamically reevaluates the path as more accurate information about the storms becomes available. In their work, Nilim et al. assume that the airplane has constant speed, consequently reducing the problem to a shortest path-finding problem among stochastic obstacles.

Unmanned aerial vehicles (UAVs) have become widely employed in civilian and military applications over the past few years. The problem of optimal path finding for mini UAVs subjected to wind is similar in nature to the vessel routing problems and has been extensively studied in recent years. The direction dependence of the speed function is introduced as a uniform wind vector field, which is added to a constant isotropic "wind-free" velocity of the airplane (Bakolas and Tsiotras 2010; McGee et al. 2006; McNeely et al. 2007; Osborne and Rysdyk 2005; Techy and Woolsey 2009). It is important to note that the resulting speed function and the minimum turning radius of a feasible path have very distinct structures, and more specifically, the properties of a convex speed polar plot and trochoidal path (Rysdyk 2007). In our work, we observe that direction-dependent speed often implies the direction-dependent nature of the minimum turning radius, and we address such problems for the generalized direction-dependent speed functions and path curvature restrictions.

10.1.3 Chapter Outline

This chapter is organized as follows. We begin our analysis by studying optimal path-finding problems in a direction-dependent, time and space homogeneous environment, which is presented in Sect. 10.2. First, we find closed form solutions for the problems with obstacle-free domain while neglecting the minimum turning radius constraint (Sect. 10.2.1). Then, we employ our findings and adapt a *visibility graph search* method of computational geometry to an anisotropic environment, delivering an algorithm that finds an optimal obstacle-avoiding path in a direction-dependent medium.

Section 10.2.2 extends our analysis of path finding in an anisotropic, time and space homogeneous environment to a set of problems where path curvature is constrained by a very general direction-dependent minimum turning radius function. We demonstrate the problem's controllability, prove existence of an optimal path, and invoke techniques from optimal control theory to derive a necessary condition for optimality. Further analysis characterizes an optimal path and delivers an algorithm that facilitates the implementation of the presented results.

The assumption of time and space homogeneity is relaxed in Sect. 10.3, where we develop a dynamic programming model to find an optimal path in a location-, direction- and time-dependent environment. The results from the preceding section are integrated into the model to improve its accuracy, efficiency, and run-time. The path finding model addresses limited information availability (Sect. 10.3.1), control-feasibility (Sect. 10.3.2), and computational demands of a time-dependent environment (Sect. 10.3.3). The step-by-step path-finding algorithm (Sect. 10.3.4) and its application to the Optimum Vessel Performance in Evolving Nonlinear Wavefields project (Sect. 10.3.5) are also presented in this section. In Sect. 10.4 we discuss how to extend our analysis to optimal path-finding problems for cost functions other than travel time. The chapter concludes with Sect. 10.5 summarizing the results, contributions and future directions of our work.

10.2 Optimal Path Finding in Direction-Dependent, Time and Space Homogeneous Environment

We begin our study of dynamic navigation in direction-dependent (anisotropic) environment by focusing our attention on the effects the direction-dependence has on path optimality. As discussed in the introduction, the anisotropic cost function is not a metric, since traveling along the straight line path from a to b does not necessarily incur the same cost as traveling along the reversed straight line path from b to a. In addition, the triangle inequality might not hold true, and the straight line path is not always optimal. In this section, we assume time and space homogeneity of the environment and extend existing results from computational geometry and control theory to the direction-dependent medium. In the following Sect. 10.3, these results are integrated into the optimal path finding algorithm for a direction-, location- and time-dependent environment.

10.2.1 Optimal Path Finding Without Turning Radius Constraint

We address the fastest-path-finding problems with anisotropic environment where the speed function, $V(\theta)$, is direction dependent, such as in the case when ocean waves, winds, or slope of the terrain affect agent's motions. Our objective is to find a path from a given starting location to a given target point that minimizes total travel time of the mobile agent. We first solve this problem in an obstacle-free domain and then integrate polygonal obstacles restricting the set of feasible paths. It is important to note that while we demonstrate our analysis and results for the fastest-path-finding problem, the discussion can be easily extended to other additive direction-dependent cost functions (e.g., fuel consumption). This section presents a brief overview of our joint work with R.L. Smith; for more detailed analysis and results, see Dolinskaya (2009), and Dolinskaya and Smith (2012).

Let $\mathcal{S}(V(\theta)) \subseteq \Re^2$ denote the set of points enclosed by the polar plot of a speed function $V(\theta)$ centered in the origin point $O = (0,0) \in \Re^2$. From the definition it follows that $\mathcal{S}(V)$ contains all the points that the mobile agent can reach from point O along the straight line path within a single unit of time. We let $\tau(V,x)$ for $x \in \Re^2$ denote the travel time along the straight line path from O to x for the speed function $V(\theta)$. Note, for all $x \in \mathcal{S}(V)$, $\tau(V,x) \leq 1$.

First, consider a special case when $\mathcal{S}(V(\theta))$ is a convex set. Then, properties of the Minkowski functional (Luenberger 1969) establish that the straight line path is the fastest path between a pair of points for the speed function $V(\theta)$, and that the triangle inequality holds true. In the case when the speed polar plot for $V(\theta)$ is not convex, we consider the augmented speed function $V_a(\theta)$, such that its polar plot is equal to the convex hull of the original speed polar plot, i.e., $\mathcal{S}(V_a(\theta)) = Conv(\mathcal{S}(V(\theta)))$ (see Fig. 10.1). Then, the speed function V_a corresponds to a convex

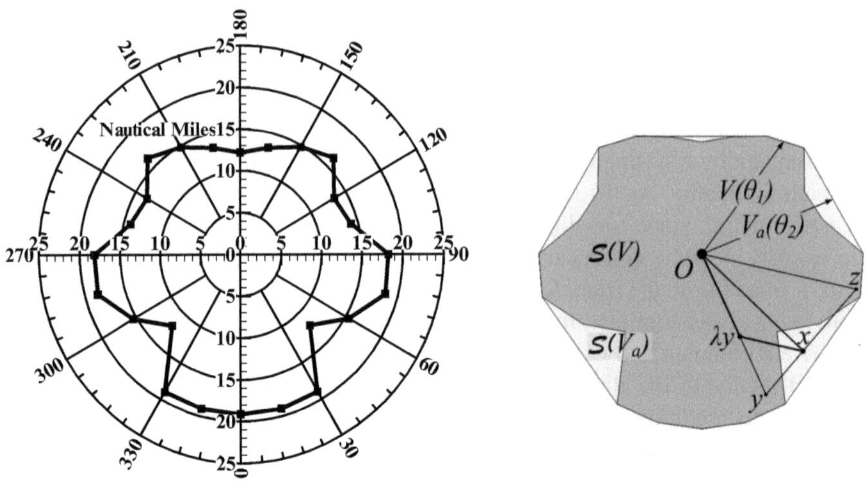

Fig. 10.1 An example of speed polar plot for the S-175 containership (Dolinskaya et al. 2009) and corresponding application of our analysis

speed polar plot and the straight line optimal paths. Furthermore, $V(\theta) \leq V_a(\theta)$ for all θ, and the travel time along an optimal (straight line) path with the speed function V_a is a lower bound on the optimal travel time between the same two points with the speed function $V(\theta)$. That is, $\tau(V_a, x)$ is the lowest possible travel time between points O and x for a mobile agent with the speed function $V(\theta)$.

Observe that, since $\mathcal{S}(V_a)$ is the convex hull of $\mathcal{S}(V)$, every point in $\mathcal{S}(V_a)$ can be expressed as a convex combination of two points in $\mathcal{S}(V)$. In other words, $\forall x \in \mathcal{S}(V_a)$, there exist $y, z \in \mathcal{S}(V)$ and $\lambda \in [0,1]$ such that $x = \lambda y + (1-\lambda)z$. Then, every point $x \in \mathcal{S}(V_a)$ can be reached following a piecewise linear path from O to λy and then from λy to x. Finally, observe that $\tau(V,y) = \tau(V_a,y)$, $\tau(V,z) = \tau(V_a,z)$ and the travel time along the described piecewise linear path with speed function $V(\theta)$ is equal to $\tau(V_a,x)$, thus establishing its optimality. See Fig. 10.1 for an illustration of the presented analysis for the test vessel used in the Optimum Vessel Performance in Evolving Nonlinear Wavefields project.

We have established that in an obstacle-free domain, an optimal path for an arbitrary anisotropic speed function is piecewise linear with at most one way-point. Next, we employ these findings to the problems that consider the presence of polygonal obstacles. For the case when the speed function corresponds to a convex polar plot, the straight line path is a fastest path in $\Re^2$. Therefore, fastest-path finding in the presence of polygonal obstacles can be restricted to a modified visibility graph, similar to Euclidian shortest path-finding problems (Alt and Welzl 1988; Lozano-Pérez and Wesley 1979) (which essentially searches for fastest path among taut-string paths between starting and target points). We construct the visibility graph by defining the set of vertices to be all the vertices of the polygonal obstacles, as well as the starting and target points. The set of edges of our visibility graph

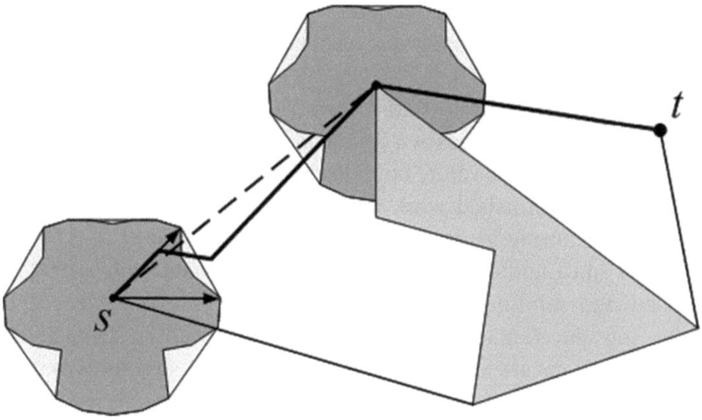

Fig. 10.2 An example of an obstacle-avoiding fastest path for the S-175 containership

consists of all the straight line edges interconnecting the defined vertices that do not intersect any of the obstacles. And finally, the cost function for each edge is set to be the travel time along the link characterized by $\tau(V,.)$ function. Then, a minimum cost path in the constructed graph corresponds to the fastest obstacle-avoiding path for our original problem.

The triangle inequality might not hold true for a general direction-dependent speed function $V(\theta)$. In such case, an augmented speed function, $V_a(\theta)$, corresponding to the convex hull of the original speed polar plot is used to find a lower bound on the minimum travel time for our problem. We construct a visibility graph as discussed above for the augmented speed function, $V_a(\theta)$. We then implement the piecewise linear results presented for the obstacle-free domain along each edge of the optimal path in this graph (see Fig. 10.2). Thus, we construct an obstacle-avoiding path that achieves this lower bound, implying its optimality.

10.2.2 Optimal Path with Bounded Curvature in an Anisotropic Medium

Our initial analysis described above shows that an optimal path in an anisotropic time and space homogeneous environment has a piecewise-linear structure. Unfortunately, the instantaneous heading change required to follow a piecewise-linear path is infeasible for most applications, including the navigation of aerial, ground, and naval vehicles. For these problems, the control system of an agent constrains the set of feasible paths and these restrictions must be integrated into the optimal path finding process. We introduce a direction-dependent minimum turning radius function $R(\theta)$ that constrains the curvature of a feasible path for a vehicle with the heading direction θ and find an optimal path with bounded curvature in an anisotropic time and space homogeneous environment. More detailed analysis and results of the work presented here can be found in Dolinskaya (2009), and Dolinskaya and Maggiar (2012).

The problem's objective is to find a fastest path that starts at the initial point (x_s, y_s) and heading angle θ_s, ends at a destination point (x_t, y_t) with a predetermined final heading θ_t, and has a curvature bounded by a specified minimum turning radius function $R(\theta)$. Most published work that discusses fastest-path-finding problems with bounded curvature (e.g., Boissonnat et al. (1994), Bui et al. (1994), Dubins (1957), Souères and Laumond (1996), and Sussmann and Tang (1991)) assumes constant speed and minimum turning radius. When the presence of direction-dependence is introduced in the existing literature, the resulting speed and minimum turning radius functions are assumed to maintain specific structures and properties. We analyze the problems in the anisotropic media where both the agent's speed and minimum turning radius are described by generalized direction-dependent functions. The direction-dependent nature of this problem implies the same asymmetry of a travel time function as discussed in preceding sections. Additionally, the non-constant turning radius results in complex sharpest turn curves, as opposed to a circle, which is an essential part of an optimal path for the isotropic problems. These facts make the task of extending the problem of optimal path finding with minimum curvature to the direction-dependent case a significant challenge.

Let $(x(t), y(t), \theta(t)) \in \Re^2 \times S^1$ denote the vehicle configuration at time $t \in [0, T]$, where $(x(t), y(t))$ are the coordinates of the mobile agent position and $\theta(t)$ is its heading angle with respect to the x axis. We set the system steering controller $u(t)$: $[0, T] \to [-1, 1]$ to represent the rate of change of the vehicle heading at time t. Then, we can write our fastest path finding problem as the following differential system:

$$\begin{aligned} & \min_u T \\ \text{subject to} \quad \dot{x} &= V(\theta)\cos(\theta), \\ \dot{y} &= V(\theta)\sin(\theta), \\ \dot{\theta} &= \frac{V(\theta)}{R(\theta)} u, \end{aligned}$$

with the boundary conditions:

$$\begin{aligned} (x(0), y(0), \theta(0)) &= (x_s, y_s, \theta_s), \\ (x(T), y(T), \theta(T)) &= (x_t, y_t, \theta_t). \end{aligned}$$

We demonstrate the problem's controllability by reducing our problem to Dubins car problem (Dubins 1957), which has been established to be controllable (Sussmann and Tang 1991). Next, we prove the existence of an optimal path using Filippov's Theorem (Filippov 1962). Then, we employ optimal control theory and

Pontryagin's principle (Pontryagin et al. 1962) and derive the necessary condition for optimality that states that any optimal path is the concatenation of the arcs with minimum turning radius $R(\theta)$ and the straight line segments. Note that this condition for optimality holds true for the very generalized anisotropic speed function $V(\theta)$ and minimum turning radius function $R(\theta)$. Further analysis delivers the detailed characterization of an optimal path structure.

Theorem 1 (From Dolinskaya and Maggiar (2012)). *There exists an optimal path from an initial configuration* (x_s, y_s, θ_s) *to a target configuration* (x_t, y_t, θ_t) *such that it is a portion of a path of type* $\mathcal{CSCSC}$ *where* $\mathcal{C}$ *denotes a sharpest turn and* $\mathcal{S}$ *a straight line.*

For the case when the speed polar plot is convex, we have more specific characterization.

Theorem 2 (From Dolinskaya and Maggiar (2012)). *When the movement along a path is characterized by a speed function with a convex polar plot, an optimal path from* (x_s, y_s, θ_s) *to* (x_t, y_t, θ_t) *is of the form* $\{\mathcal{C}, \mathcal{C}, \mathcal{C}\}$, *or* $\{\mathcal{C}, S, \mathcal{C}\}$, *where* $\mathcal{C}$ *denotes a sharpest turn curve and S denotes a straight line segment. It is implied that a path of the form* $\{\mathcal{C}, \mathcal{C}, \mathcal{C}\}$ *alternatively switches between left-hand and right-hand sharpest turn curves.*

See Dolinskaya and Maggiar (2012) for the detailed proofs of these theorems and an algorithm that facilitates the implementation of the results.

10.3 Dynamic Programming Modeling for Optimal Path Finding in a Direction-, Location- and Time-Dependent Environment

In the preceding section, an optimal path in a time and space homogeneous direction-dependent environment was found; in other words, the cost function and constraints are assumed to be independent of the time and location of an agent. Here, we relax the assumption of homogeneity and discuss a dynamic programming model for optimal path finding in a direction-, location- and time-dependent environment. This section is a brief overview of our work discussed in Dolinskaya (2009, 2012).

10.3.1 Limited Visibility Horizon

Current technological advancements in real-time data collection and forecasting call for an explicit incorporation of the available information into the decision-making process. Innovative on-board sensors, such as a Doppler radar in the

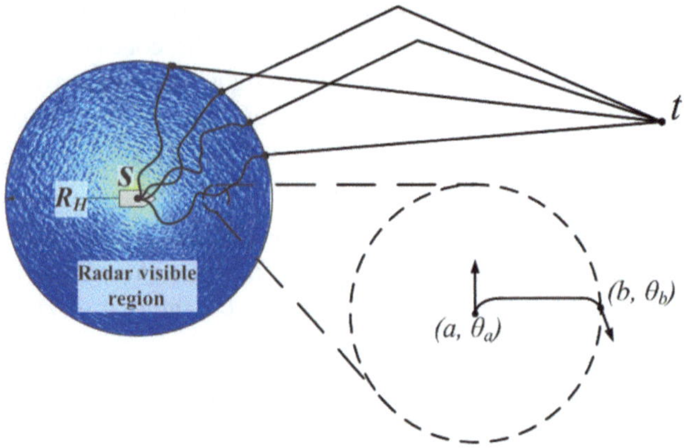

Fig. 10.3 Dynamic Programming model finds the fastest paths to the points on the visibility horizon, then results for time and space homogeneous medium are used to find the best paths to continue

Optimum Vessel Performance in Evolving Nonlinear Wavefields project, collect information about the surrounding environment in real time. The optimal path-finding model presented here makes use of the gathered information. It is important to acknowledge that physical sensors have a limited visibility horizon and cannot gather information about the medium beyond a specified distance, which is often closer than the location of the target point. To address this restriction, our model presumes to have complete information about the environment (i.e., cost function and constraints) within the radar visibility horizon (denoted by R_H) and limited information beyond that horizon (see Fig. 10.3).

Inside the radar visible region (i.e., region within R_H distance from the current location of the mobile agent) the medium is completely characterized by the available information. This allows us to construct a detailed dynamic programming model (to be described below) to evaluate optimal paths to all the points on the boundary of the visible region. Due to the limited information available to us about the environment beyond R_H, we approximate that region of the medium by a stationary distribution characterized by a global parameter. For example, in the case of vessel routing, a parameter called *sea state* characterizes the distribution of waves to be encountered for a period of few hours. As a result, we assume the environment beyond R_H is time and space homogeneous for the duration of the trip. This assumption facilitates the use of our earlier results (see Sect. 10.2) to find an optimal path from each point on the boundary of the radar visible region to the target point. Note that as the mobile agent travels along a path, the radar visible region moves with it, and the optimal path is reevaluated in real time incorporating the new information.

10.3.2 System Dynamics Restrictions

The traditional dynamic programming (DP) path-finding models discretize the path domain into a set of way-points and find an optimal ordered set of way-points to traverse from the starting point to a target location. Since the paths between a consecutive pair of way-points are assumed to be a straight line segment, the resulting optimal path is piecewise linear. However, in a number of real-life applications, such as navigation of aerial, ground, and naval vehicles, piecewise linear paths are not feasible, since the curvature of a feasible path is constrained by a minimum turning radius function. Instead of the traditional approach of addressing the optimal path-finding and path-following aspects of the problem separately, we integrate the system's operability and dynamics constraints into an optimization model resulting in a control-feasible solution.

We develop a dynamic programming model to be implemented inside the visible region to find an optimal path to all the points on the boundary of this region (i.e., R_H distance away from the agent's current location). We discretize the path domain into a set of way-points l distance away from each other, and the environment of the l-radius neighborhood surrounding each way-point is assumed to be time and space homogeneous. We set our dynamic programming state to include the location (i.e., way-point), as well as the heading angle of the mobile agent at the way-point. Then, at each state, the dynamic programming model decides on the next way-point and the heading angle with which to arrive at that point. Since the environment between the two consecutive way-points is assumed to be time and space homogeneous, we use our results for optimal path with bounded curvature presented in Sect. 10.2.2 to find the fastest path and corresponding travel time between the DP states (see Fig. 10.3).

10.3.3 Computational Demand of a Time-Dependent Environment

Due to time dependency of the environment and cost function, it is not necessarily optimal to arrive at each intermediate way-point of a path as soon as possible. To account for this fact, a time variable is traditionally added to the dynamic programming state in order to keep track of all possible times at which we might arrive, and consequently leave, a given point. This additional variable increases the number of DP states to be considered by orders of magnitude. At the same time, computational demand and run-time of the optimal path-finding model is of particular significance to timely utilization of the available real-time information in a decision-making process. We developed an alternative formulation of the dynamic programming functional equation, which allows us to eliminate the time variable from the state space, thus significantly reducing the computation time of our algorithm.

Our approach is built on work by Dreyfus (1969), who was the first one to demonstrate that in the case when unlimited waiting in the nodes is allowed, the dynamic fastest path problem can be solved using Dijkstra's algorithm as efficiently as in the case of a static network. Dreyfus redefines the cost function $d_{ij}(t_i)$ (the cost of traveling from node i to node j when leaving node i at time t_i), so that "if travel schedules are such that a delay before departure decreases the time of arrival, $d_{ij}(t_i)$ represents the elapsed time between time t_i and the earliest possible time of arrival." This alternative definition of $d_{ij}(t_i)$ ensures the validity of the consistency condition (established in Kaufman and Smith (1993)) and facilitates a straightforward dynamic programming formulation of the problem, where DP state only stores the current location in the network. However, stopping at the way-points is impractical or infeasible for many applications. For example, an airplane cannot stop in mid-air to wait for a storm to pass by. Similarly, it is not practical for a large vessel to come to a complete stop before continuing its travel. Consequently, we do not allow stopping or waiting in our model. Instead, we extend Dreyfus' approach to a path-finding model permitting voluntarily speed reduction (i.e., slow down) along a path.

Let $\tau(a,\theta_a,b,\theta_b,t_a)$ denote the vessel travel time along the fastest path with bounded curvature from point a to point b starting at heading angle θ_a at time t_a and arriving at heading angle θ_b. Define $g(a,\theta_a)$ to be the minimum travel time from the starting position (s,θ_s) to point a, arriving at a with the heading angle θ_a. Recall that the distance between two consecutive way-points is denoted by a fixed parameter l. Then, we formulate the following DP functional equation,

$$g(b,\theta_b) = \left\{ \min_{\{a,\theta_a:\|b-a\|=l\}} \{g(a,\theta_a) + \min_{\Delta_t}(\Delta_t + \tau(a,\theta_a,b,\theta_b,g(a,\theta_a)+\Delta_t))\}, \right. \tag{10.1}$$

where $\|.\|$ is Euclidean norm. By setting the initial condition $g(s,\theta_s) = 0$ and iteratively applying the functional equation (10.1), we find fastest paths to all the points on the boundary of the visible region. Note that Δ_t denotes the vessel "delay" at a given way-point before continuing its travel. Thus, the solution of our dynamic programming model returns an ordered set of optimal way-points and optimal delay time at each of those points. Since we assume that waiting (i.e., delay) is not permitted in the intermediate points of a path, we instead intentionally slow down the mobile agent to guarantee that its arrival time to a way-point coincides with the optimal time to depart it.

10.3.4 Fastest Path Finding Algorithm (Adopted from Dolinskaya (2012))

Step 1. Apply results from Sect. 10.2.2 to compute the values of $\tau(a,\theta_a,b,\theta_b,t_a)$ for all inputs where $\|a-s\| \leq R_H, \|b-s\| \leq R_H$ and $\|a-b\| = l$.

Step 2. Apply Dijkstra's algorithm to the DP recursive equation (10.1) to find the fastest paths from (s, θ_s) to a discretized set of points on the visibility horizon R_H.

Step 3. Apply results presented in Sect. 10.2.1 or Sect. 10.2.2 to find the fastest paths from the points on the visibility horizon to the target state (t, θ_t).

Step 4. Find the discretized point on the visibility horizon that has the smallest sum of the corresponding travel times found in Step 2 and Step 3. A fastest path passing through such point is the optimal path.

Step 5. For the optimal path found in Step 4, adjust the speed for each arc as discussed in Sect. 10.3.3 to ensure optimal arrival to each intermediate waypoint.

10.3.5 Numerical Results

To demonstrate the applicability and performance of our path-finding methods, we conducted computational experiments for the Optimum Vessel Performance in Evolving Wavefield project. The test runs are simulated for an S-175 containership in the Sea State 6.5 wavefield (corresponding to the mean wave height of the one third highest waves of 7 m). We compared the travel time of the found optimal path to those of the straight line path and one way-point path, which we established to be optimal for time and space homogeneous environment. We observed the average savings varying between 4% and 6%, with up to 9.7% saving. However, these estimations are very conservative due to a number of data limitations and restricted maneuverability of the 175-m long vessel. When we reduce the minimum turning radius of the test vessel by half, we report an average improvement in travel time of 12.5%. See Dolinskaya (2009, 2012) for complete discussion of our numerical results.

10.4 Optimal Path Finding for a Cost Function Other Than Travel Time

Throughout this chapter, we discuss an optimal path-finding problem with the objective of minimizing the agent's travel time. We often mention that our analysis and results can be directly extended to problems minimizing other cost functions. In many applications, we face an optimal path finding problem with alternative objective functions. For example, in the case of the Optimum Vessel Performance in Evolving Wavefield project, in addition to finding a fastest path, we are interested in minimizing root mean squared (RMS) motions, such as roll (10.2) and other measures of the path "quality."

$$\mathrm{RMS}_{\mathrm{Roll}} = \sqrt{\frac{\phi_1^2 + \ldots + \phi_n^2}{n}}. \tag{10.2}$$

However, the extension of our analysis and presented path-finding model is not straightforward for dynamic networks and path finding in a time-dependent environment. In this section we discuss how an optimal path-finding algorithm changes when the objective function is different from travel time.

The problem of minimizing cost in a dynamic network is briefly discussed in the literature. For example, Chabini (1998) looks at the minimum cost functions where travel time functions $d_{ij}(t_i)$ and cost functions $c_{ij}(t_i)$ are time dependent. He extends a backward DP formulation of the fastest path problems to this minimum cost path-finding problem. Chabini assumes that $d_{ij}(t_i)$ and $c_{ij}(t_i)$ are constant for any time greater than some specified value, resulting in a static problem. This static problem solution is then used as the boundary condition for the dynamic programming formulation of the problem.

The difference between our earlier analysis of the fastest-path-finding problems and modeling of a problem with a general cost function is that we cannot eliminate the time variable from the dynamic programming state space [see Sect. 10.3.3 and Eq. (10.1)]. Therefore, we have to set the DP state to be (a, θ_a, t_a) and consider all possible times of arrival at a given waypoint. Consequently, the resulting DP model delivers a classical functional equation and a straightforward application of Dijkstra's method or Chabini's approach to find an optimal path.

We also note that the cost function has to be additive to apply the standard dynamic programming recursive equation. However, the model can be adjusted to other objective functions. For example, the averaging measures of path quality, like RMS roll, can be implemented by fixing a constant number of arcs for all considered feasible paths, or by adding a variable keeping record of the number of arcs traveled to the dynamic programming state space. Alternately, since in the Optimum Vessel Performance in Evolving Wavefield project we are interested in minimizing roll experienced by a vessel without significant increase in trave time, we set our dynamic programming model to minimize the additive function $\phi_1^2 + \ldots + \phi_n^2$ instead of the $\mathrm{RMS}_{\mathrm{Roll}}$ defined in Eq. (10.2). This, in turn, allows the model to capture the trade-off between travel time and RMS roll of a given path. In our forthcoming work we further explore this and other similar problems.

10.5 Conclusion

This chapter discusses optimal path finding in a direction-, location- and time-dependent environment. We deliver a computationally efficient path-finding algorithm with a sufficiently small run-time for real-time implementation. A traditional dynamic programming path finding model makes a number of restrictive assumptions that jeopardize its applicability to real-life problems. Alternatively, we present a model that integrates and addresses a set of limiting aspects previously neglected in the literature:

- Our dynamic programming (DP) path-finding model integrates a limited visibility horizon and accounts for the lack of detailed information about a medium beyond a certain distance from the mobile agent's current location.
- The presented DP model finds a smooth and control-feasible fastest path by integrating the systems dynamics into the optimization process.
- By integrating the agent's controller (speed) into the decision space of the algorithm, the resulting model eliminates a time variable from the dynamic programming state space and improves efficiency and run-time of our model.

A number of special case problems corresponding to the assumption of a time and space homogeneous environment are solved analytically. These results deliver a significant contribution to the study of anisotropic (direction-dependent) problems.

We are currently working on integrating the additional system constraints, such as bounded acceleration and deceleration, into the optimal path-finding model described in Sect. 10.2.2 in order to improve its accuracy and applicability. We also plan to integrate uncertainty associated with data-collection and forecasting errors of the future environment. The goal of our ongoing work is to continue the study of integration of real-time data collection into the optimization models, especially with application to unmanned systems.

Acknowledgements The author would like to thank Robert L. Smith from the University of Michigan for his helpful guidance and discussions. This work was supported in part by the Office of Naval Research through the Multidisciplinary University Research Initiative (MURI) Optimum Vessel Performance in Evolving Nonlinear Wave Fields grant (N00014-05-1-0537) and through the Autonomous Vehicle Dynamic Navigation System grant (N00014-11-1-0516).

References

Allsopp T, Mason A, Philpott A. Optimal sailing routes with uncertain weather. In: Proceedings of The 35th Annual Conference of the Operational Research Society of New Zealand (December 2000), pp. 65–74.

Alt H, Welzl E. Visibility graphs and obstacle-avoiding shortest paths. Z Oper Res. 1988; 32(3–4): 145–164.

Bakolas E, Tsiotras P. Time-optimal synthesis for the Zermelo-Markov-Dubins problem: The constant wind case. In: Proceeding of 2010 American Control Conference (Baltimore, MD, June 30–July 2, 2010).

Boissonnat J-D, Cérézo A, Leblond J. Shortest paths of bounded curvature in the plane. J Intell Robotic Syst. 1994;11(1–2):5–20.

Bui X-N, Souères P, Boissonnat J-D, Laumond J-P. Shortest path synthesis for Dubins non-holonomic robot. In: Proceedings of the 11th IEEE Internatational Conference on Robotics Automation (1994), pp. 2–7.

Chabini I. Discrete dynamic shortest path problems in transportation applications: Complexity and algorithms with optimal run time. Transport Res Rec. 1998;1645:170–175.

Cheng S-W, Na H-S, Vigneron A, Wang Y. Approximate shortest paths in anisotropic regions. SIAM J Comput. 2008;38(3):802–824.

Chew L P, Drysdale III, RLS. Voronoi diagrams based on convex distance functions. In: SCG '85: Proceedings of the first annual symposium on Computational geometry (1985), pp. 235–244.

Collischonn W, Pilar JV. A direction dependent least-cost-path algorithm for roads and canals. Int J Geogr Inf Sci. 2000;14(4):397–406.

de Floriani L, Magillo P, Puppo E. Applications of computational geometry to geographic information systems. Handbook of computational geometry. Amsterdam: North-Holland; 2000. p. 333–388.

Dolinskaya IS. Optimal path finding in direction, location and time dependent environments. PhD thesis, University of Michigan, 2009.

Dolinskaya IS. Optimal path finding in direction, location and time dependent environments. Nav Res Logist. 2012;59(5):325–339.

Dolinskaya IS, Kotinis M, Parsons MG, Smith RL. Optimal short-range routing of vessels in a seaway. J Ship Res. 2009;53(3):121–129.

Dolinskaya IS, Maggiar A. Time-optimal trajectories with bounded curvature in anisotropic media. Forthcoming in Int J Robot Res. The International Journal of Robotics Research, 2012;31(14):1761–1793.

Dolinskaya IS, Smith RL. Fastest-Path Planning for Direction-Dependent Speed Functions: Optimization Theory and Applications. Tech. Rep. 12-01, 2012. Northwestern University, available at http://www.iems.northwestern.edu/research/papers.html. doi.10.1007/s10957-012-0248-6.

Dreyfus SE. An appraisal of some shortest-path algorithms. Oper Res. 1969;17(3):395–412.

Dubins LE. On curves of minimal length with a constraint on average curvature, and with prescribed initial and terminal positions and tangents. Am J Math. 1957;79:497–516.

Faulkner FD. A general numerical method for determining optimum ship routes. Navigation 1963;10(2):143–148.

Faulkner FD. Numerical methods for determining optimum ship routes. J Institute Navigation 1963;10(4):351–367.

Filippov AF. On certain questions in the theory of optimal control. J Soc Ind Appl Math A Contr. 1962;1(1):76–84.

Kapoor S, Maheshwari SN, Mitchell JSB. An efficient algorithm for Euclidean shortest paths among polygonal obstacles in the plane. Discrete Comput Geom. 1997;18(4):377–383.

Kaufman DE, Smith RL. Fastest paths in time-dependent networks for intelligent vehicle-highway systems application. IVHS J. 1993;1(1):1–11.

Lanthier M, Maheshwari A, Sack J-R. Shortest anisotropic paths on terrains. In: Automata, languages and programming (Prague, 1999), vol. 1644 of Lecture Notes in Compututer Science. Berlin: Springer; 1999. pp. 524–533.

Lozano-Pérez T, Wesley MA. An algorithm for planning collision-free paths among polyhedral obstacles. Commun ACM 1979;22(10):560–570.

Luenberger DG. Optimization by vector space methods. New York: Wiley; 1969.

Marks W, Goodman TR, Pierson WJ, Tick LJ, Vassilopoulos LA. An automated system for optimum ship routing. Trans Soc Naval Architects Marine Eng. 1968;76:22–55.

McGee TG, Spry S, Hedrick JK. Optimal path planning in a constant wind with a bounded turning rate. In: Proceedings of the AIAA Conference on Guidance, Navigation and Control (Keystone, Colorado, August 2006).

McNeely RL, Iver RV, Chandler PR. Tour planning for an unmanned air vehicle under wind conditions. J Guid Contr Dynam. 2007;30(5):1299–1306.

Mitchell JSB. A new algorithm for shortest paths among obstacles in the plane. Ann Math Artif Intell. 1991;3(1):83–105.

Mitchell JSB. L_1 shortest paths among polygonal obstacles in the plane. Algorithmica 1992;8(1):55–88.

Mitchell JSB. Shortest paths among obstacles in the plane. Int J Comput Geom Appl. 1996;6(3):309–332.

Mitchell JSB. Geometric shortest paths and network optimization. In: Handbook of computational geometry. Amsterdam: North-Holland; 2000. pp. 633–701.

Mitchell JSB, Papadimitriou CH. The weighted region problem: finding shortest paths through a weighted planar subdivision. J Assoc Comput Mach. 1991;38(1):18–73.

Nilim A, Ghaoui LE. Algorithms for air traffic flow management under stochastic environments. In: Proceedings of American Control Conference (July 2004), vol. 4, pp. 3429–3434.
Nilim A, Ghaoui LE, Hansen M, Duong V. Trajectory-based air trafic management (tb-atm) under weather uncertainty. In: Proceedings of the Fourth International Air Traffic Management R&D Seminar ATM (Santa Fe, New Mexico, December 2001).
Osborne J, Rysdyk R. Waypoint guidance for small UAVs in wind. In: Proceedings of the American Institute of Aeronautics and Astronautics Infotech@Aerospace Conference (Arlington, VA, 2005).
Papadakis NA, Perakis AN. Deterministic minimal time vessel routing. Oper Res. 1990;38(3): 426–438.
Perakis AN, Papadakis NA. Minimal time vessel routing in a time-dependent environment. Transport Sci. 1989;23(4):266–276.
Philpott AB. Stochastic optimization and yacht racing. In: Applications of stochastic programming, vol. 5 of MPS/SIAM Ser. Optim. Philadelphia, PA: SIAM; 2005. pp. 315–336.
Philpott AB, Mason A. Optimising yacht routes under uncertainty. In: The 15th Cheasapeake Sailing Yacht Symposium (2001).
Philpott AB, Sullivan RM, Jackson PS. Yacht velocity prediction using mathematical programming. Eur J Oper Res. 1993;67(1):13–24.
Pontryagin LS, Boltyanskii VG, Gamkrelidze RV, Mishchenko EF. The mathematical theory of optimal processes. Translated from the Russian by K. N. Trirogoff. In: Neustadt LW editor. New York: Wiley; 1962.
Reif JH, Sun Z. Movement planning in the presence of flows. Algorithmica 2004;39(2):127–153.
Rowe NC. Obtaining optimal mobile-robot paths with nonsmooth anisotropic cost functions using qualitative-state reasoning. Int J Robot Res. 1997;16(3):375–399.
Rowe NC, Ross RS. Optimal grid-free path planning across arbitrarily contoured terrain with anisotropic friction and gravity effects. IEEE Trans Robot Autom. 1990;6(5):540–553.
Rysdyk R. Course and heading changes in significant wind. J Guid Contr Dynam. 2007;30(4):1168–1171.
Souères P, Laumond J-P. Shortest paths synthesis for a car-like robot. IEEE Trans Automat Contr. 1996;41(5):672–688.
Stefanakis E, Kavouras M. On the determination of the optimum path in space. In: Frank AU, Kuhn W, editors. Spatial information theory - a theoretical basis for GIS (COSIT'95). Berlin: Springer; 1995. pp. 241–257.
Sun Z, Reif JH. On finding approximate optimal paths in weighted regions. J Algorithm. 2006;58(1):1–32.
Sun Z, Rief JH. On finding energy-minimizing paths on terrains. IEEE Trans Robot. 2005;21(1):102–114.
Sussmann HJ, Tang G. Shortest path for the Reeds-Shepp car: A worked out example of the use of geometric techniques in nonlinear optimal control. Tech. Rep. SYCON-91-10, Rutgers Center for Systems and Control, September 1991.
Techy L, Woolsey CA. Minimum-time path planning for unmanned aerial vehicles in steady uniform winds. J Guid Contr Dynam. 2009;32(6):1736–1746.
Widmayer P, Wu YF, Wong CK. On some distance problems in fixed orientations. SIAM J Comput. 1987;16(4):728–746.
Yu C, Lee J, Munro-Stasiuk MJ. Extensions to least-cost path algorithms for roadway planning. Int J Geogr Inf Sci. 2003;17(4):361–376.
Zermelo E. Über das navigationsproblem bei ruhender oder veränderlicher windverteilung. Z Angew Math Mechanik 1931;11(2):114–124.

Chapter 11
An Approach to Assess the Impact of Dynamic Congestion in Vehicle Routing Problems

H.M. Abdul Aziz and Satish V. Ukkusuri

Abstract This research proposes an integrated framework of capacitated vehicle routing problems (CVRP) and traffic flow model (cell transmission model in this research) to assess the effect of time-varying congestion. We develop a framework consisting sequence of mixed integer programs solving the CVRP with updated cost obtained from the traffic flow model. A real-world network with 15 cities and towns is tested with the framework and results show significant travel time reduction from the case where time-varying congestion is not considered. In addition, we consider system optimal type of route choice behavior within the traffic flow model.

11.1 Introduction

Vehicle routing problem (VRP) is commonly encountered in freight distribution systems and plays an important role in effective management of logistics systems. In general, VRP seeks the solution for allocating resources (vehicles or crews) with minimal cost such that all nodes in the network are visited at most only once. The cost function in VRP primarily includes travel time (from one location to all other locations) as observed in the transportation network. Evidently, the travel times on the transportation network are a function of the time-dependent flow in the network. The time-varying nature of flows is well recognized in previous works in traffic routing and network assignment. Accommodating the time-varying nature of the travel time adds congestion effects that can lead to a better design of a just-in time

H.M.A. Aziz • S.V. Ukkusuri (✉)
School of Civil Engineering, Purdue University, 550 Stadium Mall Drive,
West Lafayette, IN 47907, USA
e-mail: aziz.husain.nexus@gmail.com; sukkusur@purdue.edu

S.V. Ukkusuri and K. Ozbay (eds.), *Advances in Dynamic Network Modeling in Complex Transportation Systems*, Complex Networks and Dynamic Systems 2,
DOI 10.1007/978-1-4614-6243-9_11,

type of system (Stickel et al. 2005). Thus, it is necessary to take into account the time-varying congestion while solving the VRP problem.

Majority of the past research works [see (Toth and Vigo 2002) for an extensive review] focus on solving the VRP with non-varying travel time information. However in most real-world problems the assumptions made in the static VRP do not hold due to continuous change in traffic conditions in the network. In this research, we limit our focus only on the aspects of traffic congestion (time dependent travel time cost) and its impact on VRP solution mechanism. The travel time on different links of a road network can vary for different reasons and affect the cost function used in the VRP problem (Figliozzi 2007). One observes capacity reduction for road segments due to bad weather, incidents, work zone, etc. at different time of the day, different day of the week, or different month of a year. To obtain robust and useful solutions from VRP problem, it is important to consider of time-varying nature of travel time on road segments.

One way to account for congestion is to use a time-varying travel time function [see (Larsen 2000) for different methods used by researchers]. However, this only captures recurrent congestion to a certain degree and realistic traffic phenomenon such as spillback effects and shockwave propagation are not considered. Another approach to deal with the problem is to use the real-time information for analysis. However, using real-time traffic information to analyze the VRP problem requires installation of infrastructure and makes the computational complexity much higher in most cases (Chen et al. 2006). Information on fluctuating travel time can be acquired from traffic management center. Although the main purpose is to influence and control traffic in real time, a traffic management center also provides useful data for routing and dispatching a particular vehicle fleet (Fleischmann et al. 2004). This concerns two types of travel time information: first, there are forecasts of the variation of the travel times during the day, e.g., due to the regular rush hour congestion in certain streets. These data are not dynamic in the strong sense, because they are available in advance for the whole day. Second, there is online information on changes in travel times due to unforeseen events such as accidents, information that leads to an update of the travel time forecast. Hu et al. (2003) proposed model that uses real-time information from commercial vehicle operations (CVO) and accounts for time-sensitive demand and current traffic conditions. However, the authors mentioned that the execution time increases exponentially with the increased number of nodes. Similar computational and operational issues can also be found for conceptually similar frameworks [see (Toth and Vigo 2002) and (Larsen 2000)]. This implies that integrating the components of real-time travel time of road segments into the VRP adds computational complexity which often makes the problem intractable especially for large-scale networks.

This research a framework that accounts for the time-varying travel time obtained from a mesoscopic traffic flow simulator. Using the simulator we solve a sequence of VRP problems with updated link travel time. In addition, the proposed framework incorporates the route choice attribute of the road users which is an essential element in traffic dynamics. The rest of the paper is organized as follows: Sect. 11.2 discusses some related works from previous literature, Sect. 11.3 describes the proposed framework, and Sect. 11.4 illustrated the results for test networks. Finally Sect. 11.5 gives a short summary on the research contribution and future research direction are discussed.

11.2 Background and Related Works

The VRP has been studied with much interest in the past few decades. Initially Dantzig and Ramser (1959) proposed a linear programming-based heuristic for the VRP. Later Schrage (1981) discussed many different variants of the basic problem (frequency, time windows, varying costs, number of vehicles, etc.). Afterwards, many researchers worked on different aspects of the basic VRP problem and developed both exact (branch-and-bound, branch-and-cut, branch-and-pricing, etc.) and heuristics (tabu search, simulated annealing, genetic algorithms, ant-colony system) to get applicable solutions [for details, see (Augerat et al. 1995; Chang et al. 2003; Cordeau et al. 2001; Fukasawa et al. 2006; Laporte 1992; Lysgaard et al. 2004; Shieh and May 1998; Toth and Vigo 2002)].

Most of the previous research works are in the context of static vehicle routing problem. This implies that the travel time or the link costs are fixed for the analysis period. The concept of dynamic vehicle routing problem is relatively recent and only a few of the researchers have focused on the dynamic aspects of the vehicle routing problem. Note that, throughout this chapter dynamic aspects only refer to the time varying travel time due to congestion on road networks in context of vehicle routing problems. In the static problems all information about customers and travel times are known beforehand. On the contrary, in dynamic vehicle routing problem (DVRP) the relevant information are not deterministic and exhibit time-varying nature. A survey of the DVRP can be found in Psaraftis (1995). Time-dependent travel time, in addition to other real-time attributes (customer demand, location, time-window, etc.) is another important component in dynamic vehicle routing problem (Haghani and Jung 2005). Very few research works comprise VRP problem that accounts for time varying travel time [see (Fleischmann et al. 2004; Haghani and Jung 2005; Ichoua et al. 2003; Malandraki and Daskin 1992; Woensel et al. 2008) for recent works]. Ichoua et al. (2003) considered the time-dependent nature of travel time through time-varying travel speeds satisfying first-in-first-out principle. A restricted dynamic programming method was proposed by Malandraki and Dial (1996) for solving a time-dependent TSP which can be applicable to VRP. Chen et al. (2006) developed a series of mixed integer programming models for the time-dependent vehicle routing problems. A heuristic for route construction and improvement is proposed and rolling horizon approach is applied to predict travel time.

Most of these research works have computational issues when applied to larger networks. Further, inclusion of real time information update makes the problem more difficult (e.g., frequency of analysis, updating cost interval, and time window limitations). Only a few researchers deploy simulation tools to avoid the complexity of real-time information systems. Taniguchi and Thompson (2002) described the application of intelligent transportation systems to deal with dynamic vehicle routing problem considering time-dependent travel time deploying traffic simulation. Conrad and Figliozzi (2010) used Google data and archived historical data to assess the impact of congestion on VRP solutions using traffic simulation to represent the congestion scenarios.

Evidently, the existing literature lacks research works that integrate traffic simulation and VRP solution techniques to incorporate the time-dependent congestion effect. Besides this, the few research works related to traffic simulation and VRP do not consider any kind of route choice behavior (to the best knowledge of the authors). In this research, we seek to integrate a mesoscopic traffic flow model within the framework of vehicle routing problem. We adopt the cell transmission model (CTM) proposed by Daganzo (1994; 1995) which is a discrete approximation of Lighthill, Whitham and Richards' (LWR) hydrodynamic model (Lighthill and Whitham 1955) that assumes a piecewise linear relationship between traffic flow and density for each segment on the road. The proposed framework solves a sequence of VRP problems with updated cost from the embedded traffic flow model and thus considers the time-varying nature of travel time along with route choice behavior of road users within VRP.

11.3 Proposed Framework

In this section, the integrated VRP-CTM framework is described along with the proposed algorithm. First, we discuss primary components of the framework: the capacitated vehicle routing (CVRP) problem and cell transmission model (CTM). Afterwards, we illustrate the framework with a detailed algorithm.

11.3.1 Capacitated Vehicle Routing Problem

The vehicle routing problem (VRP) is a generalization of the traveling salesman problem (TSP) in that the VRP finds *m* vehicle routes, where each route is a tour that begins at the depot, visits a subset of the customers in a given order, and returns to the depot. The vehicle must visit each customer or node exactly once. When the total vehicle capacity is imposed as constraints on satisfying the customer demand, the variant is called capacitated vehicle routing problem (CVRP). In this research, we adopt the multi-vehicle capacitated vehicle routing problem formulation (a mixed integer program) as developed in Ahuja et al. (1993). In a CVRP, we have a fleet of vehicles at a common depot and a set of customers with demand specified. The CVRP problem seeks to determine the set of minimum cost routes for delivering (or picking up) the goods to customer sites. This research uses the simplest version of CVRP problem with the assumption of a homogenous vehicle fleet (same capacity).

11.3.2 Cell Transmission Model

The developed formulation has an embedded mesoscopic traffic flow model, namely the cell transmission model (CTM). This model was proposed by Daganzo (1994; 1995) and later modified by other researchers. CTM (cell transmission model) divides each link of the network into cells (however, cell size can be varied throughout the network). CTM can represent congestion and queue spillover effects on a road link and accordingly can serve as an ideal choice to capture traffic dynamics due to desirable properties such as the link spillover and shockwave propagation (Ukkusuri and Waller 2008; Ukkusuri et al. 2010; Ziliaskopoulos 2000). However, CTM represented as a linear program has few drawbacks such as the vehicle holding back problem, representation of merging and diverging, and First-In-First-Out(FIFO) violation for multiple origin-destination network [see (Ziliaskopoulos 2000; Ukkusuri and Waller 2008; Zheng and Chiu 2011) for details]. The flow propagation conditions in original CTM constitute a feasible region of non-convex set (Daganzo 1994; Zheng and Chiu 2011). Ziliaskopoulos (2000) relaxed the formulation and proposed a linear programming formulation of system optimal objective with embedded CTM for single destination. Later Ukkusuri and Waller (2008) and Ukkusuri et al. (2010) modified the formulation for the multi-destination system optimal objective. In this paper, for traffic simulation without any route choice behavior we adopt the Daganzo's (1995) original network-based model. To implement system optimal-based CTM simulation, we adopt the formulation by Ukkusuri and Waller (2008) and Ukkusuri et al. (2010).

11.3.3 Integrated Framework and Solution Algorithm

The integrated framework can be illustrated with the help of the following algorithm:

Step 0. Initialization:

(a) Path generation for all O–D pairs in the network. Dijkstra's shortest path algorithm can provide with shortest paths between any two nodes of the road network.
(b) Travel time matrix is developed either using free flow travel time or best available historical data (either from direct observation or possible simulation).
(c) Cost matrix for the CVRP is prepared from the initial travel time matrix.
(d) Solve the CVRP using initial cost matrix. We will call it the basic solution.

Step 1. Tour initiation:

(a) From the basic solution, start from depot O and visit the first node i_O as determined in the basic solution.
(b) Set i_O as the current node i_{current} and O as the previous node i_{previous}.

(c) Obtain the travel time for $i_{previous} \rightarrow i_{current}$ and set it as current time $T_{current}$ (e.g., if the vehicle starts at 8 AM and reaches the first node at 9 AM, then current time is 9 AM).

Step 2. Update traffic status:

(a) Obtain the traffic status from traffic simulation (CTM in our case) at $T_{current}$. Traffic status includes link travel time, average speed, queue length (if any), and density, etc.
(b) Update the travel time matrix based on current traffic status (only if travel times are changed).
(c) Update the cost matrix for CVRP.
(d) Within the CVRP formulation, set condition that the vehicle already has traversed $i_{previous} \rightarrow i_{current}$.
(e) If the cost matrix is updated, then solve the CVRP with updated cost matrix and condition defined as in (d).

Step 3. Tour update:

(a) Get the next node j from $i_{current}$ from the updated solution obtained in step 2.
(b) Set $i_{current}$ as the previous node $i_{previous}$.
(c) Set j as the current node $i_{current}$.
(d) Set $i_{previous} \rightarrow i_{current}$ as current time $T_{current}$.

Step 4. Tour Termination:

(a) If current node $i_{current}$ is the depot, tour is complete and exit the algorithm.
(b) Otherwise go back to step 2.

Figure 11.1 shows the flow chart of the proposed algorithm.

11.4 Numerical Results and Discussions

The proposed framework is tested with two networks: Test-Network-1 (toy network with seven nodes) and Test-Network-2 (real-world network containing 15 cities within the state of Indiana and the state of Illinois in the U.S.). For Test-Network-1 we apply CTM simulation with route choice behavior of the road users (system optimal formulation) and for Test-Network-2 we apply CTM simulation with and without user behavior consideration. In addition, we also experiment the networks with low and high level of demands at different locations in the network. As expected, the low demand profiles do not produce any significant changes in the basic solution of the CVRP. Therefore, we only include the results from high demand profiles for the test networks in this paper. For the test networks, we have made the following assumptions:

1. No traffic control characteristics in the network such as signal control and ramp meter are considered.

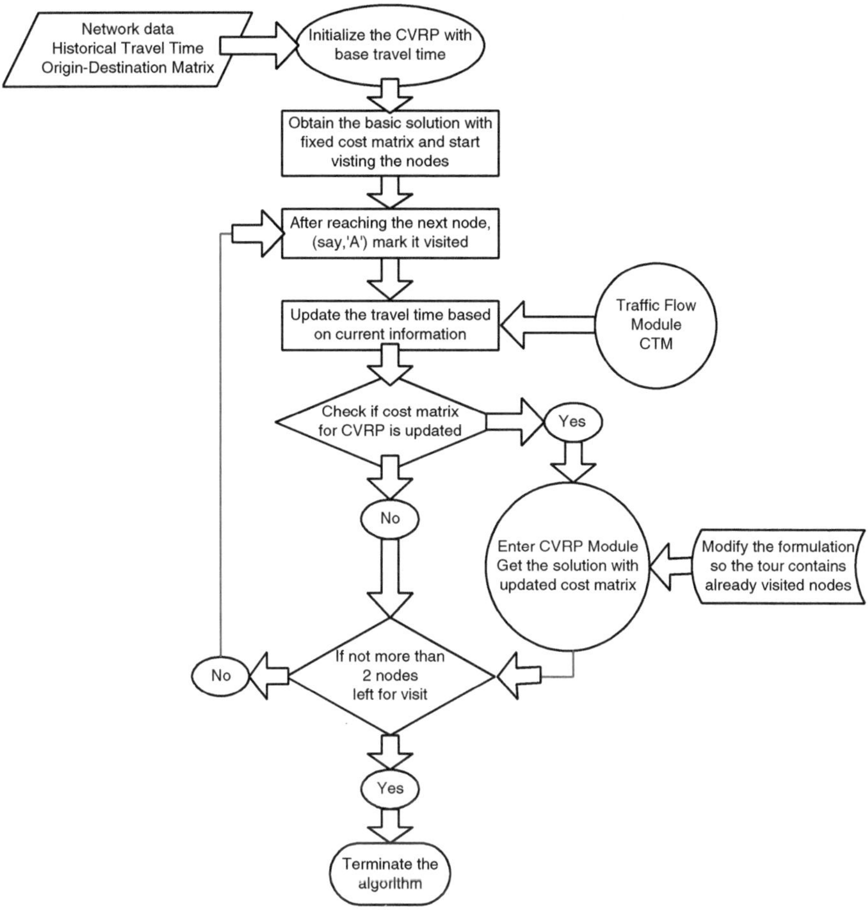

Fig. 11.1 Flowchart of the proposed algorithm

2. Demand profiles for the network are known to the analyst.
3. The CVRP costs are symmetric and we only consider travel time (since operating cost other maintenance costs are fixed in most cases) as the cost for traversing from one node to another.

11.4.1 Test-Network-1

Table 11.1 describes the input parameters and geometric characteristics for the Test-Network-1 (second column). Figure 11.2 shows the Test Network-1 with its nodes and arcs. Node 1 is the depot and nodes from 2 to 7 are the customer site required to be served in the CVRP. This simple network is used to illustrate the algorithm.

Table 11.1 Geometric attributes and input parameters for Test-Network-1 and Test Network-2

Attributes	Test Network-1	Test Network-2 (15 City Network)
Total number of nodes in the traffic network	12	19
Total number of road segments in the network	17	26
Total number of customers/cities in the CVRP	7	15
Demand range at origins (vphpl[a])	1,600–2,100	1,800–2,400
Saturation flow(vphpl)	1,900	2,160
Free flow speed or posted speed limit (mph[b])	40	60
Backward propagation speed(mph)	40	45
Minimum cell length (miles)	2.33	10
Time step (minutes)	3.5	10
Jam density(max. no. of vehicles in a cell)	200	1,056
Max. flow allowed (no. of vehicles per time step)	110	360

[a]vehicles per hour per lane
[b]miles per hour

We only apply system optimal-based linear program (Ziliaskopoulos 2000) to simulate CTM for this network. This implies that a system optimum-based route choice behavior is considered in the traffic simulation in CTM (Figs. 11.3, 11.4).

11.4.2 Results for Test-Network-1

Table 11.3 (Part-A) shows the results obtained for Test-Network-1. The results show the effect of congestion on the CVRP solutions at different iterations. The network is heavily loaded at the beginning of the simulation. As the vehicle starts the tour, the travel times on road links begin to rise and the cost matrix for the CVRP is affected. As a result, we see the change in the VRP solution in the first iteration. Similarly in the second iteration the travel times change the CVRP cost matrix and accordingly the tour (sequence of visiting nodes). One should note that it is not necessary that the CVRP solution must change with the change of travel times in the road networks. This is shown in the results of Test-Network-2. For some iterations, the results do not show any significant changes in the travel time that can affect the CVRP cost matrix. Finally, we obtain the modified CVRP

Fig. 11.2 Test-Network-1

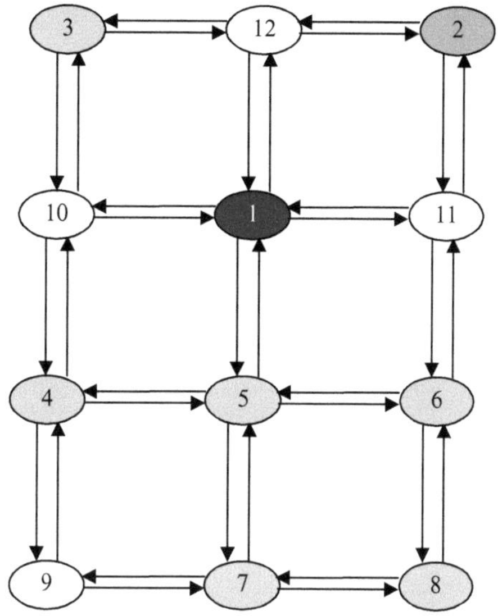

solution. If we compare the modified CVRP solution with basic CVRP solution (without considering the congestion), we see that there is a significant reduction in travel time.

The Basic CVRP solution yields 84 min of travel time with fixed travel time (fixed cost matrix). However in actual case, the travel time will change due to high congestion and the basic solution yields 146 min (this total travel time is calculated based on the varying travel time on the links). Now, the proposed framework that accounts for time-varying congestion yields a CVRP solution with total travel time of 116 min. Since the proposed framework takes into account the current traffic condition and updates CVRP cost matrix, the solution clearly shows significant reduction in travel time.

11.4.3 Test-Network-2

Table 11.1 describes the input parameters and geometric characteristics for the Test-Network-2 (third column). This network comprises of 10 cities and 5 towns in the states of Indiana and Illinois (in the U.S.). All these locations are connected with highways and interstate freeways. In the simulation we consider the on-ramp and off-ramp (exits) as the sources and sinks, respectively. Since the network mostly contains interstate freeways, without loss of generality we assume no traffic control over the network. To represent congestion we use high demand at different on-ramp nodes of the network. Further, bottlenecks are found at some exit points when many vehicles reach the sink which slows down the vehicles on the freeway. The intent of using Test-Network-2 is to present a real-world CVRP where distribution centers

Fig. 11.3 Test-Network-2 with geographical location ID (source: http://www.bing.com/maps/)

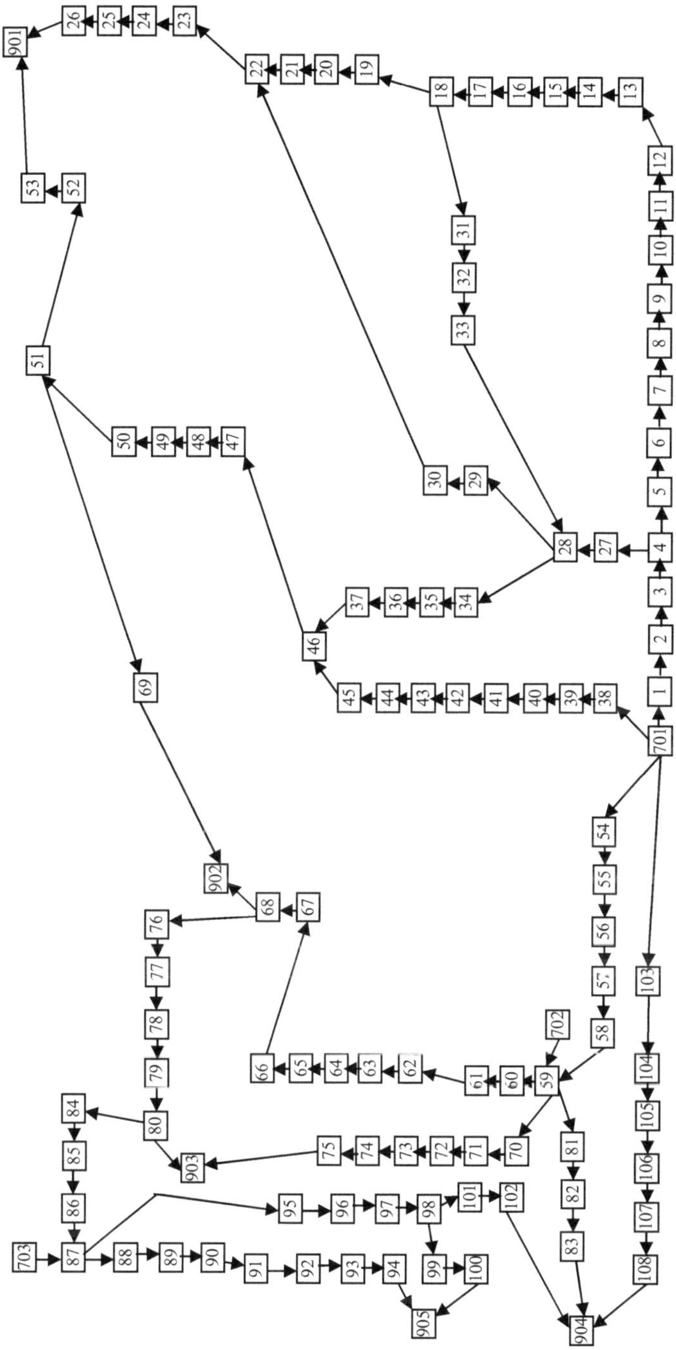

Fig. 11.4 Cell representation of Test-Network-2

Table 11.2 Geographical locations used in the real-world test network (Network-1)

Node ID	Location	Entity	State
1	Champaign	City	Illinois
2	Covington	Town	Indiana
3	Indianapolis	City	Indiana
4	Lafayette	City	Indiana
5	Rensselaer	City	Indiana
6	Merrillville	Town	Indiana
7	Manteno	Town	Illinois
8	Bloomington	City	Illinois
9	Pontiac	City	Illinois
10	Joliet	City	Illinois
11	LaSalle	City	Illinois
12	Braidwood	Town	Illinois
13	Galesburg	City	Illinois
14	Rock Island	Town	Illinois
15	Lincoln	City	Illinois

are located at different geographical locations and the network is connected with highways and freeways (Table 11.2).

11.4.4 Results for Test-Network-2

Table 11.3 (Part-B and Part-C) shows the results obtained for Test-Network-1 from traffic simulation with and without route choice behavior, respectively. Similar to Test-Network-1, we observe the effect of congestion on the cost matrix of CVRP with time. In iteration-1 of Table 11.3 (Part-B) we see changes in the travel time and corresponding change in the solution of CVRP. In iteration-2, one should notice that even though the travel times are changed, there is no effect on the solution of the CVRP. This happens because the changes in the travel times in the links do not significantly change the cost hierarchy of the CVRP nodes. The travel time on different links constitute the cost from one node to another node. Therefore, changes in few links might not change the cumulative cost from one node to another in CVRP. Further, the changes in the cost from one node to another may not change the sequence of visiting nodes in CVRP solution. This is possible because the relative cost of visiting different nodes affects the CVRP solution. Thus, the change of cost from one node to another might not change the relative cost and as a result the CVRP solution remains same.

The results in Table 11.3 (Part-B) show that there is no change in the CVRP solution up to node 15 and then the CVRP solution changes. Afterwards, there is no significant changes in the solution (even though we observe travel time changes as obtained in the traffic simulation). The total travel time for the basic solution

Table 11.3 Results obtained from the test networks (Network-1 and Network-2)

Algorithm Progress	Current CVRP Solution	Previous Node	Current Node	Travel time Update	VRP Cost Matrix Update	Updated CVRP solution	Tour Changed	Updated current Node
Test Network-1 (System Optimal Dynamic Traffic Simulation in CTM using Linear program framework) PART-A								
Iteration-1	**1–2**→2–3→3–4 →4 – 7→7–8→ 8–6→6–5→5–1	1	2	Yes	Yes	1–2→2–6→6–8 →8 – 7→7–5→ 5–4→4–3→3–1	YES	6
Iteration-2	**1–2→2–6**→6–8 →8 – 7→7–5→ 5–4→4–3→3–1	2	6	Yes	Yes	1–2→2–6→6–8→ 8–5→5–7→7–4→ 4–3→3–1	YES	8
Iteration-3	**1–2→2–6→6–8** →8 – 5→5–7→ 7–4→4–3→3–1	6	8	No	No	No	NA	5
Iteration-4	1–2→2–6→6–8 →8 – 5→5–7→ 7–4→4–3→3–1	8	5	No	No	No	NA	7
Iteration-6 (final Iteration)	**1–2→2–6→6–8** →**8** – 5→5–7→ 7–**4**→4–3→3–1	7	4	No	No	No	NA	3
Test Network-2 (System Optimal Dynamic Traffic Simulation in CTM using Linear program framework) PART-B								
Iteration-1	**1–8**→8–9→9–12 →12 – 10→10–11 →11 – 14→14–13 →13 – 15→15–2 →2 – 3→3–4→4–5 →5 – 6→6–7→7–1	1	8	Yes	Yes	1–8→8–9→9–12 →12 – 10→10–11 →11 – 14→14–13 →13 – 15→15–7 →7 – 6→6–5→5–4 →4 – 3→3–2→2–1	Yes	9
Iteration-2	**1–8→8–9**→9–12 →12 – 10→10–11 →11 – 14→14–13							

(continued)

Table 11.3 (continued)

Algorithm Progress	Current CVRP Solution	Previous Node	Current Node	Travel time Update	VRP Cost Matrix Update	Updated CVRP solution	Tour Changed	Updated Next Node
	→ 13 – 15 →15–7→7–6 → 6 – 5 →5–4→4–3→3–2	8	9	Yes	Yes	No	No	12
Iteration-3	**1–8→8–9→9–12** → 12 – 10 →10–11 → 11 – 14 →14–13 → 13 – 15 →15–7 → 7 – 6 →6–5→ 5 – 4 → 4 – 3 →3–2	9	12	Yes	Yes	No	No	10
(After reaching node-15) Iteration-4	**1–8→8–9→9–12** **→ 12-10 →10–11** **→11–14→14–13** **→ 13-15**→15–7 →7-6→6–5→5–4 →4-3→3–2	13	15	Yes	Yes	1–8→8–9→9–12 → 12 – 10 →10–11 →11–14→14–13 → 13–15→15–2 → 2-7→7–6→6–5 → 5 – 4 →4–3→3–1	Yes	2
Iteration-5	**1–8→8–9→9–12** **→ 12-10 →10–11** **→ 11-14 →14–13** **→13-15→15–2**	15	2	Yes	Yes	No	No	7

	→ 2-7→7–6→6–5 → 5 – 4 →4–3→3–1							
Iteration-6	**1–8→8–9→9–12** **→ 12-10 →10–11** **→ 11-14 →14–13** **→ 13-15 →15–2** **→2–7**→7–6→6–5 → 5 – 4 →4–3→3–1	2	7	No	No	No	No	6
Final-Iteration	**1–8→8–9→9–12** **→ 12-10 →10–11** **→ 11-14 →14–13** **→ 13-15 →15–2** **→ 2-7 →7–6→6–5** → 5 – 4 →4–3→3–1	6	5	Yes	Yes	No	No	4
Test Network-2 (Traffic Simulation in CTM using without any Route-choice Framework) PART-C								
Iteration-1	**1–8**→8–9→9–12 → 12 – 10 →10–11 → 11 – 14 →14–13 → 13 – 15 →15–2 → 2 – 3 →3–4→4–5 → 5 – 6 →6–7→7–1	1	8	Yes	Yes	No	No	9
Iteration-2	**1–8→8–9**→9–12 → 12 – 10 →10–11 → 11 – 14 →14–13					1–8→8–9→9–12 → 12 – 10 →10–11 → 11 – 14 →14–13		

(continued)

Table 11.3 (continued)

Algorithm Progress	Current CVRP Solution	Previous Node	Current Node	Travel time Update	VRP Cost Matrix Update	Updated CVRP solution	Tour Changed	Updated current Node
	→ 13 – 15 →15–2 → 2 – 3 →3–4→4–5 → 5 – 6 →6–7→7–1	8	9	Yes	Yes	→ 13 – 15 →15–3 → 3 – 4 →4–5→5–6 → 6 – 7 →7–2→2–1	Yes	12
Iteration-3	**1–8→8–9→9–12** → 12 – 10 →10–11 → 11 – 14 →14–13 → 13 – 15 →15–3 → 3 – 4 →4–5→5–6 → 6 – 7 →7–2→2–1	9	12	Yes	Yes	No	No	10
Iteration-4	**1–8→8–9→9–12** **→12–10**→10–11 → 11 – 14 →14–13 → 13 – 15 →15–3 → 3 – 4 →4–5→5–6 → 6 – 7 →7–2→2–1	12	10	No	No	No	No	11
(After reaching Node 4) Iteration-5	**1–8→8–9→9–12** **→ 12-10 →10–11** **→ 11-14 →14–13** **→ 13-15 →15–3** **→3–4**→4–5→5–6 → 6 – 7 →7–2→2–1	3	4	Yes	Yes	1–8→8–9→9–12 → 12 – 10 →10–11 → 11 – 14 →14–13 → 13 – 15 →15–3 → 3 – 4 →4–7→7–6 → 6 – 5 →5–2→2–1	Yes	7

Final Iteration	**1–8→8–9→9–12 → 12-10 →10–11 → 11-14 →14–13 → 13-15 →15–3 → 3-4 →4–7→7–6** →6–5→5–2→2–1	7	6	No	No	No	No	5

is 787 min when the travel times are assumed to be unchanged within the tour. If this solution is followed, the total travel time experienced would be 841 min due to travel time changes in the link for time-varying congestion. The proposed integrated framework in this paper offers a solution that accounts for the time-varying travel time and in that case total travel is 811 min for the solution tour in CVRP.

Next, Table 11.3 (Part-C) shows the results for Test-Network-2 without route choice consideration. We see similar trends of the results. The basic solution yields 787 min with fixed travel time assumption. The simulation suggests the basic solution would actually yield 865 min due to time-varying congestion. On the contrary, the CVRP solution obtained from the proposed framework yields 834 min for the tour. Thus, the proposed framework accounts for time-varying congestion and provides solution that saves travel time.

11.4.5 System Optimal Behavior vs. Predefined Route Choice Behavior

From the results as seen in Table 11.3 (Part-B and Part-C) we see some differences between the solution with (Part-B) and without (Part-C) route choice behavior. When we consider the system optimal behavior of all the users in the road network, the flows are assigned in such a way that total travel time of the system is minimized. This causes less delay on road links compared to simulation with predefined route choice behavior. As a result, the CVRP is less affected. This is reflected in the results shown in Table 11.3 (Part-B and Part-C). The total travel time is lesser in Part-B as compared to Part-C.

In addition, when route choice behavior is considered we see switching of paths during the solution process. For example, to reach point B from point A initially the CVRP suggests minimum cost C_1 corresponding to path P_1. However, due to high congestion in path P_1 the cost becomes higher with time and CVRP suggests minimum cost C_2 that corresponds to a different path P_2. Thus, we can include different route choice behavior within this framework. If one were to include the dynamic user equilibrium as the governing route choice within this framework, the analytical formulation will be equivalent to a mathematical program with equilibrium constraints (MPEC). This problem offers a significant potential for future study.

11.5 Conclusion

In this research, we proposed an integrated simulation-based framework for solving the capacitated vehicle routing (CVRP) problem incorporating the time-varying congestion. We adopt a density-based dynamic traffic flow model, namely the CTM, that updates the cost matrix to solve the CVRP problem. The algorithm combines a series of mixed integer programs to solve the CVRP problem. The results show the total cost (in terms of travel time) is significantly less than the basic solution when solved accounting for the time-varying congestion. In addition, we show the difference in solution process when route choice behavior is taken into account.

This research applies travel time for the links of the network from a density-based traffic flow simulation. An approach that is more appropriate would be to use the real-time travel time information to build the cost matrix to solve the CVRP problem. However, in that case we need an efficient approach that would solve the problem within reasonable time bound so that the vehicle (which is already on the road) can get the optimal node visiting sequence at the right point of time. Considering the size of practical problems and computational complexity (NP hard), it is not always feasible to go for such an efficient system. As an alternative, the system operators can apply this framework proposed here to obtain solution that is more accurate than the static CVRP approach. However, the algorithm may not yield good solutions as the real-time travel time algorithms. One can see this as a trade-off between the quality of solution and computational cost.

The proposed framework incorporates the effect of congestion and traffic dynamics into the CVRP problem. It is observed that at low levels of congestion the solution from the static VRP coincides with the time-varying congestion model proposed here. However, in cases where there is medium and heavy congestion, the results are significantly different and the proposed approach allows to model the problem more accurately.

The proposed framework has few limitations as well. The embedded traffic flow model (CTM) has some known drawbacks due to the linear nature of the constraint set (approximating the minimum operator as set of linear equations). This causes the well-known vehicle holding problem (Ukkusuri and Waller 2008; Ukkusuri et al. 2010; Zheng and Chiu 2011). Again, the formulations do not consider diverging and merging, and lane changing phenomenon within the traffic flow model explicitly. In addition, we consider only the system optimal type of route choice behavior that is more appropriate from the system operator's perspective. One might also focus on user optimal type of behavior to see the results that are appropriate from the user's perspective. Finally, no control (traffic light or ramp metering) is considered in the proposed framework that can affect the time-dependent flows in the network and accordingly the results.

Nonetheless, the proposed framework provides a sound approach to assess the impact of time-dependent flows in the context of CVRP. The approach is more realistic in the sense that it captures the time-dependent nature of congestion and accommodates its contribution to the cost function in CVRP. The framework is expected to be useful for logistics planners when dealing with CVRP in real-world context.

References

Ahuja RK, Magnanti TL, Orlin JB. Network flows: theory, algorithms and applications. Englewood Cliffs, NJ: Prentice Hall; 1993.

Augerat P, Belenguer JM, Benavent E, Corberan A, Naddef D, Rinaldi G. Computational results with a branch and cut code for the capacitated vehicle routing problem. In: Grenoble editor. France: Institut d'informatique et de mathématiques appliquées de Grenoble; 1995. p. 30.

Chang M-S, Hsueh C-F, Chen S-R. Real-time vehicle routing problem with time windows and simultaneous delivery/pickup demands. J Eastern Asia Soc Transport Stud. 2003;5:2273–2286.

Chen H-K, Hsueh C-F, Chang M-S. The real-time time-dependent vehicle routing problem. Transport Res E Logist Transport Rev. 2006;42:383–408.

Conrad R, Figliozzi M. Algorithms to quantify impact of congestion on time-dependent real-world urban freight distribution networks. Transport Res Rec J Transport Res Board 2168, Washington, D.C.: Transportation Research Board of the National Academies; 2010, p. 104–113.

Cordeau JF, Laporte G, Mercier A. A unified tabu search heuristic for vehicle routing problems with time windows. J Oper Res Soc. 2001;52:928–936.

Daganzo CF. The cell transmission model, part II: network traffic. Transport Res B Methodological 1995;29:79–93.

Daganzo CF. The cell transmission model: A dynamic representation of highway traffic consistent with the hydrodynamic theory. Transport Res B Methodological 1994;28:269–287.

Dantzig GB, Ramser JH. The truck dispatching problem. Manag Sci. 1959;6:80–91.

Donati AV, Montemanni R, Casagrande N, Rizzoli AE, Gambardella LM. Time dependent vehicle routing problem with a multi ant colony system. Eur J Oper Res. 2008;185:1174–1191.

Figliozzi MA. Analysis of the efficiency of urban commercial vehicle tours: Data collection, methodology, and policy implications. Transport Res B Methodological 2007;41:1014–1032.

Fleischmann B, Gietz M, Gnutzmann S. Time-varying travel times in vehicle routing. Transport Sci. 2004;38:160–173.

Fukasawa R, Longo H, Lysgaard J, Aragão MP, d Reis M, Uchoa E, Werneck RF. Robust branch-and-cut-and-price for the capacitated vehicle routing problem. Math Program. 2006;106:491–511.

Haghani A, Jung S. A dynamic vehicle routing problem with time-dependent travel times. Comput Oper Res. 2005;32:2959–2986.

Hu T-Y, Liao T-Y, Lu Y-C. Study of solution approach for dynamic vehicle routing problems with real-time information. Transport Res Rec J Transport Res Board 1857, Transportation Research Board of the National Academies, Washington, D.C., 2003, pp. 102–108.

Ichoua S, Gendreau M, Potvin J-Y. Vehicle dispatching with time-dependent travel times. Eur J Oper Res. 2003;144:379–396.

Laporte G. The vehicle routing problem: An overview of exact and approximate algorithms. Eur J Oper Res. 1992;59:345–358.

Larsen A. The dynamic vehicle routing problem. PhD dissertation, Department of Mathematical Modeling, Technical University of Denmark, 2000.

Lighthill MJ, Whitham GB. On kinematic waves II: A theory of traffic flow on long crowded roads. In Proceedings of the royal society of london series a-mathematical and physical sciences, 1955, p. 317–345.

Lysgaard J, Letchford AN, Eglese RW. A new branch-and-cut algorithm for the capacitated vehicle routing problem. Math Program. 2004;100:423–445.

Malandraki C, Daskin MS. Time dependent vehicle routing problems: formulations, properties and heuristic algorithms. Transport Sci. 1992;26:185–200.

Malandraki C, Dial RB. A restricted dynamic programming heuristic algorithm for the time dependent traveling salesman problem. Eur J Oper Res. 1996;90:45–55.

Psaraftis HN. Dynamic vehicle routing: status and prospects. Ann Oper Res. 1995;61:143–164.

Schrage L. Formulation and structure of more complex/realistic routing and scheduling problems. Networks 1981;11:229–232.

Shieh H-M, May M-D. On-line vehicle routing with time windows: optimization-based heuristics approach for freight demands requested in real-time. Transport Res Rec J Transport Res Board 1617, Washington, D.C.: Transportation Research Board of the National Academies; 1998, p. 171–178.

Stickel M, Darger J, Furmans K. Vehicle routing with regard to traffic prognosis and congestion probabilities. Adv OR AI Method Transport. 2005;1:780–786.

Taniguchi E, Thompson R. Modeling city logistics. Transport Res Rec J Transport Res Board 1790, Washington, D.C.: Transportation Research Board of the National Academies; 2002. p. 45–51.

Toth P, Vigo D (eds.) The vehicle routing problem. SIAM monographs on discrete mathematics and applications. Philadelphia: SIAM; 2002.

Ukkusuri SV, Waller ST. Linear programming models for the user and system optimal dynamic network design problem: formulations, comparisons and extensions. Network Spatial Econ. 2008;8:383–406.

Ukkusuri SV, Ramadurai G, Patil G. A robust transportation signal control problem accounting for traffic dynamics. Comput Oper Res. 2010;37:869–879.

Woensel TV, Kerbache L, Peremans H, Vandaele N. Vehicle routing with dynamic travel times: A queueing approach. Eur J Oper Res. 2008;186:990–1007.

Zheng H, Chiu Y-C. A network flow algorithm for the cell-based single-destination system optimal dynamic traffic assignment problem. Transport Sci. 2011;45:121–137.

Ziliaskopoulos AK. A linear programming model for the single destination system optimum dynamic traffic assignment problem. Transport Sci. 2000;34:37–49.

Chapter 12
Incident Duration Prediction with Hybrid Tree-based Quantile Regression

Qing He, Yiannis Kamarianakis, Klayut Jintanakul, and Laura Wynter

Abstract Accurate prediction of incident duration is critical for efficient incident management which aims to minimize the impact of non-recurrent congestion. In this chapter, a hybrid tree-based quantile regression method is proposed for incident duration prediction and quantification of the effects of various incident and traffic characteristics that determine duration. Hybrid tree-based quantile regression incorporates the merits of both quantile regression modeling and tree-structured modeling: robustness to outliers, simple interpretation, flexibility in combining categorical covariates, and capturing nonlinear associations. The predictive models presented here are based on variables associated with incident characteristics as well as the traffic conditions before and after incident occurrence. Compared to previous approaches, the hybrid tree-based quantile regression offers higher predictive accuracy.

12.1 Introduction

Incidents, including accidents, vehicle breakdowns, spilled loads, or other random events, reduce the capacity of the road and cause congestion when traffic demand exceeds the reduced capacity at the incident location. Oak Ridge National Laboratory estimates that 55% of motorist delays on freeways are incident related (Chin et al. 2004). Effective management is essential for mitigating the negative effects of incidents on congested urban freeways. Various studies have been undertaken to

Q. He • Y. Kamarianakis • K. Jintanakul • L. Wynter (✉)
IBM T. J. Watson Research Center, P.O. Box 218, Yorktown Heights, NY 10598, USA
e-mail: qhe@us.ibm.com; yiannis@us.ibm.com; kjintan@us.ibm.com; lwynter@us.ibm.com

S.V. Ukkusuri and K. Ozbay (eds.), *Advances in Dynamic Network Modeling in Complex Transportation Systems*, Complex Networks and Dynamic Systems 2,
DOI 10.1007/978-1-4614-6243-9_12,

develop mitigation measures that minimize non-recurrent congestion due to freeway incidents. A typical example of such efforts is the development of various types of incident management systems that aim to clear traffic incidents quickly to minimize its impact on traffic flow.

In existing incident management systems, an ability to anticipate incident characteristics allows traffic managers to make better decisions on how to use management and control resources, such as advanced traveler information system (ATIS) and route guidance systems (RGS). Incident duration is an essential characteristic since it highly determines both the magnitude and the extent of congestion. Therefore, it is important to understand which factors can affect the incident duration. This study explores these critical factors and develops statistical models for incident duration prediction.

Incident duration can be defined as the duration between the instances of incident occurrence and of departure of the response vehicles from the accident scene (Garib et al. 1997; Nam and Mannering 2000; Smith and Smith 2001). As indicated in previous studies, an incident is composed of the following four phases: (a) incident detection and reporting time, (b) response time, (c) clearance time, and (d) recovery time. Traditionally incident duration is defined as the sum of first three phases.

Incident duration prediction models can be used as a means to improve incident management systems under non-recurrent traffic congestion. Incident management systems generally encompass three main modules, including incident detection technology, incident impact prediction, and incident-responsive traffic management and control. Incident duration prediction models are essential components in such a system, especially in the last two modules. Travelers and traffic management entities can generally realize the impact by the forecasted incident duration. In general, the impact of an incident in terms of both magnitude and extent of the congestion is significantly affected by incident duration. Virtually all existing impact prediction models developed in the literature require knowing the incident duration before producing a prediction. Since duration of an incident is usually not known until the incident is cleared, an accurate estimate is needed for accurate real-time prediction of incident impacts.

Likewise, an accurate estimate of the incident duration is also required in deriving effective response management and control strategies. Effective traffic control strategies are supposed to alleviate impacted traffic without unnecessarily interrupting the normal traffic or creating a secondary bottleneck. Rerouting factors such as diversion and merge points, diversion percentages, and diversion duration need to be derived on the basis of accurate magnitude and extent of the congestion as well as the duration of congestion (Lee et al. 2003; Srinivasan and Krishnamurthy 2003). In other words, the ability to redistribute flows over time is important for effective incident management (Oh and Jayakrishnan 2000). For instance, the projected incident duration will enable responsible traffic agencies to notify the en-route drivers of traffic congestion in a timely manner with VMS, and assess if any detour operators or control actions are needed. Drivers with better traffic information when encountering an incident can then make a proper route choice decision with less anxiety, which may consequently increase their compliance to suggestions or guidance by responsible traffic agencies (Garib et al. 1997).

The vast majority of previously related studies focused on predicting incident duration solely from incident characteristics. This work incorporates not only incident characteristics but also traffic data from both before and after the incident occurrence. Traffic data collected prior to incidents act as spatial and temporal indicators. Spatially, sequential traffic measurements indicate if the location of an incident is a bottleneck in the network. Temporally, time of day and day of week are associated with different levels of traffic variables and consequently, with different effects for incidents of the same type. Furthermore, the levels of traffic variables after an incident are associated with incident severity and hence with clearance times.

In addition, this work uses hybrid tree-based quantile regression. Other tree approaches have been used in the literature previously, including (Ozbay and Kachroo 1999; Smith and Smith 2001). A critical feature of the method used here is that it is designed to overcome the fundamental problems of previous trees such as over-fitting and selection bias towards predictors with many possible splits or missing values.

This chapter is organized as follows. Section 12.2 is devoted to a literature review which discusses different methods for online prediction of incident duration. In Sect. 12.3, we present the methodology of hybrid tree-based quantile regression, which combines conditional inference trees with quantile regression. Data description and preliminary data analysis are illustrated in Sect. 12.4. Section 12.5 describes the calibration of the statistical models, which is followed by an evaluation of their predictive accuracy. Finally, Sect. 12.6 presents some concluding remarks.

12.2 Literature Review

Incident duration is one of the essential characteristics of incidents that determine the magnitude and extent of the congestion. Thus, it has been extensively studied over the last few decades. Different approaches proposed in the literature can be grouped into the following categories:

- Linear regression: Garib et al. (1997) performed duration prediction using regression models, to provide real-time incident information to travelers. As the empirical distribution of incident duration is skewed (Golob et al. 1987; Giuliano 1989), linear models are based on its logarithmic transformation. Khattak et al. (1994) used a series of truncated regression models to predict incident duration, which account for the fact that incident information at a Traffic Operations Center is obtained sequentially.
- Tree models: Ozbay and Kachroo (1999) constructed decision trees which do not require knowledge of all observable incident characteristics. A similar approach was followed in Smith and Smith (2001) where classification trees

were applied to predict incident duration, defined as a categorical variable. The classification tree was shown to be well suited for forecasting the phases of incident duration with reliable and informative characteristics. Recently, Kim et al. (2008) constructed a rule-based tree model coupled with a discrete choice model, aiming at improved predictive ability. However, these models do not use any traffic data nor do they consider the tails of the distributions.

- K-nearest neighbor (KNN): Smith and Smith (2001) investigated incident duration prediction with KNN methods. Qi and Smith (2004) developed a distance metric that can be effectively used with categorical data. They argued that KNN outperformed parametric forecasting models significantly. Again, these methods did not leverage traffic data.
- Survival analysis: Incident duration can be viewed as the time period an incident can survive before being cleared. However, to implement survival analysis, selecting an appropriate probability distribution for incidence duration can be a challenging task (Jones et al. 1991; Nam and Mannering 2000; Qi and Teng 2008; Chung 2010; Chung et al. 2010).
- Artificial intelligence: Wei and Lee (2007) applied artificial neural network (ANN)-based models and data fusion techniques to forecast incident duration. Recently they employed genetic algorithms (GA) and ANNs to construct two models that forecast accident duration from the moment of accident notification to accident clearance (Lee and Wei 2010). Demiroluk and Ozbay (2011) developed three structure learning algorithms to construct Bayesian network (BN) structures. They demonstrated that BNs were very useful in uncovering important relationships among predictors, using the concept of strength of links.

12.3 Methodology

The adopted methodology, proposed recently in Hothorn et al. (2006), combines unbiased recursive partitioning (URP) with piecewise constant fitting using permutation tests. The conditional distribution of statistics measuring the association between incident duration and its predictors is the basis for an unbiased selection of the predictors in the model. Multiple tests are applied to determine whether no significant association between any predictor and duration can be stated and the recursion needs to stop. The above framework aims to solve both the over-fitting and the variable selection problems of older recursive partitioning methods (a detailed overview is provided in Murthy 1998).

Our implementation was based on the software provided by the developers of the method (Hothorn et al. 2011). Significance levels for the test statistics were set to conventional levels (0.05) and a Bonferroni correction was applied in multiple testing procedures, in accordance with the suggestion in Hothorn et al. (2006).

Predictions from conventional tree models are compared to the ones derived from a hybrid approach that combines regression trees based on the incident characteristics with quantile regression models that use traffic variables as predictors.

The latter are robust to outliers and skewed response distributions (Koenker 2005) and are widely used in applications instead of conventional least-squares regression during the last decade. It is worth noting that our hybrid method is similar to the one adopted in GUIDE (Loh 2008).

In Sect. 12.5 we display predictive models for the 0.5 (median regression) and the 0.9 quantiles of log-duration. Median regression models can be used as conventional incident duration predictors, while models for the 0.9 conditional quantile quantify the uncertainty associated with each prediction and can also be viewed as predictors of worst-case scenarios.

Finally, URP (prediction from only tree models) and hybrid tree-based quantile regression models are compared with the well-known older approach known as Classification and Regression Tree[1] (CART) (Breiman et al. 1984) as well as with the classic K-nearest-neighbor (KNN) methods without using traffic data as predictors, as was done in previous work found in the literature. Overall prediction accuracy is measured by mean absolute error (MAE1), median absolute error (MAE2), mean absolute percentage error (MAPE1), and median absolute percentage error (MAPE2). We also present percentages of predictions that are within a certain tolerance of their actual duration times, as suggested by Smith and Smith (2001).

12.4 Data Description

We examine incidents that occurred in 17 major freeways in Bay area, California, from April to June, 2010. The freeway network, shown in Fig. 12.1, connects ten cities. Incident data were obtained from the California Highway Patrol computer-aided dispatch (CHP/CAD) system (CHP 2011). Incident information was collected from two sources: the first source provided the incident type and the corresponding spatio-temporal information, while the second source provided further details on incident characteristics, such as number of vehicles involved.[2] Original incident types were classified into three groups: collision, disabled vehicle, and traffic hazard. In total, 1,245 incidents with valid data were analyzed. Table 12.1 contains the basic summary statistics of the dataset. The empirical probability distribution of incident duration has a long tail, which is in accordance with observations from previous studies (Chung 2010). The average incident duration is 20.61 min, while

[1]CART is implemented in R (R Development Core Team 2009), using rpart (Therneau and Atkinson 2011).

[2]Incidents associated with scheduled road closures or without any log were excluded from the analysis. Duplicated incidents were identified by incident reporting time and location and were excluded as well while their logs were reviewed and merged. An automatic text recognition program was developed to parse incident logs.

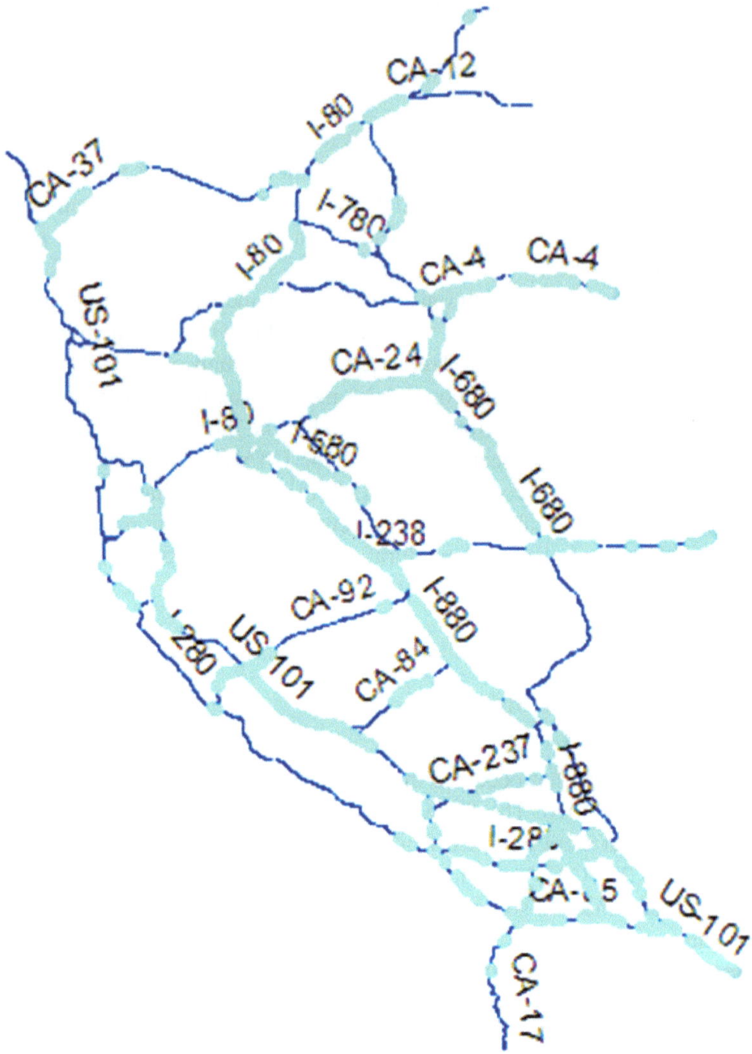

Fig. 12.1 Bay area freeway network with detectors in highlighted links

the median incident duration is 15.5 min. The set of incidents was randomly cut into a training dataset (60% of data) and test dataset (40% of data).

Traffic data were obtained from the Caltrans Performance Measurement System (PeMS). PeMS is a system designed to maintain California freeway traffic data and compute annual congestion for facilities with surveillance systems in place, typically loop detectors spaced approximately 0.5 mile apart on each freeway lane (Choe et al. 2002). There are around 850 detectors in Bay area freeways, shown as highlighted links in Fig. 12.1. The analysis that follows uses 5-min aggregated

Table 12.1 Summary statistics

Incident data	
Number of incidents	1,245
Median incident duration (min)	15.5
Average incident duration (min)	20.61
Proportion of incidents in "Collision"	0.52
Proportion of incidents in "Disabled"	0.26
Proportion of incidents in "Hazard"	0.22
Proportion of incidents with injuries	0.08
Average number of vehicle involved	1.30
Traffic data	
Average historical speed across all incident sites (mph)	52.84
Average historical volume across all incident sites (veh/hr/ln)	1,361
Average historical occupancy across all incident sites	0.139
Average speed before incident (mph)	48.75
Average speed after incident (mph)	29.27
Average volume before incident (veh/hr/ln)	1,303
Average volume after incident (veh/hr/ln)	1,300
Average occupancy before incident	0.147
Average occupancy after incident	0.316

volume, speed, and occupancy data. Each incident was associated with traffic data spatially and temporally:

- Spatially, each incident was matched with the closest link, which satisfied the incident location descriptions. Upstream and downstream traffic detectors were also identified accordingly.
- Temporally, a modified incident detection algorithm based on the DELOS (also called Minnesota) algorithm (Chassiakos and Stephanedes 1993) was developed to trace differences in occupancy between adjacent detectors through time, and to detect an incident when these differences change significantly in a short time period. This incident detection algorithm associates incident data with upstream and downstream traffic data, locates the time stamp when the shockwave hits the nearest upstream detector, and records traffic data before and after the incident's time of occurrence.

Table 12.1 depicts summary statistics of traffic data before and after the incident. Prior knowledge suggests that incidents will cause congestion on an upstream detector whereas traffic conditions will become less congested at downstream stations (Payne and Tignor 1978). In our study, it is found that speed and occupancy are affected dramatically by incidents, while volume remains relatively stable. On average, speed drops 40% after an incident, while occupancy increases by 115%. The impact of incidents on traffic data is illustrated in Fig. 12.2. Speeds, volumes, and occupancies at the first upstream detector before and after an incident's time of occurrence are normalized and plotted in the same graph. Points on the 45° line correspond to data that are not affected by an incident. One notes that speeds tend to

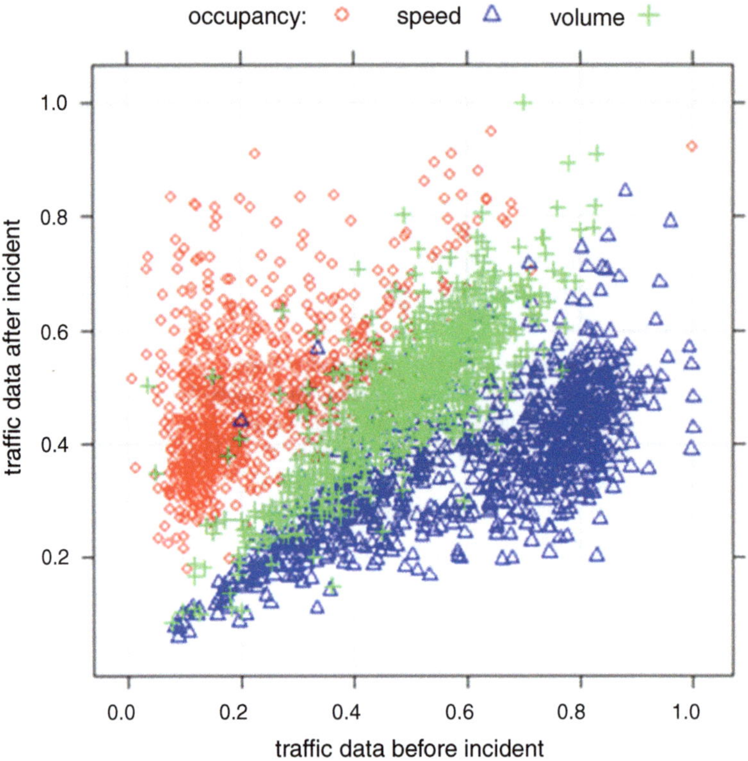

Fig. 12.2 Scatter-plots of normalized traffic data (speed, volume, and occupancy) at the first upstream detector before and after incident occurrence

decrease while occupancies tend to increase after the incident, in accordance with prior expectations. On the other hand, volumes may increase or decrease, depending on the levels of traffic congestion before and after the incident.

The candidate predictor variables are displayed in Table 12.2. All incident-related variables are categorical except for the number of vehicles involved in the incident. Figure 12.3 depicts a heteroscedastic relationship between incident duration and occupancy range. The latter is measured by occupancy differences, i.e., $Occ(s,t) - Occ(s,t-1)$, where section s indicates the first upstream detector, and t is the time when the incident-induced impact is observed, with $t-1$ being the preceding time period to t. Each violin-type plot represents the empirical probability density of incident duration at different ranges of occupancy. Clearly, the variability of incident duration increases as occupancy increment increases. Low occupancy ranges are associated with short incident durations, while high occupancy increments may be related to both short and long incident durations. This suggests that traffic data may provide significant predictive power for incident duration. An increasing relationship between incident duration and the number of vehicles involved in an incident can be observed in Fig. 12.4.

Table 12.2 Candidate independent variables

Information type	Independent variables	Notation
Weather characteristics	Rainy	Rain
	Snowy	Snow
Temporal characteristics	Time of day (AM, PM, Mid, Off-peak)	t_am, t_pm, t_mid, t_off
	Day of week (Weekday or not)	Weekday
Incident characteristics	Incident type (collision, disabled, or hazard)	Type
	Num of vehicles involved	num_veh
	Lanes blocked (binary)	lane_block
	Truck involved (binary)	Truck
	Person injured (binary)	Injured
	CHP officer assigned (binary)	CHP
Geometric characteristics	Freeway (CA-17, CA-237, CA-24, CA-242, CA-4, CA-84, CA-85, CA-87, CA-92, I-238, I-280, I-580, I-680, I-80, I-880, I-980, US-101)	freeway1~freeway17
	City (Castro Valley, Contra Costa, Dublin, Hayward, Marin, Oakland, Redwood City, San Francisco, San Jose, Solano)	city1~city10
	Interstate highway	Interstate
	Ramp exists near incident location (upstream/downstream on-ramp/off-ramp; binary)	uponramp, upofframp, downonramp, downofframp
	Upstream off-ramp and a downstream on-ramp exist near incident location (binary)	Junction
	Upstream on-ramp and/or downstream off-ramp exist near incident location (binary)	Junctionbwt
	number of lanes (2 or 3, 4, 5+)	ln23, ln4, ln5
Traffic characteristics	Historical mean of traffic data (speed, volume, and occupancy) at the time of incident	v_mean, q_mean, o_mean
	Traffic data at the first upstream detector before incident detection	v_prior, q_prior, o_prior
	Traffic data at the first upstream detector after incident detection	v_inc, q_inc, o_inc
	Traffic data after incident occurrence divided by measurements collected before incident occurrence	v_ratio, q_ratio, o_ratio
	Traffic data increments after incident occurrence	v_diff, q_diff, o_diff

Note: v_ratio = v_inc/v_prior; v_diff = v_inc – v_prior

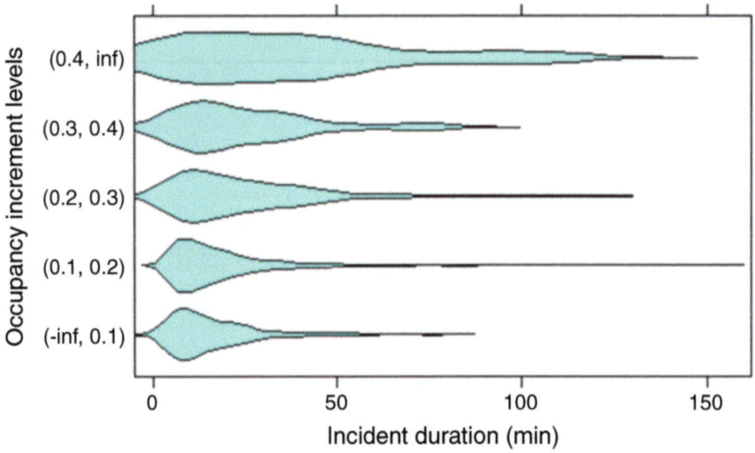

Fig. 12.3 Empirical distributions of incident duration for different occupancy increment levels after incident detection at the first upstream detector

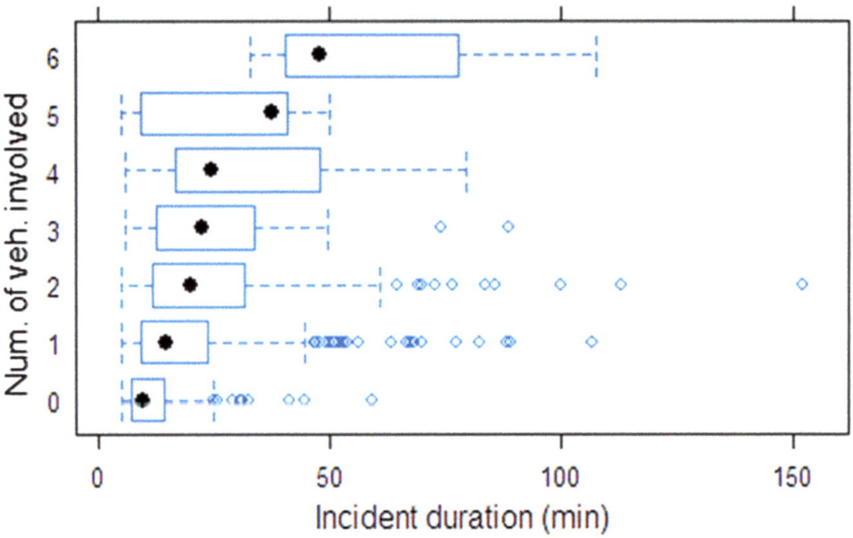

Fig. 12.4 Box-plots of incident duration for different numbers of vehicles involved

12.5 Model Estimation and Validation

Two URP trees were built based on the different sets of predictors. The first one, called URP tree1, shown in Fig. 12.5, was created using all candidate variables in Table 12.2. The second one (URP) was obtained using all but traffic variables and is depicted in Fig. 12.6. The decision path of the tree model is followed by answering a yes or no question at each node. Eventually, at each terminal node, a prediction is made based on the mean of incident duration of the data in that category.

Specifically, URP tree2 is a subset of URP tree1 that does not contain traffic data variables. According to the *p*-values in each node in both URP tree1 and URP tree2, the most significant predictor variables are incident characteristics (type, injured, num_veh and lane_block). As shown below, URP tree1 turns out to yield improved prediction accuracy than URP tree2, demonstrating that the incorporation of traffic data provides increased predictive power to the model.

In both URP trees the first node separates incidents according to type. This finding is in accordance with earlier studies which suggest that the empirical distribution of incident duration depends significantly on incident type (Kim et al. 2008). In the case of traffic collisions, the second node divides incidents according to the presence or not of an injury. In the URP tree with traffic data, URP tree1, if both collision and injury occur, v_prior, the level of speed prior to an incident divides the dataset further. Hence, collisions with injuries and high prior speeds (>48.8 mph) cause the longest incident durations in URP tree. High speeds prior to the incident are usually associated with off-peak periods; severe off-peak incidents are expected to last longer due to fewer available response units.

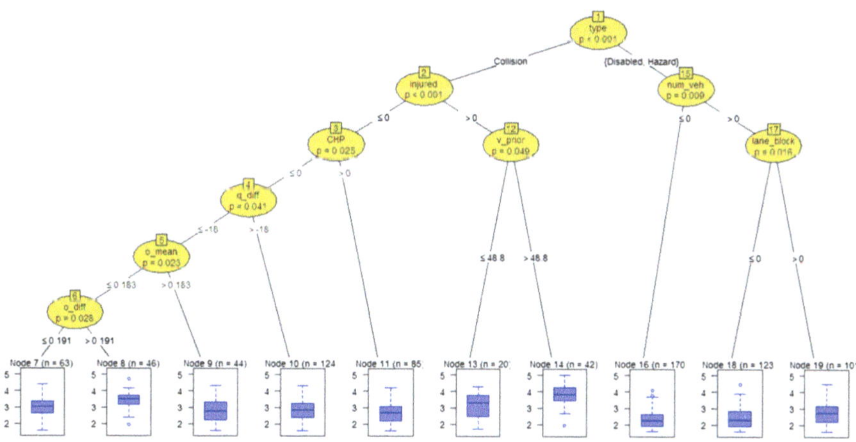

Fig. 12.5 URP tree1 (with traffic data): Unbiased recursive partition tree using all candidate predictors in training data. For each inner node, the Bonferroni-adjusted p-values are given. A box-plot of the log of incident duration is displayed in each terminal node

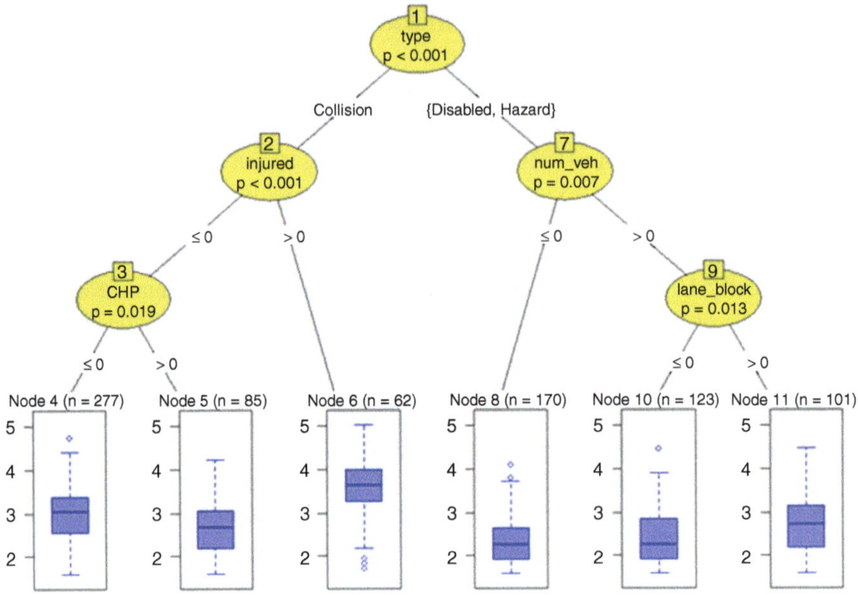

Fig. 12.6 URP tree2 (without traffic data): Unbiased recursive partition tree using only categorical variables in training data. For each inner node, the Bonferroni-adjusted p-values are shown. A box-plot of the log of incident duration is displayed in each terminal node

Node 3 indicates that CHP officer involvement reduces incident duration while node 4 indicates that a large reduction in traffic volume is associated with elevated incident duration. Node 5 uses historical occupancy to split incidents. Large values correspond to peak-periods, which tend to have short incident duration. Information from node 6 is consistent with the observations from Fig. 12.3: a larger occupancy increment is associated with larger incident duration. Incidents with disabled vehicles (num_veh = 1) have longer expected duration than traffic hazard (num_veh = 0). Node 17 shows that information on blocked lanes is significant for duration prediction with disabled vehicles.

To gain better prediction accuracy, quantile regression models are built for each terminal node in URP tree2. Unlike least-squared regression trees, which concentrate on modeling the relationship between the response and the covariates at the center of the response distribution, quantile regression can provide insight into the nature of that relationship at the center as well as the tails of the response distribution. Table 12.3 shows the coefficients of the six estimated regression models for the 0.5 (median) and 0.9 quantiles of the logarithm of incident duration. Besides traffic characteristics, geometric characteristics appear in most of the estimated regression models, such as ramp, city, and freeway junction information. This implies that incident duration varies significantly for different geometry factors, as well as different jurisdictions. For example, incidents that happen in freeway

Table 12.3 Coefficients of regression models estimated at each terminal node of URPtree2

0.5 quantile				0.9 quantile			
Regressor	Value	Std. Error	t value	Regressor	Value	Std. Error	t value
Node 4				Node 4			
constant	−2.7325	0.1266	−21.5818	constant	−3.3950	0.1967	−17.2611
city2	−0.2881	0.1692	−1.7025	interstate	−0.1924	0.1353	−1.4215
q_diff	−0.0007	0.0003	−2.3113	junctionbwt	−0.2772	0.1076	−2.5768
o_diff	−1.8049	0.6023	−2.9968	o_prior	−1.3523	0.6291	−2.1495
Node 5				o_diff	−1.8274	0.7205	−2.5364
constant	−2.6672	0.0910	−29.2972	Node 5			
Node 6				lane_block	−0.3862	0.1603	−2.4096
v_diff	−0.0175	0.0144	−1.2180	city2	−0.9949	0.4525	−2.1985
v_prior	−0.0566	0.0074	−7.6151	city6	−0.7597	0.3037	−2.5012
o_prior	−6.6888	1.3426	−4.9818	city7	−0.3904	0.2175	−1.7954
city8	−0.8139	0.3257	−2.4987	city9	−0.6949	0.1800	−3.8601
city2	−0.5440	0.2358	−2.3071	q_mean	−0.0006	0.0003	−1.8679
Node 8				v_prior	−0.0441	0.0108	−4.0738
constant	−1.7954	0.1688	−10.6352	o_prior	−4.3970	1.2343	−3.5624
junction	−0.6095	0.1287	−4.7376	v_diff	−0.0631	0.0182	−3.4759
uponramp	−0.2327	0.1087	−2.1402	o_diff	−8.0788	2.2776	−3.5470
upofframp	−0.4717	0.1057	−4.4644	Node 6			
v_prior	−0.0060	0.0028	−2.1330	interstate	−0.4334	0.3388	−1.2792
Node 10				o_diff	−3.1075	1.5401	−2.0177
constant	−1.7715	0.2757	−6.4263	v_prior	−0.0426	0.0071	−5.9916
t_mid	−0.4259	0.1787	−2.3832	o_prior	−7.9935	1.7221	−4.6418
t_off	−0.6355	0.2892	−2.1977	city8	−0.8302	0.3984	−2.0837
t_pm	−0.2677	0.1373	−1.9501	Node 8			
junction	−0.2560	0.1296	−1.9751	constant	−2.8679	0.1357	−21.1301
CHP	−0.3997	0.1244	−3.2133	junction	−0.8737	0.2749	−3.1787
downonramp	−0.2272	0.1599	−1.4208	junctionbwt	−0.5531	0.2895	−1.9109
city6	−0.3466	0.2346	−1.4772	upofframp	−1.1451	0.2178	−5.2583
q_mean	−0.0002	0.0002	−1.5193	downonramp	−0.3267	0.1980	−1.6497
o_diff	−1.7876	1.0319	−1.7324	Node 10			
Node 11				constant	−3.7102	0.3994	−9.2896
constant	−1.6724	0.6992	−2.3918	junction	−1.2756	0.4000	−3.1891
weekday	−0.5759	0.2481	−2.3209	upofframp	−1.1815	0.3823	−3.0906
interstate	−0.6921	0.2970	−2.3303	downonramp	−0.8234	0.3498	−2.3537
truck	0.4832	0.2702	−1.7882	downofframp	−0.8556	0.4051	−2.1121
uponramp	−0.2953	0.1660	−1.7790	city6	−0.3029	0.1863	−1.6260
freeway12	−0.3397	0.2504	−1.3564	v_diff	−0.0110	0.0057	−1.9181
v_prior	−0.0198	0.0087	−2.2799	q_diff	−0.0007	0.0004	−2.0189
o_prior	−4.8327	1.7361	−2.7836	o_diff	−2.4608	1.4465	−1.7012
o_diff	−2.5521	1.3578	−1.8796	Node 11			
				constant	−3.6738	0.1235	−29.7529
				uponramp	−0.4549	0.1742	−2.6114

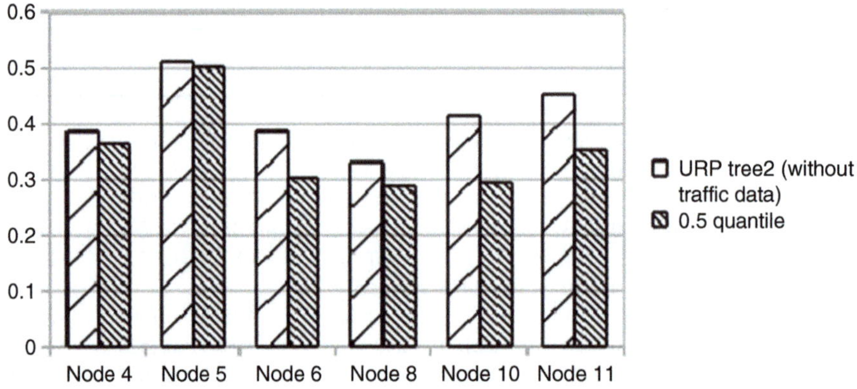

Fig. 12.7 Comparisons of median absolute percentage error for URPtree2 and 0.5 quantile regression in each terminal node

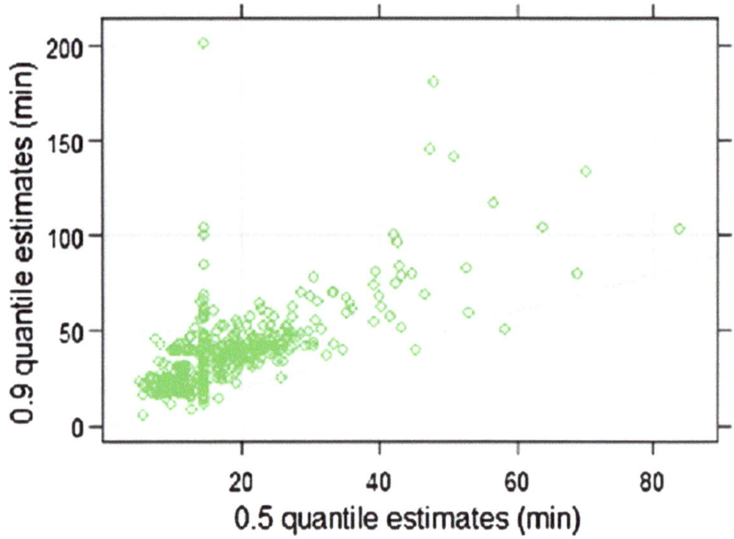

Fig. 12.8 Comparisons of 0.5 and 0.9 quantile estimates

junctions are related to increased clearance times, while the presence of an upstream off-ramp may decrease incident duration.

By replacing the mean in the final nodes of tree2 by quantile regression models, the forecasting accuracy (measured by median absolute percentage error) on each terminal node was improved on average by 15%, as can be observed in Fig. 12.7. To better visualize the difference between 0.5 and 0.9 quantile estimates, the corresponding predictions are plotted in Fig. 12.8; the average ratio of 0.9 and 0.5 quantile estimates is 2.29.

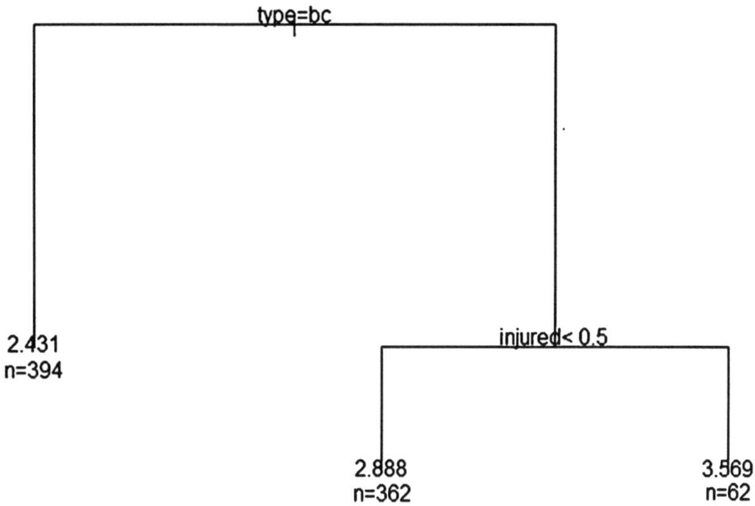

Fig. 12.9 Regression tree from CART for the training data. The split is beneath each intermediate node. Types a,b, and c represent collision, disabled vehicle and hazard, respectively. The number beneath each terminal node is the predicted logarithm of incident duration

Finally, the proposed hybrid tree-based quantile regression model is compared with the well-known Classification and Regression Tree[3] (CART) (Breiman et al. 1984) and K-nearest-neighbor (KNN) methods that do not use traffic data, as was reported in earlier studies in the literature. Hence, we consider the CART and the KNN approaches as performed here to be benchmarks for this research. The tree model from CART is depicted in Fig. 12.9. Again, the first nodes are incident type and the presence of an injury. KNN selects k past incidents that are closest to the current one and takes the mean or median of incidents in the neighborhood. A similar KNN approach implemented for incident prediction was reported in previous studies (Qi and Smith 2004; Smith and Smith 2001). The predictors in URPtree2 (Fig. 12.6) were used as the set of descriptors for each incident in KNN. The distance metric of Qi and Smith (2004) was adopted for measuring similarity between current and past incidents.

Table 12.4 reports measures of predictive accuracy for all examined methods. Overall prediction accuracy was measured by mean absolute error (denoted as MAE1), in minutes, median absolute error (denoted as MAE2), in minutes, mean absolute percentage error (denoted as MAPE1), and median absolute percentage error (denoted as MAPE2). As can be observed from the table, the URP tree approaches, and specifically the hybrid tree-based quantile regression, reduced error across the board as compared to the KNN and CART approaches used in the literature.

[3]CART is implemented in R (R Development Core Team 2009), using rpart (Therneau and Atkinson 2011).

Table 12.4 Evaluation of predictive error with different methods

	KNN	CART	URP tree2 (without traffic data)	URP tree1 (with traffic data)	Hybrid tree-based quantile reg.
MAE1 (min)	9.77	9.62	9.39	9.15	8.54
MAPE1 (%)	59.2	57.1	55.1	53.2	49.1
MAE2 (min)	6.2	6.02	5.81	5.78	4.99
MAPE2 (%)	42.2	42.2	40.4	39.8	34.5

Table 12.5 Comparisons of percentage of test samples in different prediction tolerances

	KNN (%)	CART (%)	URP tree2 (without traffic data) (%)	URP tree1 (with traffic data) (%)	Hybrid tree-based quantile reg. (%)
Prediction error <= 5 min	42.8	42.6	43.4	43.96	50.1
Prediction error <= 10 min	69.1	70.1	71.2	72.1	72.3
Prediction error <= 15 min	82.5	82.4	84.8	83	84.6
Prediction error <= 30 min	94.2	94.6	94.7	94.5	94.3
Prediction error <= 60 min	98.8	99.2	99	99.2	99.1

An alternative measure of effectiveness is related to a certain tolerance of the prediction error. As suggested by Smith and Smith (2001), it is useful to know the percentage of predictions that are within a certain tolerance of their actual duration times. Table 12.5 reports accuracy in terms of tolerance levels. Five tolerance values were used: 5, 10, 15, 30, and 60 min.

Over 50% of incidents have been predicted with less than 5 min prediction error with hybrid tree-based quantile regression, while other published methods reached at most 44%. For the ranges of prediction error under 5, 10, and 15 min, there were clear advantages to using the URP approaches proposed here. Note that for thresholds of 30 and 60 min, the benefits of the proposed approach decrease. That is not surprising for this dataset in that the average and median incident durations were 20 and 15 min, respectively. Hence relatively few incidents fall into the range of 30 min or more, and the benefits of predicting better those durations is therefore not as visible. Nonetheless, the most important tolerance levels for the incidents in the dataset used in this study, namely the 5, 10, and 15 min thresholds, all demonstrated significant improvements via the use of the URP techniques developed here.

12.6 Concluding Remarks

In this chapter, the use of unbiased recursive partitioning (URP) on both incident characteristic data and traffic data is proposed for incident duration prediction. In particular, a hybrid tree-based quantile regression method was developed; hybrid tree-based quantile regression modeling incorporates the merits of both quantile regression modeling and tree-structured modeling. Its merits include simple interpretation and ease of handling categorical covariates, robustness, and flexibility for nonlinearity. Given a URP tree, the hybrid method works by obtaining quantile regression models for each terminal node. With both 0.5 and 0.9 quantile estimates, traffic operators may understand not only the actual prediction but also the worst case results, and visualize the prediction range easily. Compared with the classic classification and regression tree (CART) approach, as well as a K-nearest neighbor approach, the URP trees and hybrid tree-based quantile regression proposed here appear to offer higher prediction accuracy.

The overall findings of this chapter can be summarized as follows:

- Incident characteristics (type, injuries, blocked lanes, number of vehicle involved, etc.) are the most significant predictors of incident duration.
- Traffic data can provide additional information that improves forecasting accuracy. Incidents with high prior speeds (occurring for instance during the night or during off-peak hours) generally last longer than those in daytime due to the lack of sufficient response units for incident clearance operations. Incidents with large occupancy increment tend to have longer duration than those with small occupancy changes.
- Incident location matters. Different geometry factors and jurisdiction may result in different incident duration.

In summary, it is essential to forecast the spatial-temporal incident impact based on both incident duration prediction and traffic conditions. Spatial-temporal incident impact aims to capture how congestion propagates over space and time. Future work in this area should leverage not only the model structure developed here, but in an online decision support system would incorporate real-time traffic predictions as predictor variables, in addition to the incident characteristics as they become available. Together, such a system can provide traffic operators with important components of an optimized control strategy for non-recurrent congestion.

References

Breiman L, et al. Classification and regression trees. New York: Chapman & Hall; 1984.

Chassiakos AP, Stephanedes YJ. Smoothing algorithms for incident detection. Transport Res Rec J Tranport Res Board 1993;1394:8–16.

Chin SM, et al. Temporary loss of highway capacity and impacts on performance: phase 2. Oak Ridge, Tennessee: Oak Ridge National Laboratory; 2004.

Choe T, Skabardonis A, Varaiya P. Freeway performance measurement system: operational analysis tool. Transport Res Rec J Transport Res Board c;1811:67–75.

CHP. 2011. CHP traffic incident information page. Available at: http://cad.chp.ca.gov/ [AccessedApril27,2011].

Chung Y. Development of an accident duration prediction model on the Korean Freeway Systems. Accid Anal Prev. 2010;42:282–289.

Chung Y, Walubita LF, Choi K. Modeling accident duration and its mitigation strategies on South Korean Freeway Systems. Transport Res Rec J Transport Res Board 2010;2178:49–57.

Demiroluk S, Ozbay K. Structure learning for the estimation of non-parametric incident duration prediction models. In: Proceedings 90th Annual Meeting of TRB (CD-ROM). Washington D.C.: 2011.

Garib A, Radwan AE, Al-Deek H. Estimating magnitude and durationof incident delays. J Transport Eng. 1997;123(6):459–466.

Giuliano G. Incident characteristics, frequency, and duration on a high volume urban freeway. Transport Res. 1989;23A:387–396.

Golob TF, Recker WW, Leonard ID. An analysis of truck involved freeway accidents. Accid Anal Prev. 1987;19:375–395.

Hothorn T, et al. 2011. CRAN - Package party. party: A Laboratory for Recursive Partytioning. Available at: http://cran.r-project.org/web/packages/party/[AccessedMay3,2011].

Hothorn T, Hornik K, Zeileis A. Unbiased recursive partitioning: a conditional inference framework. J Comput Graph Stat. 2006;15:651–674.

Jones B, Jassen L, Mannering FL. Analysis of the frequency and duration of freeway accidents in Seattle. Accid Anal Prev. 1991;23:239–255.

Khattak AJ, Schofer JL, Wang M-H. A simple time sequential procedure for predicting freeway incident duration. IVHS J 1994;1:1–26.

Kim W, Natarajan S, Chang G. Empirical analysis and modeling of freeway incident duration. In: Proceedings of the 11th International IEEE Conference on Intelligent Transportation Systems. Beijing, China:2008.

Koenker R. Quantile regression. Cambridge: Cambridge University Press; 2005.

Lee D-H, Jeng S, Ng M. Defining the incident impact area for traffic diversion: knowledge discovery via a data mining approach. In: Proceedings 82th Annual Meeting of TRB (CD-ROM). Washington D.C. 2003.

Lee Y, Wei C-H. A computerized feature selection method using genetic algorithms to forecast freeway accident duration times. Comput Aided Civ Infrastruct Eng. 2010;25(2):132–148.

Loh W-Y. Classification and regression tree methods. In: Ruggeri F, Kenett R, Faltin FW, editors. Encyclopedia of statistics in quality and reliability. Chichester: Wiley; 2008. p. 315–323.

Murthy SK. Automatic construction of decision trees from data: a multi-disciplinary survey. Data Min Knowl Discov. 1998;2:345–389.

Nam D, Mannering FL. An exploratory hazard-based analysis of highway incident duration. Transport Res A 2000;34:85–102

Oh J, Jayakrishnan R. Temporal control of variable message signs toward achieving dynamic system optimum. In: Proceedings 79th Annual Meeting of TRB (CD-ROM) 2000.

Ozbay K, Kachroo P. Incident management in intelligent transportation systems. Boston: Artech House; 1999.

Payne HJ, Tignor SC. Freeway incident-detection algorithmsbased on decision trees with states. Transport Res Rec J Tranport Res Board 1978;682:30–37.

Qi Y, Smith BL. Identifying nearest neighbors in a large-scale incident data archive. Transport Res Rec J Transport Res Board 2004;1879:89–98.

Qi Y, Teng H. An information-based time sequential approach to online incident duration prediction. J Intell Transport Syst. 2008;12(1):1–12.

R Development Core Team, R: a language and environment for statistical computing. Vienna, Austria: R Foundation for Statistical Computing; 2009.

Smith KW, Smith BL. Forecasting the clearance time of freeway accidents. Charlottesville, VA: Center for Transportation Studies, University of Virginia; 2001.

Srinivasan K, Krishnamurthy A. Roles of spatial and temporal factors in variable message sign effectiveness under nonrecurrent congestion. Transport Res Rec J Tranport Res Board 2003;1854:124–134.
Therneau TM, Atkinson B. 2011. CRAN - Package rpart. rpart: Recursive Partitioning. Available at: http://cran.r-project.org/web/packages/rpart/index.html[AccessedMay4,2011].
Wei CH, Lee Y. Sequential forecast of incidentduration using artificial neural network models. Accid Anal Prev. 2007;39(5):944–54.

Index

A
ABM. *See* Activity-based models (ABM)
Activity-based models (ABM), 226
Activity rescheduling. *See* Within-day activity
Activity travel network (ATN), 227
Adaptive signal control, 44–45
Advanced traveler information systems (ATIS)
  behavioral response data, 219
  description, 194
  driver response
    data collection methods, 200–201
    day-to-day learning, 202–204
    dynamic and static information, 200
    modeling efforts and data analysis, 209
    personal experience and exogenous information, 199–200
    real-time (*see* Real-time information, ATIS)
    WTP, 201–202
  economic benefits, 210
  evaluation frameworks
    demand prediction error, 215, 216
    guided *vs.* unguided mean travel time, 215, 216
    laboratory evaluation, 212
    microscopic traffic simulator, 214
    real-time traffic data, 211
    real-world evaluations, 212–213
    route guidance, 218
    simulation methods, 213
    update frequency, 214–215
    usage level and update frequency, 213, 214
    VMS, 218
    Westchester county network and incidents, 217, 218
  parameters, 211
  route guidance
    attributes, 195–196
    features and technological characteristics, 199
    "message," 195
    network performance overall, 194
    predicted conditions, 197–199
    prevailing conditions, 196–197
  technological advances, 219
  traffic simulation tools, 210
Agent for utility-driven rescheduling of routinized activities (AURORA), 226–227
Airplane path planning, 249
Anisotropic environment, 250, 251
Anisotropic medium, 253–256
Arterial travel times analysis
  data
    from/to times observation, 121
    interpretation, 121
    reader deployment locations, 118–119
    records, 119
    tag-reads observation, 120
    trip times creation, 121
  description, 115–116
  estimation, 138
  incident detection and characterization, 139
  long trip times, 139
  monitor trip times, 116
  prior work (*see* Travel times)
  trends within individual days
    histogram, mean trip rates, 136–137, 136–138
    minimum trip rate, τ-CHs, 133–134
    over time, 131

S.V. Ukkusuri and K. Ozbay (eds.), *Advances in Dynamic Network Modeling in Complex Transportation Systems*, Complex Networks and Dynamic Systems 2,
DOI 10.1007/978-1-4614-6243-9,

Arterial travel times analysis (*cont.*)
spot trip rates, speed data, 138
τ–CHs, snow day, 134–136
trip rate, single week, 133
trip rates ≤1,200 s/mile, 134–136
two-percentile point, τ-CH, 132
trip times
analysis, 122–123
creation, 121
cumulative histogram, 125–126
density functions, 139–140
diurnal pattern, "to" time, 124
multi-day patterns, 123
northbound trips, τ-CHs, 126, 127
observations, 139
percentile travel time graphs, 131
percentile trip rates, 127–128
rates, 124–125
shorter, 124
southbound trips, τ-CHs, 126–127
statistics, typical weekday and weekend, 129, 130
trip rate statistics, southbound OD pairs, 131, 132
variation, 122
Artificial neural network (ANN) models, 290
Assignment
load, 117
traffic, 116
ATIS. *See* Advanced traveler information systems (ATIS)
ATN. *See* Activity travel network (ATN)
AURORA. *See* Agent for utility-driven rescheduling of routinized activities (AURORA)
Automatic vehicle identification (AVI)
technology, 116, 118
travel time, 118
trip times, 122
Automatic vehicle location (AVL)
equipment, travel time, 117
technology, 116, 118
AVI. *See* Automatic vehicle identification (AVI)
AVL. *See* Automatic vehicle location (AVL)

C

Capacitated vehicle routing problems (CVRP)
CTM, 269
demand profiles, test networks, 270–271
DVRP, 267
integrated framework and solution algorithm
initialization, 269
tour initiation, 269–270
tour update and termination, 270, 271
update traffic status, 270
integrated simulation-based framework, 282
linear programming-based heuristic, 267
mesoscopic traffic flow model, 266, 268
optimal node visiting sequence, 283
real-time information systems, 267
static problems, 267
system optimal behavior *vs.* predefined route choice behavior, 277–282
test-network-1
geometric attributes and input parameters, 271, 272
nodes and arcs, 271, 273
system optimal dynamic traffic simulation, CTM, 272–273, 277–281
traffic simulation, CTM, 272, 274, 275
travel time, 272–273
test-network-2
input parameters and geometric characteristics, 272, 273
real-world test network, 273, 276
system optimal dynamic traffic simulation, CTM, 276–281
time-varying congestion, 282
travel time, 276, 282
time-varying nature, 265
travel time information, 266
TSP, 268
CART. *See* Classification and regression tree (CART)
Cell-based dynamic equilibrium models
calibration issues, 188
CTM, 164–165
DTA models, 163–164
efficient and convergent solution methods, 186
formulation and algorithmic approaches
FPP, 182, 183
NCP, 181–183
VIP, 181
lane changing traffic behavior and moving bottlenecks, 186–187
large time-dependent path set, 184–185
nonexistence and nonuniqueness, RSDUO solutions, 183–184
quality, OD matrix, 187
realistic travel and traffic behaviors, 165
route choice component (*see* Route choice component)

teal-time, large-scale applications, 187
time-consuming Monte Carlo simulation, 185
traffic flow component (*see* Traffic flow component)
Cell-transmission model (CTM)
control
minimizing delay, corridor, 97–98
physical constraints, traffic, 98–99
flow dynamics, corridor roadway
LWR model, 92–93
metered freeway onramp, 95–96
roadway junctions, 93
signalized diverges, 93–94
signalized merges, 94–95
stop and yield merge, 96–97
urban intersections, 93
linear program, 269
LWR model, 144
multi-destination system, 269
traffic demand input and vehicle routing, 97
traffic flow model, 166–168
CEMDEP. *See* Comprehensive econometric micro-simulator for daily activity-travel patterns (CEMDEP)
CFP. *See* Cyclic flow profiles (CFP)
Classification and regression tree (CART)
and KNN, 291, 301
training data, 301
C-MIXCROS and D-MIXCROS, feedback control
algorithm
ALINEA, 71, 72
Ball Aerospace/FHWA, ARMS, 71
Fuzzy logic, 70
helper ramp and bottleneck, 69–70
linear programming, 70
METALINE, 70–71, 73
Sperry ramp metering, 70
SWARM, 71
traffic-responsive metering (FLOW), 70, 71
Zonc, 68–69
control law
coupled, 81–83
decoupled, 80–81
design, 80
system dynamics, 78–79
time-discretizing, 78
traffic-responsive ramp metering control system, 77–78
description, 73, 74
freeway control, ramp metering, 68
freeway traffic model, 74–76
local and coordinated ramp metering, 68
macroscopic testing, RMSE, 87
"*n*" freeway sections with 1 on-ramp, 71–72
objective
absolute and appropriate values, 77
definition, 73
design, 76
error function, 76, 77
fundamental image, 76, 77
"*n*" freeway sections with 1 on-ramp, 72, 76
performance, 87
ramp metering controls, 87
ramp metering strategy, 73
simulation model
macroscopic, 83–84
microscopic, 84–86
traffic demand, 68
Comprehensive econometric micro-simulator for daily activity-travel patterns (CEMDEP), 226
CTM. *See* Cell-transmission model (CTM)
Cyclic flow profiles (CFP), 90

D
Dallas Network
DNL, 107
offset optimization, 108
procedure, 107
real-value coding scheme, 107
SPSA-*vs*. GA-based optimization, network II, 107–108
SPSA-*vs*. hill-climbing method, 108–109
TNTT, 107
Dijkstra's method/Chabini's approach, 260
Direction-dependent environments
airline industry, 249
asymmetric cost function, 246
DP modeling, 255–259
Euclidean distance, 246
GIS, 249
mathematical programming methods, 249
mobile robot, 249
optimal path finding (*See* Optimal path finding)
optimum vessel performance
dynamic real-time path optimization and vessel control, 247
path-finding problem, 248
real-time measurement, ocean wavefields, 247

Direction-dependent environments (*cont.*)
short-term forecasts, nonlinear wavefields, 247
time-domain computation, nonlinear ship motions, 247
wavefield forecast, 248
time and space homogeneous environment, 261
travel time, 259–260
UAVs, 250
vessel routing, 249
DNL. *See* Dynamic network loading (DNL)
Driver response
ATIS, 219
network control, 217
travel information
data analysis and modeling efforts, 209
data collection methods, 200–201
day-to-day learning, 202–204
dynamic and static information, 200
personal experience and exogenous information, 199–200
real-time information (*see* Real-time information, ATIS)
valuation and WTP, 201–202
DTA. *See* Dynamic traffic assignment (DTA)
DTA models. *See* Dynamic traffic assignment (DTA) models
DUO. *See* Dynamic user optimal (DUO)
DVRP. *See* Dynamic vehicle routing problem (DVRP)
Dynamic network loading (DNL), 107
Dynamic programming (DP) modeling
computational demand, time-dependent environment, 257–258
fastest path finding algorithm, 258–259
limited visibility horizon, 255–256
numerical results, 259
system dynamics restrictions, 256
Dynamic traffic assignment (DTA)
continuous time, 13–14
description, 2
discrete time, 13
feedback control, 19
incident management, 2
macroscopic model (*see* Macroscopic model, DTA)
macroscopic traffic network, 18–20
mathematical programming
numerical schemes (*See* User equilibrium)
system optimum (*See* System optimal solution)
user equilibrium (*See* User equilibrium)
modal split analysis, 2
projected dynamics (*See* Projected dynamics)
real-time traffic operations, 18–19
scalar initial-boundary data, 18
simulation based (*See* Simulation)
travel time and FIFO (*See* Travel times)
trip distribution, 2
trip generation models, 1
variational inequality (*See* Variational inequality)
Dynamic traffic assignment (DTA) models, 163–165, 227, 237, 242
Dynamic user equilibrium. *See* User equilibrium
Dynamic user optimal (DUO)
departure time choice principle, 176–177
NCP, 182
route choice principle, 176
Dynamic vehicle routing problem (DVRP), 267
DynaSmart-P network, 104

E
Euclidean distance, 246

F
Feedback control
C-MIXCROS and D-MIXCROS (*See* C-MIXCROS and D-MIXCROS, feedback control)
DTA, 19
max-pressure (*See* Max-pressure controller, arbitrary networks)
Fixed-cycle controller
performance, 48–50
queue size, delay, and busy period, 37, 38
stabilizing, 54
traffic bursts, 29
work-conserving (*See* Work-conserving controllers)
Fixed-point problem (FPP), 182, 183
FPP. *See* Fixed-point problem (FPP)
Frank-Wolfe algorithm, 6
Freeway traffic model, MIXCROS control law design, 74
flow and speed relationship, 75–76
ramp dynamics, 75
system dynamics, 75
traffic density, 74–75

G
Geographic information systems (GIS)
computational geometry, 249
travel time, 117
Geometric shortest path, 248
GIS. *See* Geographic information systems (GIS)
Global positioning system (GPS)
large time-dependent path set, 184
navigation units, 219
technologies, 206, 241
travel time, 117
GPS. *See* Global positioning system (GPS)

H
Hybrid tree-based quantile regression
artificial intelligence, 290
ATIS and RGS, 288
CART, 303
characteristics, 289
data description
box-plots, 296
candidate independent variables, 294, 295
empirical distributions, 294, 296
freeway network, 291, 292
PeMS, 292–293
Scatter-plots, normalized traffic data, 293–294
summary statistics, dataset, 291, 293
traffic data, 292
description, 287–288
KNN, 290, 303
linear regression, 289
management systems, non-recurrent traffic congestion, 288
model estimation and validation, 297–302
survival analysis, 290
system model, 288
traffic control strategy, 288
travelers and traffic management entity, 288
tree models, 289–290
URP (*See* Unbiased recursive partitioning (URP))

I
Incident duration prediction. *See* Hybrid tree-based quantile regression
Incident management systems
management and control resources, 288
modules, 288
Integrated corridor control
algorithm, network traffic, 102–103
constrained optimization, 101–102
minimizing delay, corridor, 97–98
network traffic loads, 109–111
optimization techniques, 99
physical constraints, corridor, 98–99
procedure, 99–100
regularity conditions, 100–101
Iteration
CVRP, 272, 276–281
travel time and vehicle hours traveled, 218
user-equilibrium
network loading, 22
path assignment adjustment, 22
path set update, 22
process, 21

K
Kinematic wave traffic flow model, 143
Kirchoff's law, 19
K-nearest neighbor (KNN), 290, 291, 301
KNN. *See* K-nearest neighbor (KNN)
Kuhn-Tucker conditions, 10

L
Lighthill–Whitham–Richards (LWR) model
road networks (*See* Multibuffer, LWR road networks)
traffic control, 90
LWR model. *See* Lighthill–Whitham–Richards (LWR) model

M
MAC. *See* Media access control (MAC)
Macroscopic model, DTA
Greenshield, 15–16
LWR model
generalized solutions, 16–17
traffic characteristics, 16, 17
weak solutions, 17
traffic network, 18–20
Macroscopic traffic network
Coclite/Piccoli formulation, 19
distribution, 19
Kirchoff's law, 19
traffic node, incoming and outgoing links, 18–19
Wardrop condition, 20
MAE1. *See* Mean absolute error (MAE1)
MAE2. *See* Median absolute error (MAE2)

MAPE1. *See* Mean absolute percentage error (MAPE1)
MAPE2. *See* Median absolute percentage error (MAPE2)
Max-pressure controller, arbitrary networks
  advantages, 57
  cumulative departures, 58–59
  description, 27–28
  designs, traffic-responsive controllers, 55–56
  fixed-time controllers, 57
  intuition, 54
  limitations, "store and forward" (SF) model, 56
  multiphase isolated intersection permit, 56
  network calculus (*See* Network calculus)
  network performance (*See* Network intersections performance)
  over-saturated networks, 57
  policies, 27
  queue size, arrivals and departures, 57–58
  single intersection performance
    all movement analysis, 36–38
    assumptions, 33
    fixed-cycle, 32
    network calculus, 32
    phase movement, 31
    single movement analysis, 33–36
    standard intersection, 32
  stabilizing, 57
  store and forward (SF) queuing system, 27
  work-conserving controllers, 59–60
MC-SCTM. *See* Monte-Carlo-based stochastic cell transmission model (MC-SCTM)
MCTM. *See* Modified cell transmission model (MCTM)
Mean absolute error (MAE1), 291
Mean absolute percentage error (MAPE1), 291
Media access control (MAC), 118
Median absolute error (MAE2), 291
Median absolute percentage error (MAPE2), 291
Minnesota, 293
Modified cell transmission model (MCTM), 169–170
Monte-Carlo-based stochastic cell transmission model (MC-SCTM), 171–173, 186
Multibuffer, LWR road networks
  CTM, 144
  definitions and notations, 149–151
  dynamics junctions, 144
  justification
    buffer capacity and load, 146–147
    dynamics, traffic distribution matrix, 147–149
    RS_{CGP} induction, 145–146
    solution, junctions, 145
  kinematic wave theory, 143
  literature, 144
  ODE-PDE model, 158–159
  Riemann problem (*See* Riemann solvers)
  Riemann solvers (*See* Riemann solvers)
  transportation engineering community, 143

N

NCP. *See* Nonlinear Complementarity Problem (NCP)
Network calculus
  definition, arrival and service process, 30, 31
  discrete time, 29–30
  maximum queue size and delay, 30–31
  queue size, 30
  queuing system, 29, 30
Network control. *See* Max-pressure controller, arbitrary networks
Network intersections performance
  fixed-cycle controller
    counterpart, 49
    network stage matrix, 48–49
    stabilizing, 50
  max-pressure controller
    definition, 50–51
    extensions, 52–53
    external arrivals, 51
    stabilizing, 52
    stage selection, 51
  network model
    routers, 46–47
    routing matrix, 48
    vector-matrix notation, 47–48
Nonlinear Complementarity Problem (NCP), 181–182

O

Obstacle-avoiding path, 250, 253
O-D-T journey. *See* Origin-destination-departure time (O-D-T) journey
Optimal control theory, 249, 250, 254–255
Optimal path finding
  DP modeling
    computational demand, time-dependent environment, 257–258
    fastest path finding algorithm, 258–259
    limited visibility horizon, 255–256

numerical results, 259
system dynamics restrictions, 256
time and space homogeneous environment
anisotropic medium, 253–255
turning radius constraint, 251–253
travel time
Dijkstra's method/Chabini's approach, 260
dynamic network, 260
RMS motions, 259
Origin-destination-departure time (O-D-T) journey, 227
Over-reaction, 213, 215

P
PeMS. *See* Performance measurement system (PeMS)
Performance measurement system (PeMS), 292–293
Prediction-based guidance
advantages, 198
consistency, 197–198
disadvantages, 198
technology and models, 199
Projected dynamics
dynamic route choice, 11–12
equilibrium point, 11
stability, equilibrium point, 11
variational inequality, fixed point problem, 11

Q
Quantile regression. *See* Hybrid tree-based quantile regression
Queue size
arrivals and departures, 57–58
maximum and delay, 30–31

R
Real-time information, ATIS
long-term response
habitual trip-making behavior adjustments, 208–209
residence/employment location choice, 209
short-term response
daily activity schedule adjustment, 207–208
departure time choice, 204–205
destination choice and trip cancellation, 206–207
mode choice, 206
route choice, 205–206
stress and anxiety reduction, 208
travel demand, 204
Regression trees. *See* Hybrid tree-based quantile regression
Reliability-based stochastic dynamic user optimal (RSDUO)
MC-SCTM, 183, 188
nonexistence and nonuniqueness, 183–184
route choice principle, 179–181
SAM, 182
Rescheduling decision
activity attribute change, 231–232
decision context, 229–230
mathematical model, 232–234
modeling considerations, 228–229
network condition change, 230–231
RGS. *See* Route guidance systems (RGS)
Riemann solvers
buffer capacity and load, 146–147
buffer dynamics, 147–149
collection, functions, 152
construction, 153
definition, 156–157
difference, 149
function, 151–152
incoming and outgoing roads, 154–155
index buffer, 155
junctions, 155
multibuffer solutions, 157–158
properties, 152–153
RS_{CGP} induction, 145–146
Root mean squared (RMS) motions, 259, 260
Route choice component
cell-based dynamic equilibrium models, 175–176
departure time choice principle/DUO route, 176–177
departure time choice principle/SDUO route, 178–179
DUO, 176
RSDUO, 179–181
SDUO, 178
Route guidance systems (RGS), 288

S
SAM. *See* Self-regulated averaging method (SAM)
SCTM. *See* Stochastic cell transmission model (SCTM)
SDUO. *See* Stochastic dynamic user optimal (SDUO)

Self-regulated averaging method (SAM), 182, 186
Signal timing design procedure, 105–106
Simulation based DTA
  calibration, field data, 22
  iterations, user equilibrium, 21–22
  microscopic and macroscopic, 21
Simulation model
  CTM, 270
  macroscopic
    ALINEA, 83
    D-MIXCROS and C-MIXCROS, 83
    METALINE, 83–84
    RMSE, 87
    total travel time and networks, 84
  microscopic
    congestion levels, ramps, 85
    PARAMICS, 84–85
    ramp metering controls, 85–86
  real-time information systems, 267
  traffic, 268, 270, 276
Simulation of travel/activity responses to complex household interactive logistic decisions (STARCHILD), 226, 227
Simultaneous perturbation stochastic approximation (SPSA)
  corridor control, 111, 112
  dynamic network flow (*See* Cell-transmission model (CTM))
  guidelines, parameters selection, 111
  optimization methods, 91–92
  real network
    Dallas Network, 107–109
    differences, 104–105
    DynaSmart-P network, 104
    geometric layout, Dallas fort worth network, 104, 106
    integrated corridor control (*See* Integrated corridor control)
    optimal control plan, 105
    signal timing design procedure, 105–106
    SR-81 corridor, 104
  simple network
    convergence process, 104, 105
    geometric layout and demand, 103–104
  techniques, 92
  traffic flow model (*See* Traffic flow model)
  transportation corridor, 90
  urban signal control, 111
SMM. *See* Switching-mode model (SMM)
Spatial traffic flow model. *See* Traffic flow model
SPSA. *See* Simultaneous perturbation stochastic approximation (SPSA)
STARCHILD. *See* Simulation of travel/activity responses to complex household interactive logistic decisions (STARCHILD)
Stochastic approximation. *See* Simultaneous perturbation stochastic approximation (SPSA)
Stochastic cell transmission model (SCTM), 170–171
Stochastic dynamic programming models, 249
Stochastic dynamic user equilibrium, 165
Stochastic dynamic user optimal (SDUO)
  cell-based dynamic equilibrium model, 181, 182
  departure time choice principle, 178–179
  route choice principle, 178
Store and forward (SF) network, 27, 56
Switching-mode model (SMM), 169–170
System optimal solution
  equivalence, marginal user-equilibrium condition, 6
  mathematical programming formulation, 5–6

T

TASHA. *See* Travel activity scheduler for household agents (TASHA)
Time-varying network. *See* Within-day activity
Time-varying travel cost, 228, 229
TNTT. *See* Total network travel time (TNTT)
Total network travel time (TNTT), 104, 107
Traffic assignment. *See* Dynamic traffic assignment (DTA)
Traffic dynamics. *See* Dynamic traffic assignment (DTA)
Traffic equilibrium assignment models, 203
Traffic flow component
  actual route travel time, 173–175
  CTM, 166–168
  incident management systems, 288
  mapping, 166
  MC-SCTM, 171–173
  MCTM and SMM, 169–170
  SCTM, 170–171
Traffic flow model
  CTM, 91
  HCM methods, 90
  LWR model, 90
  METANET, 91

point-queue (P-Q) and spatial queue (S-Q), 90–91
TRANSYT version 8, 90
Traffic operation design, 18–19
Transportation corridor, 90
Travel activity scheduler for household agents (TASHA), 226
Traveling salesman problem (TSP), 268
Travel times. *See* also Arterial travel times analysis
ADVANCE project, 117
arterial-based urban network, 117
ATIS system, 117
AVI technology, 118
AVL equipment, 117
convexifaction method, 175
CTM simulation, 165, 171, 174
CVRP, 272, 273, 276
direction-dependent nature, 254
discontinuity and non-monotonic route, 186
dynamics, 20–21
estimation, 118
and FIFO constraint, 14–15
GPS-GIS integrated system, 117
information programs, 116
license-plate and test-car survey, 116
linear path, 252
MAC, 118
mean, 215, 216
monitor trip times, 116
optimal path finding, 259–260
people and shippers, 115–116
route, 203
safety margin, 180
surveys and techniques, 116
taxis and buses, 117–118
teal-time, large-scale applications, 187
time-dependent nature, 267
TNTT, 104, 107
traffic management center, 266
transportation systems, 115
trip frequencies, 218
variations, 91
video and machine vision analysis, 116–117
VMS network, 208
VRP, 266
work-conserving controller, 39
TSP. *See* Traveling salesman problem (TSP)

U

UAVs. *See* Unmanned aerial vehicles (UAVs)
Unbiased recursive partitioning (URP)
Bonferroni correction, 290
CART and KNN, 291
conditional distribution, statistics measurement, 290
conventional tree models, 290
MAE1 and MAE2, 291
MAPE1 and MAPE2, 291
median regression models, 291
model estimation and validation
accuracy in terms, tolerance levels., 302
CART, 301
CHP officer, 298
coefficients, six estimated regression models, 298, 299
hybrid tree-based quantile regression model, 301
KNN, 301
least-squared regression trees, 298
median absolute percentage error, 300
predictive accuracy measurement, 301–302
quantile estimates, 0.5 and 0.9, 300
tree1 (with traffic data), 297
tree2 (without traffic data), 297, 298
outliers and skewed response distributions, 291
over-fitting and variable selection, 290
Unmanned aerial vehicles (UAVs), 250
URP. *See* Unbiased recursive partitioning (URP)
User equilibrium
DTA problem, 165
FHWA method, 6–7
Frank-Wolfe algorithm, 6
incremental assignment, 7
iterations, 21–22
mathematical programming formulation, 3–4
static traffic assignment, 6
travelers, 180
Wardrop principlc, 4–5
Utility maximization, 201–202, 225, 229, 232

V

Variable message sign (VMS), 206, 217, 218
Variational inequality problem (VIP)
boundary minimizer, 8–9
formulation, 7
framework, 8
gradient relationship, 10
interior minimizer, 8
Kuhn-Tucker conditions, 10, 11

Variational inequality problem (VIP) (*cont.*)
user-equilibrium problem, 9
VI Problem, 7
Vehicle routing and traffic demand input, 97
Vehicle routing problem (VRP). *See* Capacitated vehicle routing problems (CVRP)
Vessel routing
optimal path, 249
sea state characterization, 256
VIP. *See* Variational inequality problem (VIP)
VMS. *See* Variable message sign (VMS)

W

Wardrop principle, 4–5
Willingness to pay (WTP), 201–202
Within-day activity
ABM, 226
activities, schedule set, 237, 238
ATN, 227
branch-and-cut technique, 228
branching step, 238, 239
C language, 237
consistency search step, 239
decision process, 228
dynamic traffic assignment, 242
GPS technology, 241
hypothetical traveler, 237
model and solution methodology
mathematical model, 232–234
rescheduling decision process, 230–232
solution algorithm, 234–237
modified schedule, travel time change, 240, 241
network/traffic conditions, 227
O-D-T journey, 227
relaxation step, 238, 239
rescheduling decision (*See* Rescheduling decision)
sequence connections, 238, 239
STARCHILD, 226, 227
time-varying travel times, 239, 240
traveling public makes trips, 225
trip-based approach, 226
updated schedule by solution algorithm, 238
Work-conserving controllers
adaptive signal control, 44–45
bound α, 60–61
bound β, 61–62
feed back, 59–60
fixed cycle, 39
single-phase intersections
definition, 40
description, 39–40
network calculus, 41
Poisson processes, 41
stabilizing, 41
weighted queue, 40–41
travel time, 39
two counter
actuated time, intersections, 42
construction, 44
eight phases, standard intersection, 42–43
nonempty queue, 43
vector-matrix
bound μ, 62–64
bound í, 64
WTP. *See* Willingness to pay (WTP)

MIX
Papier aus verantwortungsvollen Quellen
Paper from responsible sources
FSC® C105338

If you have any concerns about our products,
you can contact us on
ProductSafety@springernature.com

In case Publisher is established outside the EU,
the EU authorized representative is:
Springer Nature Customer Service Center GmbH
Europaplatz 3, 69115 Heidelberg, Germany

Printed by Libri Plureos GmbH
in Hamburg, Germany